UNIVERSITY OF BATH • SCIENCE 16-19

APPLIED GENETICS

Other titles in the Project

Biology Martin Rowland
Applied Ecology Geoff Hayward
Micro-organisms and Biotechnology Jane Taylor
Biochemistry and Molecular Biology Moira Sheehan

Physics Robert Hutchings
Telecommunications John Allen
Medical Physics Martin Hollins
Energy David Sang and Robert Hutchings
Nuclear Physics David Sang
Chemistry Ken Gadd and Steve Gurr

UNIVERSITY OF BATH • SCIENCE 16-19

Project Director: J. J. Thompson, CBE

APPLIED GENETICS

GEOFF HAYWARD

Nelson

Thomas Nelson and Sons Ltd
Nelson House Mayfield Road
Walton-on-Thames Surrey
KT12 5PL UK

51 York Place
Edinburgh
EH1 3JD UK

Thomas Nelson (Hong Kong) Ltd
Toppan Building 10/F
22A Westlands Road
Quarry Bay Hong Kong

Thomas Nelson Australia
102 Dodds Street
South Melbourne
Victoria 3205 Australia

Nelson Canada
1120 Birchmount Road
Scarborough Ontario
M1K 5G4 Canada

First published by Macmillan Education Ltd 1990
ISBN 0-333-46659-4

This edition published by Thomas Nelson and Sons Ltd 1992

ISBN 0-17-438511-0
NPN 9 8 7 6 5 4

Printed in Hong Kong.

Contents

The Project: an introduction

The **University of Bath · Science 16–19 Project,** grew out of a reappraisal of how far sixth form science had travelled during a period of unprecedented curriculum reform and an attempt to evaluate future development. Changes were occurring both within the constitution of 16–19 syllabuses themselves and as a result of external pressures from 16+ and below: syllabus redefinition (starting with the common cores), the introduction of AS-level and its academic recognition, the originally optimistic outcome to the Higginson enquiry; new emphasis on skills and processes, and the balance of continuous and final assessment at GCSE level.

This activity offered fertile ground for the School of Education at the University of Bath to join forces with a team of science teachers, drawn from a wide spectrum of educational experience, to create a flexible curriculum model and then develop resources to fit it. This group addressed the task of satisfying these requirements:

- the new syllabus and examination demands of A- and AS-level courses;
- the provision of materials suitable for both the core and options parts of syllabuses;
- the striking of an appropriate balance of opportunities for students to acquire knowledge and understanding, develop skills and concepts, and to appreciate the applications and implications of science;
- the encouragement of a degree of independent learning through highly interactive texts;
- the satisfaction of the needs of a wide ability range of students at this level.

Some of these objectives were easier to achieve than others. Relationships to still evolving syllabuses demand the most rigorous analysis and a sense of vision – and optimism – regarding their eventual destination. Original assumptions about AS-level, for example, as a distinct though complementary sibling to A-level, needed to be revised.

The Project, though, always regarded itself as more than a provider of materials, important as this is, and concerned itself equally with the process of provision – how material can best be written and shaped to meet the requirements of the educational market-place. This aim found expression in two principal forms: the idea of secondment at the University and the extensive trialling of early material in schools and colleges.

Most authors enjoyed a period of secondment from teaching, which not only allowed them to reflect and write more strategically (and, particularly so, in a supportive academic environment) but, equally, to engage with each other in wrestling with the issues in question.

The Project saw in the trialling a crucial test for the acceptance of its ideas and their execution. Over one hundred institutions and one thousand students participated, and responses were invited from teachers and pupils alike. The reactions generally confirmed the soundness of the model and allowed for more scrupulous textual housekeeping, as details of confusion, ambiguity or plain misunderstanding were revised and reordered.

The test of all teaching must be in the quality of the learning, and the proof of these resources will be in the understanding and ease of accessibility which they generate. The Project, ultimately, is both a collection of materials and a message of faith in the science curriculum of the future.

J.J. Thompson
January 1990

How to use this book

Genetics is an area of intense scientific research and, with the current developments in genetic engineering, one which is likely to have an increasing impact on our lives. New discoveries and new applications require a constant updating of publications about genetics and its possible applications to ensure a true picture of the subject. This book explains the mechanics of genetics and their application, but one of the main aims is to help you understand what genetic engineering, modern breeding techniques and genetic counselling involve. The book looks at the moral and social issues these new techniques pose and provides the information for informed discussion of them.

Applied Genetics is written for A- or AS-level courses in biology and in particular for options in genetics. Genetics is a subject students often find difficult and a primary objective has been to present information in an accessible and interesting way. The book assumes an understanding of biology up to GCSE level and some topics also require some understanding of topics which are covered in A-level biology courses. Where particular scientific content underlines a section this is listed in the pre-requisites at the beginning of each theme.

The book is divided into five themes and can be used in a number of different ways. For example, you could use it as a course book working your way through all the chapters in the order in which they are given. Alternatively, you might only need to study some of the chapters or you may wish to use the questions and summary assignments to help in revision.

The best way to study any subject is to take some responsibility for your own learning. This means, for example, ensuring that you make a good set of notes and answer the questions as you come to them so that you can assess your progress. The book is designed to help you organise your studies and on the page opposite you will find information about the book which will help you get the best out of it.

Reading is, on its own, usually too passive to promote effective thinking and learning and the questions within the chapters are, therefore, an important feature of this book. They are intended to help you understand what you have just read. You should write down the answers to the questions as you come to them and then check the answers either with those at the back of the book or with your teacher. If you do not understand a question or an answer, make a note of it and discuss it with your teacher at the earliest opportunity.

Practical work in applied genetics in schools and colleges is difficult because of the time and the expensive equipment needed. Consequently, this book contains no suggestions for practical work though some may be possible in some situations. Instead the book contains a variety of investigations which range from extended data analysis to topics requiring library research.

Applied genetics is a very active and exciting area of biological research which offers a multitude of careers. I hope this book will not only enable you to learn the material you need to pass an exam, but will also encourage you to investigate whether you would like to work in this area.

Learning objectives

These are given at the beginning of each chapter and they outline what you should gain from the chapter. They are statements of attainment and often link closely to statements in a course syllabus. Learning objectives can help you make notes for revision, especially if used in conjunction with the summary assignments at the end of the chapter, as well as for checking progress.

Questions

In-text questions occur at points when you should consolidate what you have just learned, or prepare for what is to follow, by thinking along the lines required by the question. Some questions can, therefore, be answered from the material covered in the previous section, others may require additional thought or information. Answers to questions are at the end of the book.

Boxes

The information in the boxes supplements the text by providing descriptions of techniques and procedures referred to in the text and explanations of key ideas with which you need to be familar.

Investigations

These raise the sorts of issues which arise during practical and experimental work in genetics, much of which is not possible in a school laboratory. Questions such as: How do you isolate an antibiotic strain of bacteria? What is the human genome mapping project? How big a risk is radiation? The investigations will help you answer these, and similar, questions.

Summary Assigments

These require you to think about the key points in the chapter and so can be used to build up your understanding of the topic. They can, therefore, be used to compile your own revision notes, or at the end of the course to test your understanding of a chapter.

Additional Reading

A list of other books, identified by the chapter to which they will be relevant, is given at the end of the book. This provides an opportunity to read around a topic in order to understand it in more depth, or to go a bit beyond what is required purely for an examination.

Acknowledgements

The author and publishers wish to thank the following who have kindly given permission for the use of copyright material:

American Association for the Advancement of Science for letter, P. Berg, 'Potential Biohazards of Recombinant DNA Molecules', *Science*, Vol. 185, p 303, 26 July 1974. Copyright © 1974 by the AAAS; Blackwell Scientific Publications Ltd. for material by Greaves and Rennison in *Mammal Review*, Vol. 3, 1973, pp 27-9; Cambridge Local Examinations Syndicate and Joint Matriculation Board for questions from past examination papers; Columbia University Press for material from *The Genetic Basis of Evolutionary Change* by Lewontin. Copyright © 1974 Columbia University Press; Mr. Fothergill's Seeds Ltd. for illustrative material of various seed packets; New Scientist Syndication for material form 'Finger-printing undermines the case for killing whales', 2.6.88; 'How Jeffreys secured a Ghanian boy's right to enter Britain', 28.1.88; 'Europe's researchers design code for gene therapy', 9.6.88; 'Cancer and the viral trigger', 19.5.88, and 'The route to gene therapy' from 'Viruses work to improve their image' by Andrew Scott, 19.5.88; from various issues of *New Scientist;* The Open University for adapted material from S101, *Science Foundation Course,* Chap. 5, p.48. Copyright © The Open University; Paul Parey Verlagsbuchhandlung for material from G. F. Sprague, P. A. Miller and B. Brimall, 'Additional studies on the effecttiveness of two systems of selection for oil content of the corn kernel', *Journal of Agrimony,* 1952, Vol. 44, pp 329-331; Thompson & Morgan for illustrative material from various seed packets; Unwin Hyman Ltd. for material from C. D. Darlington and K. Mather, *The Elements of Genetics,* 1950, Allen & Unwin.

The author and publishers wish to acknowledge, with thanks, the following photographic sources:

Action Images *p 183;* Barnaby's Picture Library *p 6 bottom centre;* Biophoto Associates *p 111;* Dr Bootsma, Erasmus University *p 182 bottom;* Dr Jennet Blake, Wye College *p 79;* Connecticut Agricultural Experiment Station *pp 49 top and bottom, 74;* Cystic Fibrosis Research Trust *p 184;* Dr Joy Delharity, University College, London *p 192;* Eli Lilly Co *p 156;* Farmers Weekly Picture Library *p 66;* Food and Agricultural Organisation of the United Nations *pp6 top left, 33, 52 left and top right, 60 left, 62, 72 left and right, 80 bottom, 101 top right;* Sally and Richard Greenhill *pp 182 top, 193;* Pat Herron/Greenpeace *p 167;* Hutchison Library *p 52 bottom right;* Professor Jeffreys, University of Leicester *pp 163, 164;* Ministry of Agriculture Fisheries and Food *pp 101 left and bottom right, 105;* Muscular Dystrophy Group of Great Britain and Northern Ireland *p135;* Natural History Photo Agency *p 95;* Panos Pictures *p 6 bottom right;* Purdue University Agricultural Communication Service, West Lafayette, Indiana *p 60 right;* Science Photo Library *pp17 left and right, 43, 80 top, 88, 109, 116, 120, 122, 128, 130, 169, 191 insert;* Scottish Society for the Prevention of Vivisection *p 6 top right;* Society for Anglo-Chinese Understanding *p 6 bottom left.*

Every effort has been made to trace all the copyright holders but if any have been inadvertently overlooked the publishers will be pleased to make the necessary arrangement at the first opportunity.

Theme 1

VARIATION – THE SPICE OF LIFE

PREREQUISITES

To complete the work in this theme successfully you will need a working knowledge of basic Mendelian genetics (mono- and dihybrid crosses), linkage, protein synthesis and the structure of DNA.

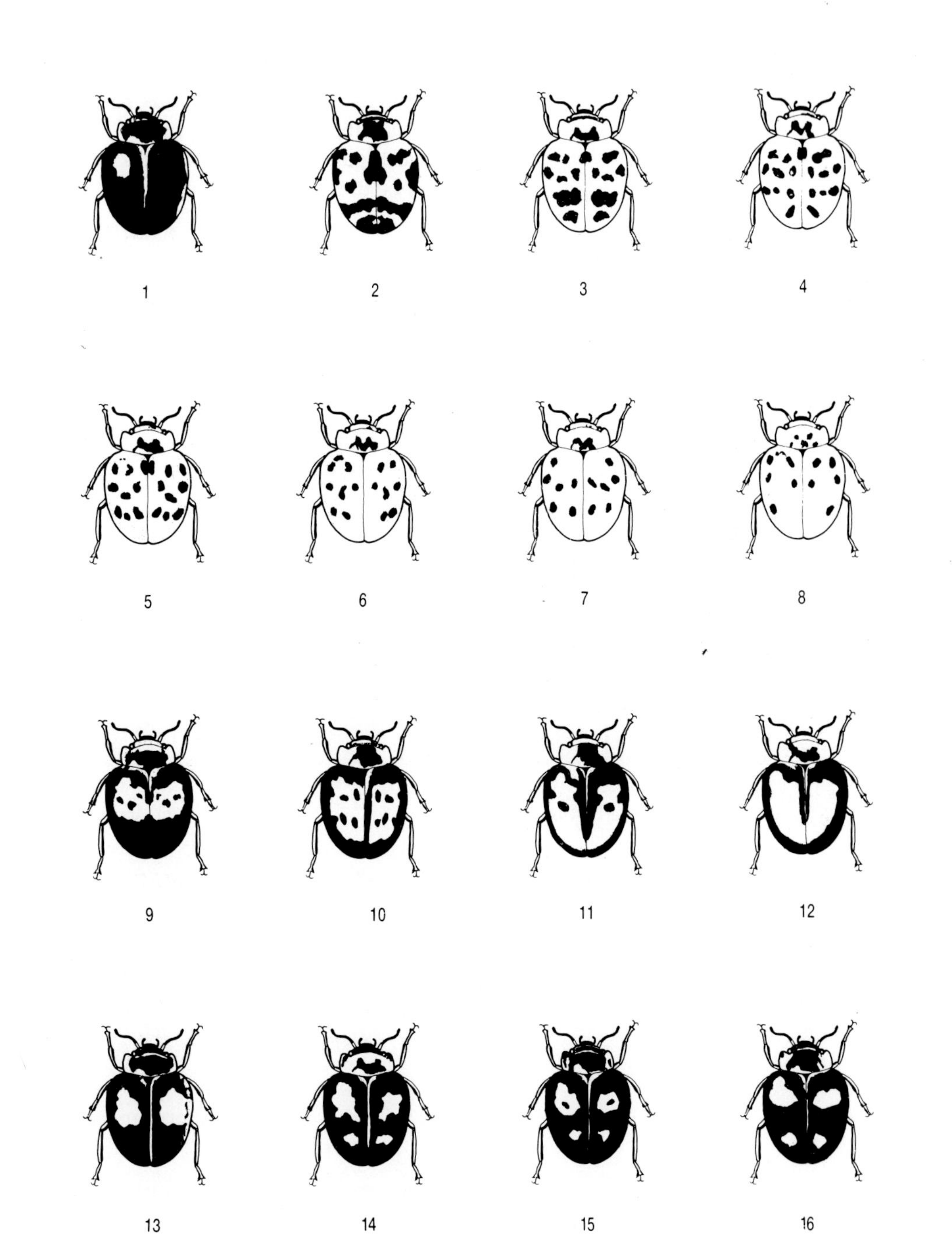

Variation in the colour pattern of *Harmonia axyridis*, a ladybird from eastern Asia.

Chapter 1

INTRODUCING VARIATION

The two graphs shown in Fig 1.1 indicate how the amount of an agriculturally important product, the **yield**, has increased over the years. Some of this increase is undoubtedly due to changes in farming techniques, modern veterinary care and developments in pest control. However, much of the increase has been produced by analysing the **genetic differences** between individuals and then using this information to breed cows which produce more milk, sheep which produce more wool or wheat plants that yield more grain. Such genetic analysis is possible because of the small differences that exist between all organisms. This basic fact of life, **organisms vary**, makes genetics possible. Since variation is so crucial to the geneticist, the whole of this first theme is concerned with analysing and accounting for the origins of variation. In subsequent themes we will investigate how the genetic knowledge gained from analysing variation can be used to help people.

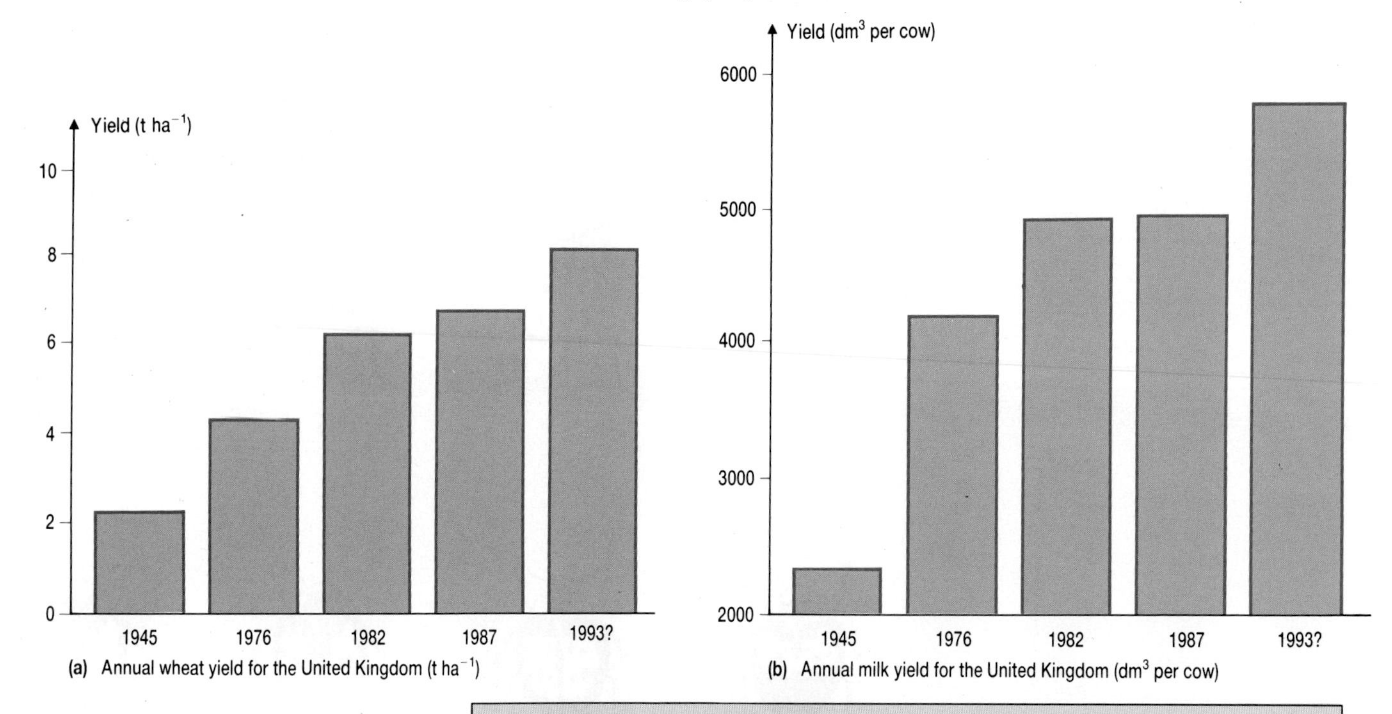

(a) Annual wheat yield for the United Kingdom (t ha^{-1})

(b) Annual milk yield for the United Kingdom (dm^3 per cow)

LEARNING OBJECTIVES

After completing the work in this chapter you will be able to:

1. explain why an organism's phenotype is the result of an interaction between its genotype and the environment;
2. distinguish between clones, pure-bred and inbred lines;
3. outline the technique of gel electrophoresis and explain its importance in determining the amount of genetic variability in a population;
4. explain the role of mutation in producing new genetic variation.

1.1 AN EXPERIMENTAL INVESTIGATION OF VARIATION

The cinquefoil, *Potentilla glandulosa*, is a small plant which grows at a range of altitudes in the Sierra Nevada mountains of California. Cinquefoils growing at different altitudes differ in their appearance, growth rate and fertility. What causes this phenotypic variation? (An organism's **phenotype** is its appearance – its morphology, physiology, behaviour and so on.) Two hypotheses suggest themselves.

1. Plants growing at different altitudes do not contain the same genetic information, that is they have different **genotypes**. Variation in the plant phenotypes is the result of this genetic variation.
2. Plants growing at different altitudes experience different environments. It is this environmental variation which produces the observed differences in the plant phenotypes.

The experiment described below tests these two hypotheses.

Method

1. Individual cinquefoils were collected from three sites at different heights above sea level – Stanford (30 m), Mather (1400 m) and Timberline (3050 m).
2. Each cinquefoil was then divided into three cuttings so producing three genetically identical Stanford plants, three genetically identical Mather plants and three genetically identical Timberline plants.
3. One cutting from each plant was then grown in an experimental garden at each collection site. Each garden therefore contained a cutting from each plant. For example, the Stanford garden at 30 m above sea level contained a cutting from the plant originally collected at the Stanford site plus cuttings from the plants collected at the Mather and Timberline sites.

Results

The results of the experiment are shown in Fig 1.2. Notice the following.

1. Looking first at the plants in each row, which have the same genotype, the results show that plants with a given genotype may have different phenotypes in different environments. This phenomenon, called **phenotypic plasticity**, is particularly pronounced in plants.
2. Comparison of plants in any column shows that in a given environment different genotypes produce different phenotypes.
3. There is no single cinquefoil genotype which survives best in all environments. For example, the plant which thrives at 30 m dies at 3050 m.

Conclusion

The results of this experiment suggest that the phenotype of an organism is the result of the interaction between the organism's genotype and the environment. An organism's genotype does not, therefore, specify absolutely the phenotype. Rather, the genotype determines the range of phenotypes that may develop. The range of phenotypes that may develop from a given genotype is called the range, or **norm of reaction**, of the genotype. Which phenotype actually develops will depend on the environment in which development takes place.

The message that you should remember from this experiment is that an individual organism has a particular phenotype because of the genes it has inherited and the environment which it has grown up in. So in order to understand what produces phenotypic variation we need to consider the effects of both the environment and the genotype on the development of the phenotype.

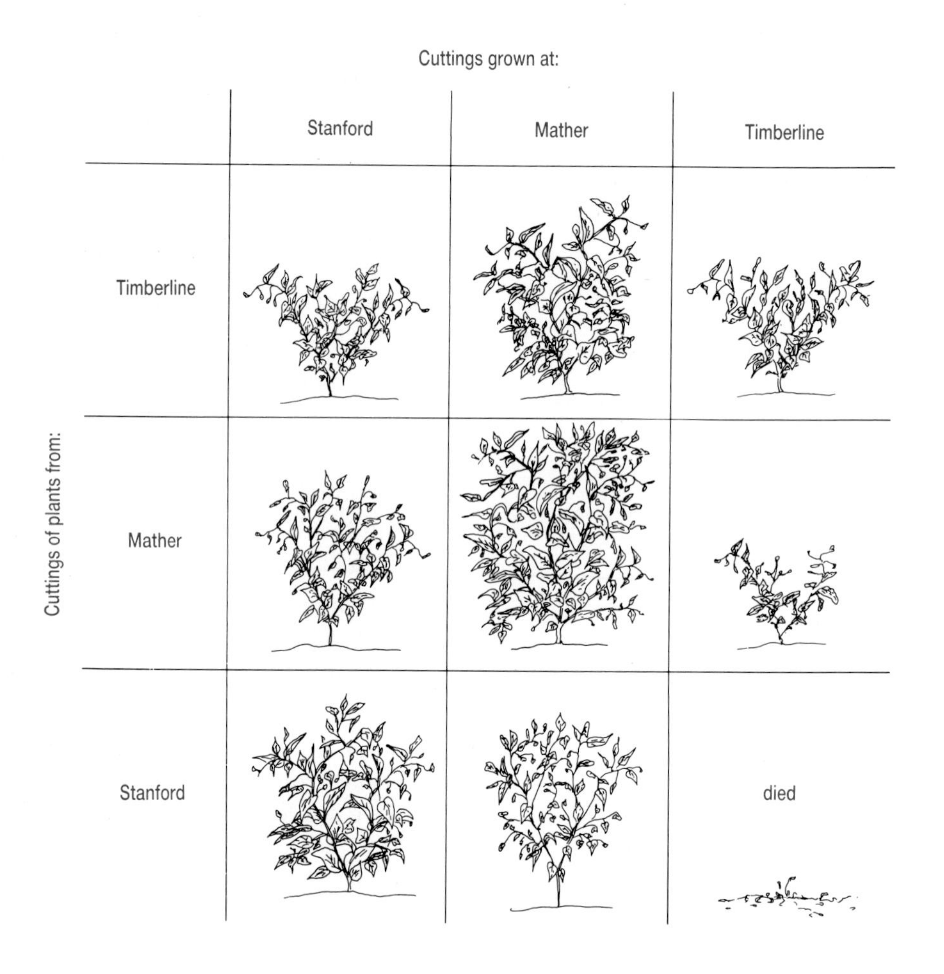

Fig 1.2 Effects of the genotype and the environment on the phenotype. Cuttings from *Potentilla glandulosa* plants collected at different altitudes were planted together in three different experimental gardens. Plants in the same row have the same genotype because they have been grown from cuttings of a single plant. Plants in the same column have different genotypes, but have been grown in the same environment.

1.2 DISTINGUISHING BETWEEN ENVIRONMENTAL AND GENETIC VARIATION

It is often necessary to discover the extent to which the observed differences between the phenotypes of different individuals are due to heredity and the extent to which they are caused by the environment. For example, breeders of agricultural plants and animals are always on the lookout for genetic variants which will give improved yields. Even slight inherited improvements in yield are desirable since they may be expected to recur again and again in the offspring (**progeny**) of an improved variety. However, yields will also be affected by environmental factors such as the quality of the soil, the amount of moisture and the quantity of food supplied. How can we dissect out the contribution made by inherited variation from this background clutter of environmental variation?

The answer is to carry out experiments in which *either* the environment *or* the genotype of the experimental organisms is strictly controlled so that observed **phenotypic differences** can be attributed *either* to variations in the environment *or* to differences in the genotype of the experimental organisms but *not* to both. In order to achieve this, two conditions must be met.

1. The environment must be controlled; experiments must be conducted in similar environments or in environments which differ in known ways.

2. The genotypes of the experimental organisms must be controlled; a supply of individuals with similar genotypes or with genotypes differing in known ways must be obtained or produced for use in experiments.

In practice it is rarely possible to satisfy both of these conditions. Even when organisms are raised under supposedly identical conditions some environmental variation is bound to creep in. For example, think about two wheat plants growing next to each other in an experimental field. They will never receive exactly the same amounts of light, moisture or nutrients. However, most of the problems posed by environmental variation can be overcome by using good experimental design and the appropriate statistical techniques. Such techniques are beyond the scope of this book but usually form part of advanced courses in applied genetics.

The challenge of controlling genetic variation is even more demanding. The best control is achieved in organisms which reproduce asexually, for example by simple fission or by runners and stolons. The progeny produced by asexual reproduction from a single individual are known as a **clone**. Members of a clone are genetically identical.

Some organisms that reproduce sexually are capable of self-fertilisation or **selfing**. Here the female gametes are fertilised by male gametes from the same individual. Self-fertilisation is, of course, only possible when an organism is hermaphrodite. This is, however, a common condition in many plants of agricultural importance, for example wheat, maize, barley, peas and tomatoes. Such organisms can be bred to produce a **pure line**. Members of a pure-bred line are genetically identical and **homozygous** at most **gene loci** (see Box 1.2 if you are unfamiliar with either of the highlighted terms). One of the consequences of this high degree of homozygosity is that all the offspring produced by selfing a pure-bred variety will inherit the same genotype. In other words, pure lines breed true. Perhaps the most famous examples of pure-bred lines were those developed by Mendel for his experiments on inheritance in peas.

QUESTION

1.1 Three varieties of plant have the genotypes
(i) *AAbbccDdee*
(ii) *AABBccDDee*
(iii) *AAbbccddee*.
Which of these plants will be pure breeding? Explain your answer.

Farm and laboratory animals like mice are capable of neither asexual reproduction nor self-fertilisation. So with these organisms it is impossible to obtain either clones or pure-bred lines. Under these circumstances the geneticist has to resort to **inbreeding** to make experimental organisms as genetically similar as possible. Inbreeding involves the mating of closely related individuals, for example brothers and sisters. After several generations of inbreeding the geneticist will have available **inbred lines** which can then be used for experimentation. Whilst the individuals in such inbred lines will not have identical genotypes they will be genetically more similar than the original crossbred lines (Fig 1.3).

QUESTIONS

1.2 The usual method of propagating some plants, for example potatoes, is by cuttings of tubers. What do you call a group of plants produced in this way? Will they necessarily form a pure line?

1.3 Why would it be important to use an inbred line of mice in a drug trial?

Fig 1.3(a) A wild house mouse (*Mus musculus*). This individual will be heterozygous at most gene loci.

Fig 1.3(b) This strain of laboratory mouse has been produced by inbreeding from the wild house mouse and will be homozygous at most gene loci.

1.3 HOW MUCH GENETIC VARIATION IS THERE IN POPULATIONS?

Even a superficial examination reveals the considerable amount of phenotypic variation present among organisms of the same species. For example, look at the ladybirds shown on page 1. This species, *Harmonia axyridis*, from eastern Asia, has at least 16 recognisably different spotting patterns. Look at any group of people (Fig 1.4) and differences in, for example, facial features, eye colour, skin colour and height are immediately apparent. However, it is not obvious to what extent the phenotypic variation we observe is due to genetic or environmental variation.

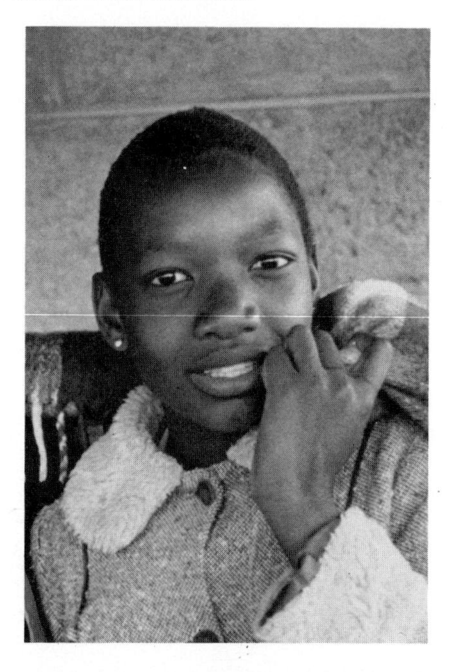

Fig 1.4 Phenotypic variation within the human species.

That at least some of the variation we see is due to genetic differences between individuals can be inferred from the results of artificial selection programmes. In artificial selection the individuals chosen to breed the next generation are those that exhibit the greatest expression of the desired characteristic. For example, if we want to increase the yield (the amount of grain produced) of a particular strain of wheat, we choose, in every generation, the wheat plants with the greatest yield and use their seed to produce the next generation. If, over the generations, the yield of wheat increases whilst the environment remains relatively constant, for example the same amount of fertiliser is used each year, then the original wheat plants had genetic variation with respect to yield.

Artificial selection has been successful for large numbers of commercially desirable **traits** (characteristics) in many domesticated species, including chickens, pigs, cows and many plants. The fact that artificial selection succeeds practically every time it is tried suggests that large amounts of genetic variation exist in populations. This view of genetically diverse populations is further supported by the discovery of a large amount of previously hidden variation among proteins.

Protein variation

The function of many genes is to specify the sequence of amino acids in a protein. So by looking at variation in protein structure we are very close to looking at variation in the structure of DNA itself. Biochemists have known since the early 1950s how to obtain the amino acid sequences of proteins but, even with modern amino acid analysers, obtaining the sequence is a time-consuming business. However, the technique of gel electrophoresis makes the study of protein variation relatively straight forward. This technique is one of the most important in modern genetics and is examined in some detail below.

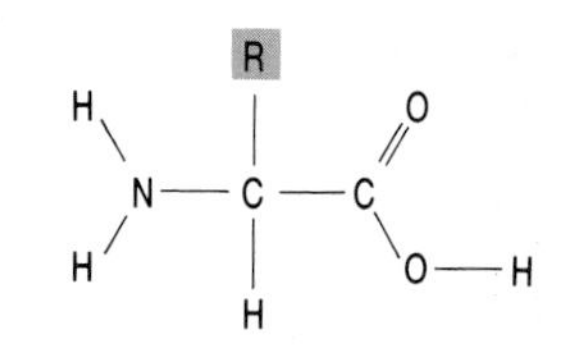

Fig 1.5 The generalised structure of an amino acid.

Gel electrophoresis. All amino acids have the same basic structure shown in Fig 1.5. The differences between the 20 amino acids commonly found in proteins are due to the nature of the R side chain. Some amino acids have side chains which carry an electrical charge at physiological pH, others do not. Some of the charges will be positive, others will be negative. The overall charge on a protein will therefore depend on the balance of amino acids that it contains. If the amino acid sequence is altered then the overall charge on the protein may also change. It is this change in charge which can be detected using electrophoresis. The apparatus and techniques used in electrophoresis are explained in Box 1.1.

BOX 1.1

Gel electrophoresis

The apparatus used in gel electrophoresis is shown in Fig 1.6. The procedure is as follows.

1. Tissue samples from individual organisms are homogenised (ground up) to release the proteins from the cells.
2. The homogenate is spun in a centrifuge and a sample of the supernatant is placed in a sample slot of a gel made of starch, agar or polyacrylamide. Note: each sample slot contains the supernatant from just one individual. The sample slots can be in the middle or at one end of the gel.
3. The gel is then subjected to an electrical field in which each protein in the gel will migrate in a direction and at a rate that will depend on the protein's net charge and its molecular size. The buffer solution

(continued)

maintains a constant pH. Altering the pH of the buffer solution will alter the charge on the protein.

4. The gel is removed from the electric field usually after a few hours and is treated with a chemical substance which stains the specific protein which you are interested in.

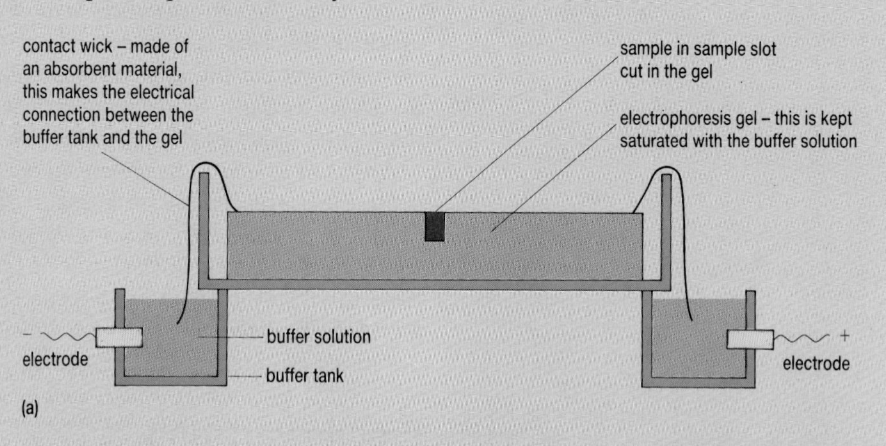

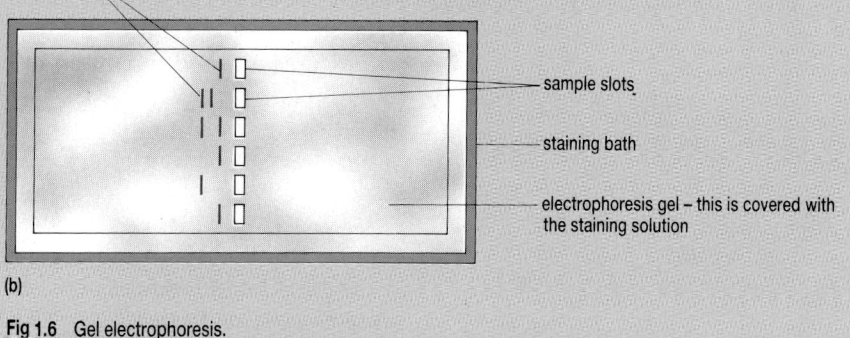

Fig 1.6 Gel electrophoresis.
(a) Running the gel. The diagram shows a cross-section through the apparatus used for the separation of proteins by gel electrophoresis. Note that the position of the sample slot will depend upon the exact nature of the procedure being used.
(b) Staining the gel. After separation the gel is transferred to a tray which is filled with a solution of a specific stain. This stain reveals the position of the protein bands on the gel.

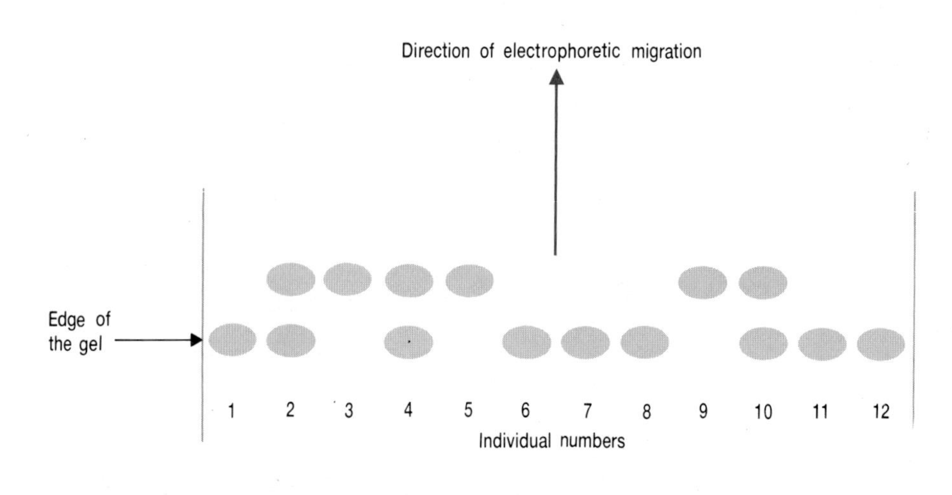

Fig 1.7 A drawing of an electrophoretic gel stained for the enzyme phosphoglucomutase. The gel contains tissue samples from each of 12 females of the fruit fly, *Drosophila pseudoobscura.*

The beauty of the method lies in the fact that the genotype at the locus coding for a protein, frequently an enzyme, can be inferred for each individual in the sample from the number and positions of the spots observed in the gels. Fig 1.7 shows a gel which has been stained to reveal the position of phosphoglucomutase, an enzyme involved in carbohydrate metabolism, from 12 fruit flies (*Drosophila pseudoobscura*). Let us call the gene coding for this protein *Pgm*. Look carefully at the gel. Individuals 1 and 3 have enzymes with different electrophoretic mobilities, in other words, they have moved different distances through the gel. They therefore have different amino acid sequences. Since the sequence of amino acids in the proteins is determined by a corresponding sequence of bases in the DNA, this implies that the proteins are coded for by different forms of the gene, that is, different alleles. If you are unsure about the distinction between genes and alleles then look at Box 1.2.

Let us represent the alleles coding for the enzyme in individuals 1 and 3 as Pgm^{100} and Pgm^{108} respectively. (The superscripts indicate that the form of phosphoglucomutase coded for by Pgm^{108} migrates 8 mm farther in the gel than the form coded for by Pgm^{100}. This is a common way of representing alleles detected using electrophoresis.) Now the first and third individuals in Fig 1.7 each produce only one coloured spot in the gel so we can infer that they are homozygotes with the genotypes $Pgm^{100/100}$ and $Pgm^{108/108}$ respectively.

QUESTION

1.4 Look at individual number 2 in Fig 1.7.

(a) How do you interpret the fact that this individual has produced two spots on the gel?

(b) How would you write its genotype?

How many alleles can a gene have?

The gel shown in Fig 1.7 indicates that the gene encoding for phosphoglucomutase has two alleles, Pgm^{108} and Pgm^{100}. However, electrophoretic studies with a wide range of organisms suggest that many, if not all, genes have multiple alleles, that is, more than two alleles. A familiar example of multiple alleles underlies the ABO blood group phenotypes. The inheritance of ABO blood types is controlled by three alleles at a single locus, as shown in Table 1.2.

The situation in which a gene or a trait exists in two or more forms in a population is called **polymorphism**. Electrophoretic studies have confirmed that most populations are highly polymorphic and that the amount of genetic variation in most natural populations is enormous.

Table 1.1 ABO blood groups in humans

Blood phenotype	Genotype
O	ii
A	I^AI^A or I^Ai
B	I^BI^B or I^Bi
AB	I^AI^B

QUESTIONS

1.5 Give four examples of polymorphism.

1.6 Produce a table which shows the number of genotypes which you can create when a gene has

(a) 2 **(b)** 3 **(c)** 4 **(d)** 5

alleles. Can you decide a mathematical relationship between the number of alleles and the number of genotypes?

BOX 1.2

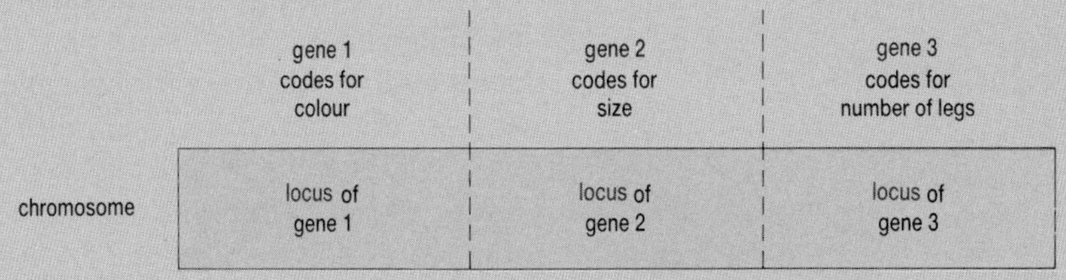

gene 1 codes for colour
gene 2 codes for size
gene 3 codes for number of legs
chromosome
locus of gene 1
locus of gene 2
locus of gene 3

Fig 1.8 The relationship between genes and loci. Each gene occupies a particular locus (place) on a chromosome.

Genes and alleles.

A simple analogy will help you to grasp the difference between genes and alleles. Fig 1.8 shows a chromosome which only carries three genes. Each gene occupies a particular place on the chromosome called a **locus**. So we have a locus for gene 1, a locus for gene 2 and a locus for gene 3. You can think of each locus as a box. Into each box you can put one form of each gene – an allele. Note that each box can only contain one allele at a time. The alleles for the three genes on our chromosome are shown in Table 1.2. So one possible combination of alleles could be:

BLUE LARGE ONE LEG

Table 1.2

Gene	Alleles
colour	blue, orange
size	large, small
legs	one, three

In diploid organisms there are two copies of each gene in a cell carried on a pair of homologous chromosomes. Homologous chromosomes are identical with respect to the gene loci they carry and their visible structure. So the partner to the chromosome shown in Fig 1.8 will also carry genes for colour, size and number of legs. However, the alleles do not have to be the same. Look at Fig 1.9. Here the alleles are different at gene locus 1 and gene locus 2 but the same at gene locus 3. So this particular organism is **heterozygous** for colour and size but **homozygous** for leg number.

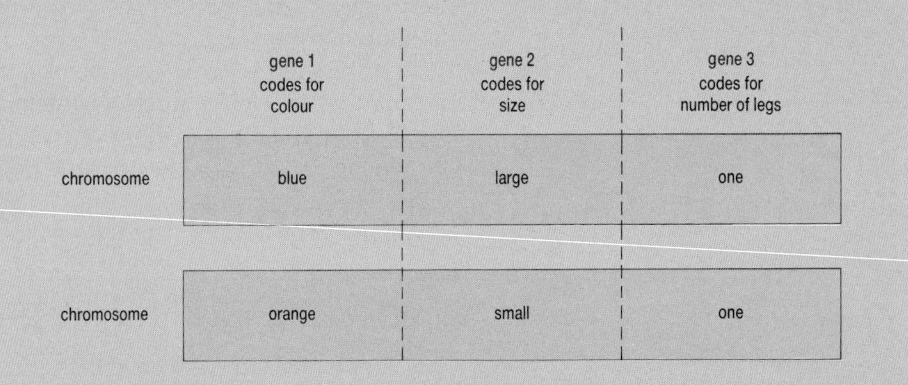

Fig 1.9 The relationship between genes and alleles. Each gene can have more than one allele. This means that a cell containing a pair of homologous chromosomes can have the same allele at both gene loci (homozygous) or different alleles (heterozygous).

1.4 THE ORIGIN OF GENETIC VARIATION – MUTATION

Clearly, an enormous amount of genetic variation exists in populations. Where did this variation come from? To answer this question let us consider only those genes which are involved in the synthesis of proteins. Such genes determine the sequence of amino acids in proteins. If the sequence of amino acids in a polypeptide is altered then its biological properties may also change. In particular, the sequence of amino acids, the primary structure of the polypeptide, determines how the polypeptide folds up and then interacts with other molecules. Enzymes, in particular, are very sensitive to changes in their shape caused by altering their primary structure, since their complex three dimensional configuration is essential in recognising the substrate molecules on which they act. Such a change in the function of a protein, caused by an alteration in the amino acid sequence, may in turn cause a change in the phenotype of the organism. For example, substitution of the amino acid glutamic acid by valine at position 6 in the ß-globin chain of haemoglobin is the cause of the inherited disease sickle-cell anaemia. The crucial question, then, is what determines the primary structure of a polypeptide?

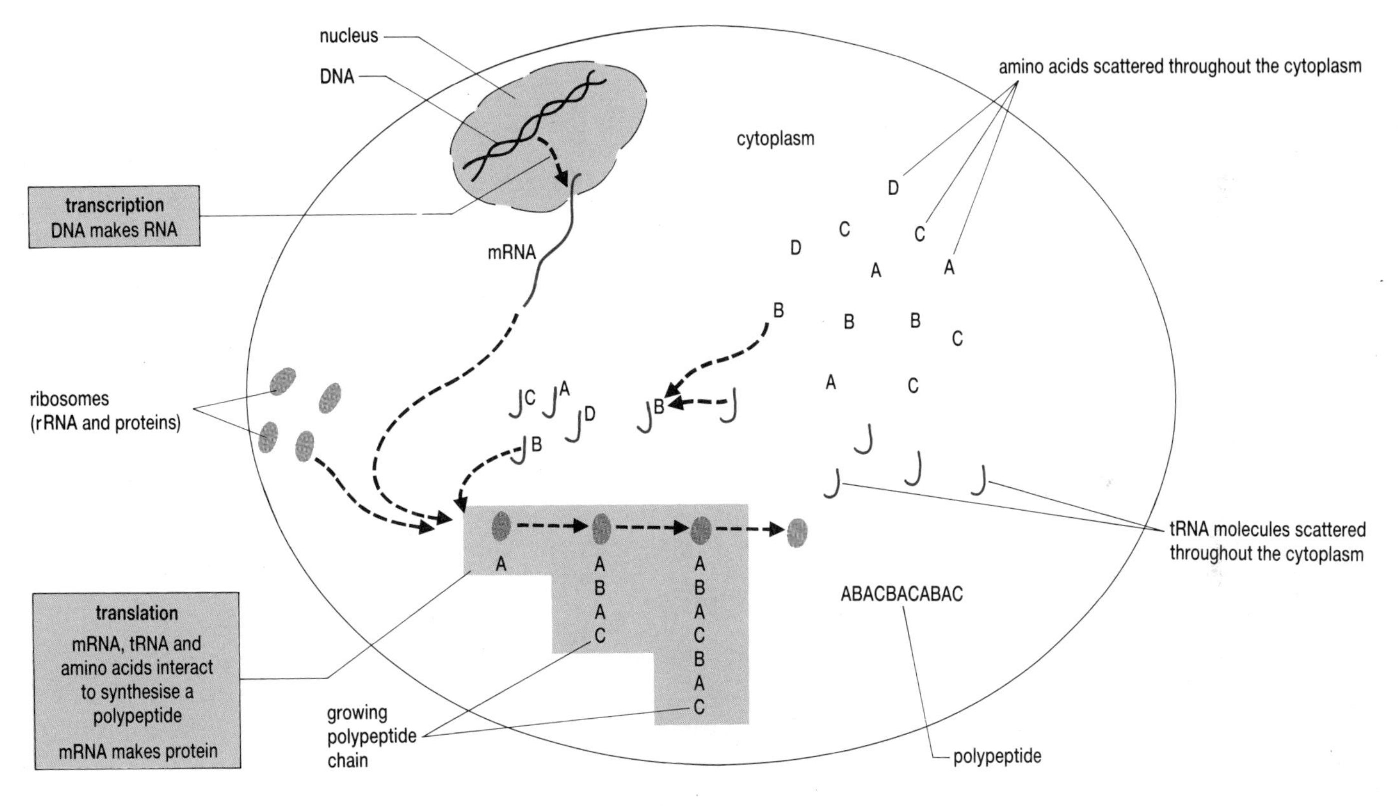

Fig 1.10 A simplified scheme of protein synthesis showing the two main stages: transcription (DNA makes RNA) and translation (mRNA makes protein).

The process of protein synthesis is shown, in outline, in Fig 1.10. Notice that the flow of information is from DNA to protein with the three types of RNA (mRNA, tRNA and rRNA) playing an intermediate role. The amino acid sequence in the protein is determined by the sequence of bases in the particular molecule of mRNA directing protein synthesis on the ribosome. Since the mRNA molecule has been transcribed from the coding strand of the DNA in the nucleus it appears that the DNA carries the code which determines the amino acid sequence in the protein.

So our gene involved in protein synthesis is, in reality, a length of DNA which contains the information required to construct a polypeptide. The information is stored as a code in the form of a sequence of the bases adenine (A), guanine (G), cytosine (C) and thymine (T). A triplet of bases specifies a particular amino acid, for example CTA = asparagine. So the sequence of triplets in the DNA will specify the sequence of amino acids in the polypeptide. If the sequence of bases in the DNA is changed by the process called **mutation**, then the sequence of amino acids may also be altered producing a change in the biological activity of the polypeptide. The effects of mutation are summarised in Fig 1.11.

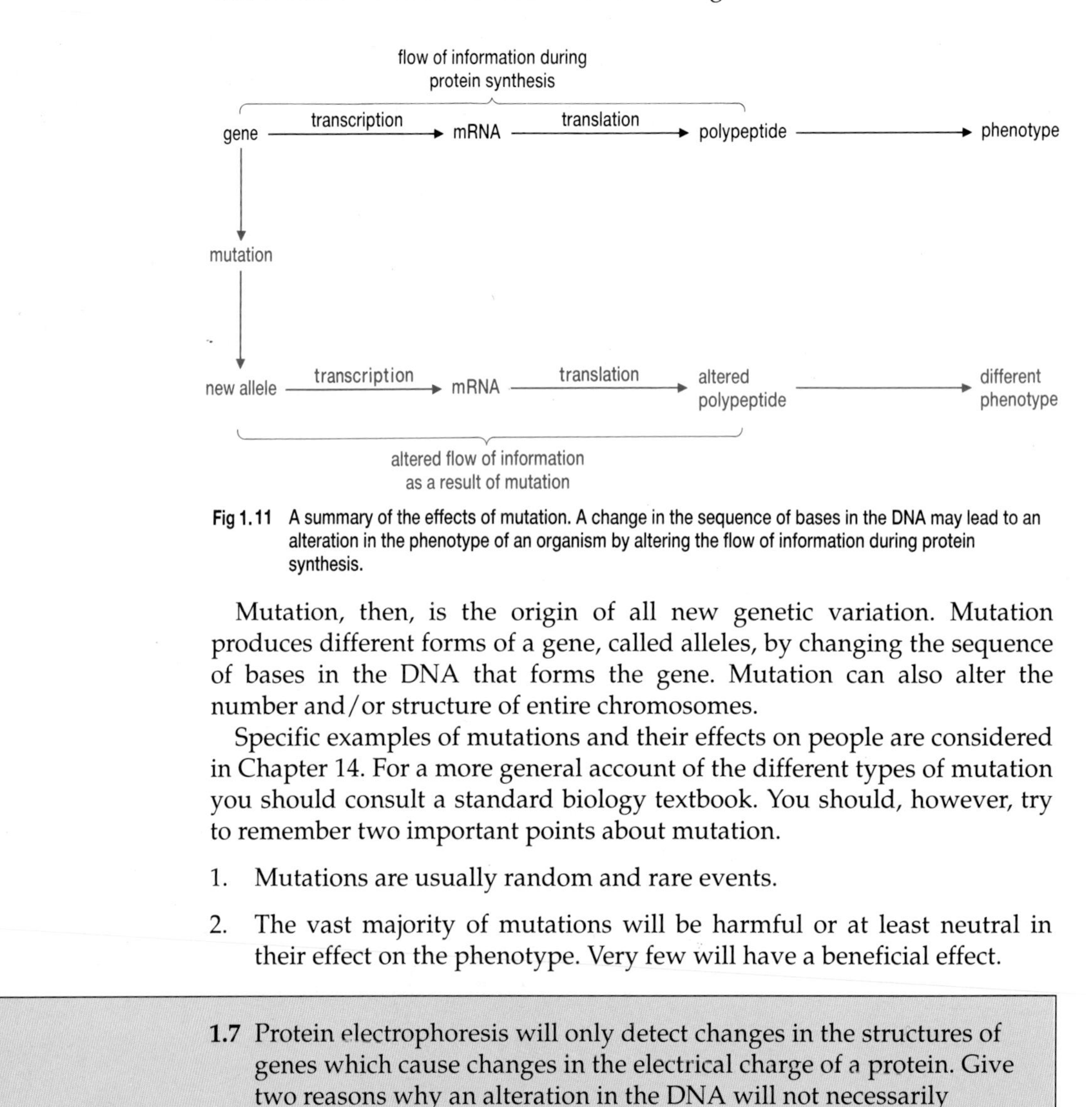

Fig 1.11 A summary of the effects of mutation. A change in the sequence of bases in the DNA may lead to an alteration in the phenotype of an organism by altering the flow of information during protein synthesis.

Mutation, then, is the origin of all new genetic variation. Mutation produces different forms of a gene, called alleles, by changing the sequence of bases in the DNA that forms the gene. Mutation can also alter the number and/or structure of entire chromosomes.

Specific examples of mutations and their effects on people are considered in Chapter 14. For a more general account of the different types of mutation you should consult a standard biology textbook. You should, however, try to remember two important points about mutation.

1. Mutations are usually random and rare events.
2. The vast majority of mutations will be harmful or at least neutral in their effect on the phenotype. Very few will have a beneficial effect.

QUESTION

1.7 Protein electrophoresis will only detect changes in the structures of genes which cause changes in the electrical charge of a protein. Give two reasons why an alteration in the DNA will not necessarily produce such a change in a protein.

SUMMARY ASSIGNMENT

1. Define the following:
   (a) gene **(b)** allele **(c)** locus **(d)** genotype **(e)** phenotype.
2. How do clones, pure-bred and inbred lines differ and why are they essential to genetic research?
3. Summarise the evidence which suggests that most natural populations are genetically variable.
4. What is mutation?

Chapter 2

ESSENTIAL GENETICS

The last chapter illustrated the fact that populations contain a lot of genetic variability. The ultimate source of this genetic variation is mutation which produces the different alleles of genes. You should also remember that most genes have two or more alleles. This means that the number of different combinations of alleles that can be made for a given species is enormous. In this chapter we are going to examine some of the patterns of phenotypic variation which this genetic variability can produce. In particular, this chapter tries to take you beyond the classic Mendelian monohybrid (3:1) and dihybrid (9:3:3:1) ratios, which you should already have met, to explain how interactions between genes can produce other phenotypic patterns. To understand the material in this chapter you will have to think at two levels: at the level of whole animals or plants, looking at their phenotypes in successive generations; and at the molecular level, thinking about what happens to the DNA during reproduction and how alleles interact to produce their effects on the phenotype.

LEARNING OBJECTIVES

After completing the work in this chapter you will be able to:

1. account for the origin of basic and modified Mendelian ratios;
2. solve problems involving dominance, complementary gene action and epistasis.

2.1 THE BASIC RULES

Look at Fig 2.1. You should already be familiar with the genetics of this monohybrid cross. If you are at all uncertain about what is going on, and how the 3:1 ratio in the F_2 has been produced, you should consult a standard biology textbook, for example Rowland's *Biology*. This cross demonstrates the first essential genetic rule.

Rule number 1

The two members of a gene pair **segregate** (separate) from each other into the gametes, so that half of the gametes carry one member of the gene pair and the other half of the gametes carry the other member of the gene pair. A gene pair refers to the pair of alleles for a particular gene present in an individual. In this case *PP* , *Pp* or *pp* are all gene pairs.

Now look at Fig 2.2. Again, you should be familiar with the genetics of this dihybrid cross, which demonstrates the second essential genetic rule.

Rule number 2

During gamete formation the segregation of one gene pair is independent of other gene pairs. In other words, if a gamete receives *R* from one locus, it could receive either *Y* or *y* from the other locus. The obvious exception to this rule is where the genes are **linked**, that is, they are on the same chromosome. This means that the alleles cannot move independently during meiosis, resulting in modified F_2 ratios.

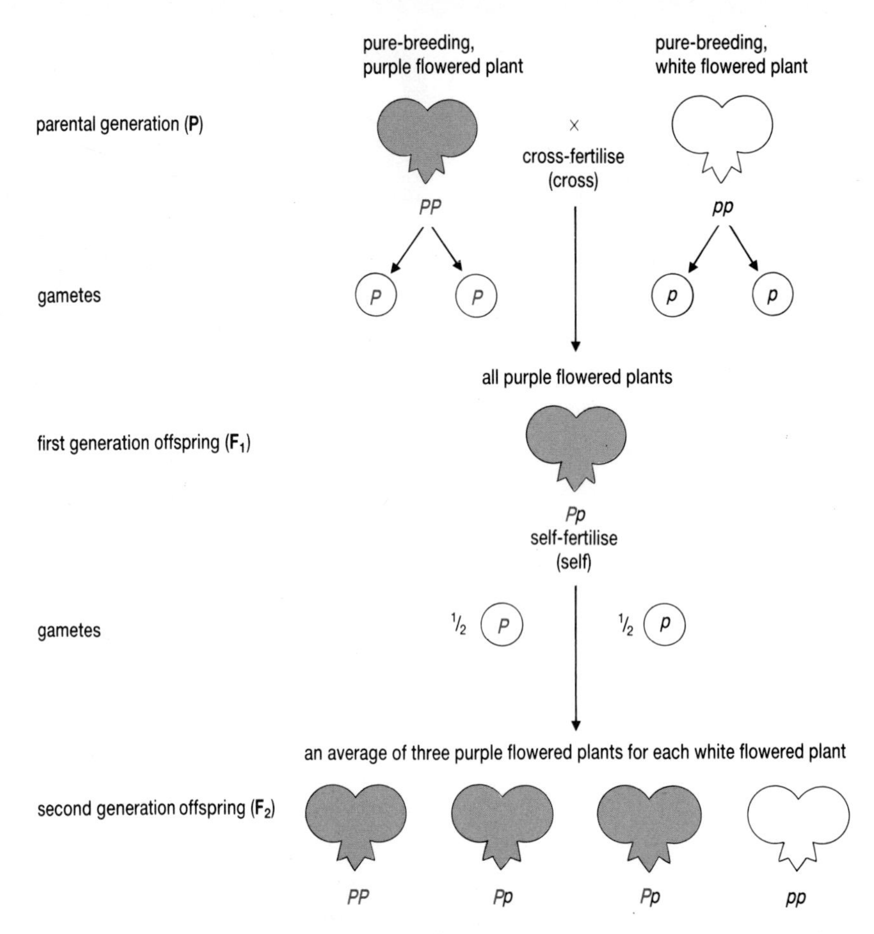

Fig 2.1 A monohybrid cross. Notice that the *P* allele is completely dominant to the *p* allele. During gamete formation the two alleles for flower colour in each parent are separated so that each gamete receives only one allele for flower colour.

You should be able to explain these two basic rules in terms of the movement of chromosomes at meiosis. Again, look at a standard biology textbook if you are unable to do this. These two rules apply across a wide range of eukaryotic organisms. Prokaryotic organisms, like bacteria, have their own genetic systems which are discussed in Theme 3. However, these two rules are only a starting point. In particular, you should realise that they are only concerned with how alleles are passed from parents to offspring. They do not tell us how alleles interact and are expressed in an individual organism's phenotype. Such interactions between genes can lead, as we will see, to modifications of the 3:1 and 9:3:3:1 F_2 ratios.

2.2 INTERACTIONS BETWEEN ALLELES AT A SINGLE LOCUS

Alleles interact constantly with each other to produce their phenotypic effects. In this section we are going to look at interactions between alleles at the same locus giving rise to **dominance** effects.

Complete dominance

Both classical Mendelian ratios (but not the two rules deduced from them in section 2.1) depend upon one allele in each gene pair of the heterozygote being completely dominant over the other allele. Complete dominance is defined as a relationship between two alleles at a single locus where if one is dominant it is expressed in the phenotype of the heterozygote. The other allele, which is not expressed in the phenotype of the heterozygote, is described as being **recessive**. Thus in Fig 2.1 the heterozygote (*Pp*) has the same phenotype as the homozygote (*PP*) because *P* is dominant to *p*. Notice that dominance is a phenotypic effect which results from the interaction between two alleles at a single locus and is *not* a property of an individual allele.

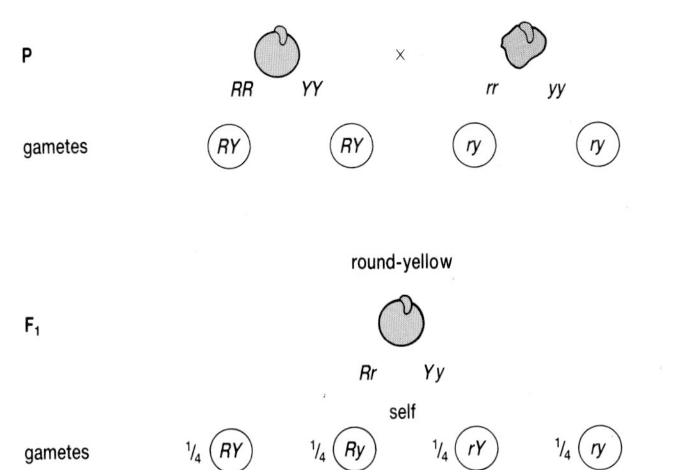

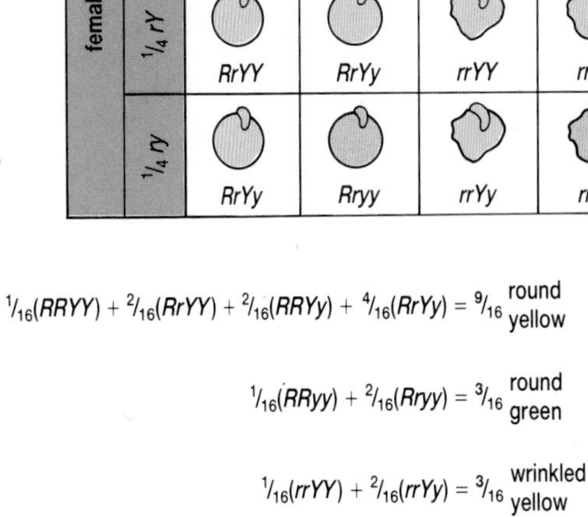

Fig 2.2 A dihybrid cross. (a) Production of the F_1 generation. (b) The F_2 produced by selfing the F_1 generation. Notice that the alleles for seed shape (*R* and *r*) assort independently of the alleles for seed colour (*Y* and *y*). This means that the F_1 plants can produce four kinds of gametes, each with a frequency of $\frac{1}{4}$. It is the random fusion of these gametes, shown in the Punnet square, which produces the characteristic 9:3:3:1 phenotypic ratio of the F_2 generation.

Incomplete dominance

Look at Fig 2.3. This shows the results of a cross between two pure-breeding strains, one red the other white, of the snapdragon, *Antirrhinum*. The F_1 generation all have pink flowers. Selfing the F_1's gives a ratio of 1 red : 2 pink : 1 white in the F_2 generation. This situation, where the heterozygote exhibits a phenotype intermediate between the two homozygous forms, is called incomplete dominance.

Co-dominance

This is a situation in which both alleles are expressed equally in the phenotype of the heterozygote. A good example is provided by the MN blood group system in humans. Three blood group phenotypes are possible – M, N and MN – which are determined by the genotypes $L^M L^M$, $L^N L^N$ and $L^M L^N$ respectively.

These blood group phenotypes indicate the presence of particular antigenic molecules, substances that stimulate the production of a specific antibody, on the surface of red blood cells. For example, people with the genotype $L^M L^M$ have only M antigens on the surface of their red blood cells so their blood group phenotype will be M. Similarly, people with the genotype $L^N L^N$ have red blood cells which carry only N antigens so their blood group phenotype will be N. By contrast, people with the $L^M L^N$

genotype have red blood cells which carry both types of antigen and they have the MN phenotype. This last example suggests that to really understand dominance we need to consider its biochemical basis.

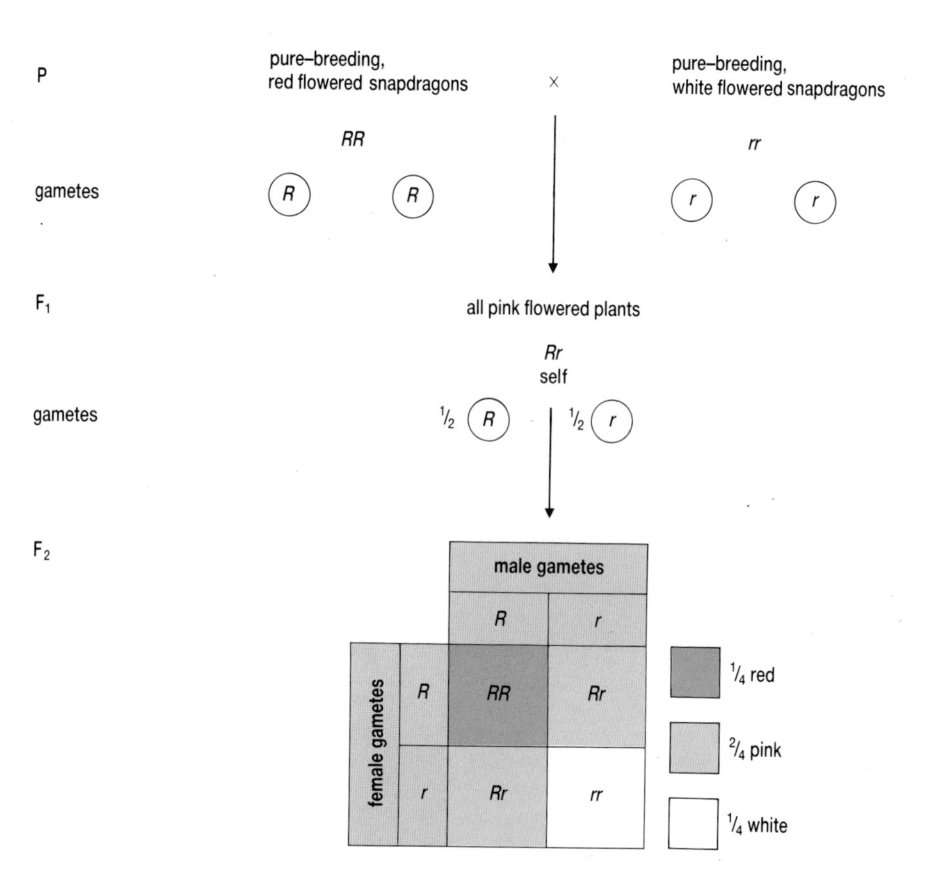

Fig 2.3 A monohybrid cross to show the phenotypic effects of incomplete dominance. Compare the F_1 and F_2 in this cross with those in the cross shown in Fig 2.1.

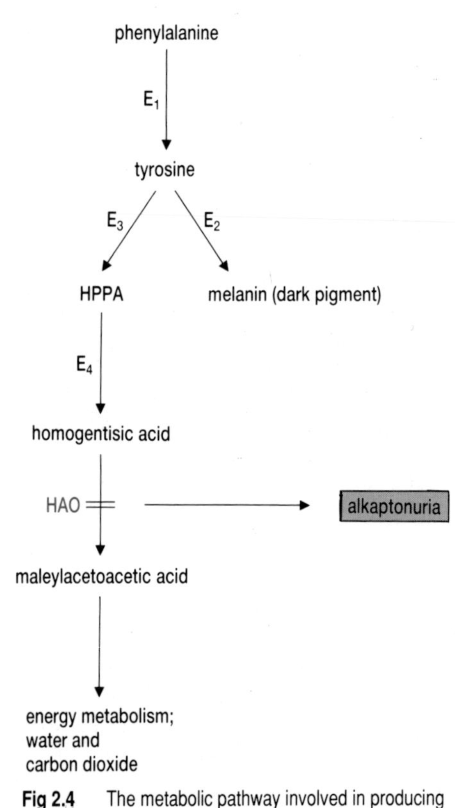

Fig 2.4 The metabolic pathway involved in producing the symptoms of alkaptonuria. E1 to E4 represent enzymes whose substrates and products are indicated above and below each enzyme respectively. HAO = homogentisic acid oxidase; HPPA = hydroxyphenylpyruvic acid.
An individual who produces nonfunctional HAO (indicated by =) will develop the symptoms of alkaptonuria.

The biochemical basis of dominance

Many genes work by controlling chemical reactions by means of enzymes. To illustrate this let us consider the inherited metabolic disorder alkaptonuria. The major symptom of this rare disorder is that the patient's urine turns black on exposure to air because it contains increased levels of homogentisic acid. The causes of alkaptonuria can be explained using Fig 2.4 (don't worry about the names!).

Imagine what would happen if the gene coding for the production of the enzyme homogentisic acid oxidase (HAO) was to undergo a mutation. The resulting allele (call it *a*) now 'produces' HAO which does not work. If you had two copies of this allele, that is, you were homozygous recessive (*aa*), then your cells would be unable to synthesise any functional HAO, the level of homogentisic acid in your urine would increase and you would be suffering from alkaptonuria.

So we now have a biochemical definition of dominant and recessive. A dominant allele is one which usually codes for a functional product, such as an enzyme, and which even when present in only one copy (the heterozygous condition) is capable of synthesising enough product to produce a phenotype identical to that produced when there are two copies of the dominant allele in the cell. In contrast, a recessive allele usually directs the synthesis of a nonfunctional product.

Now consider the situation where both alleles at a locus produce functional products. If the two alleles code for different products then both products will appear in the heterozygote. The MN blood group is an example of this – the relationship we called co-dominance.

Incomplete dominance in the heterozygote of the snapdragon can be explained by postulating that the allele *R*, when present in only one copy per cell, produces insufficient enzyme to synthesise enough red pigment. Consequently the heterozygotes are pink.

QUESTIONS

2.1 Explain why heterozygous individuals (*Aa*) do not suffer from alkaptonuria.

2.2 Table 2.1 gives details of some other inherited metabolic disorders. Using Fig 2.4, identify the defective enzyme involved in each of the disorders.

Table 2.1

Disorder	Symptoms
phenylketonuria	brain damage; increased level of phenylalanine in the blood
albinism	lack of melanin pigment in skin, eyes and hair
tyrosinosis	increased levels of HPPA

2.3 Assume that in a diploid organism hairiness is due to a threshold effect such that more than 50 units of 'hairy factor' will result in a hairy phenotype and less than 50 will result in a bald phenotype. Allele *H* is functional so that each *H* allele contributes 40 units of hairy factor, thus *HH* homozygotes will have 80 units and will be phenotypically hairy. A mutant allele, *h*, arose which was nonfunctional, contributing no hairy factor at all.
(a) Which allele will show dominance, *H* or *h*?
(b) Are functional alleles necessarily always dominant?

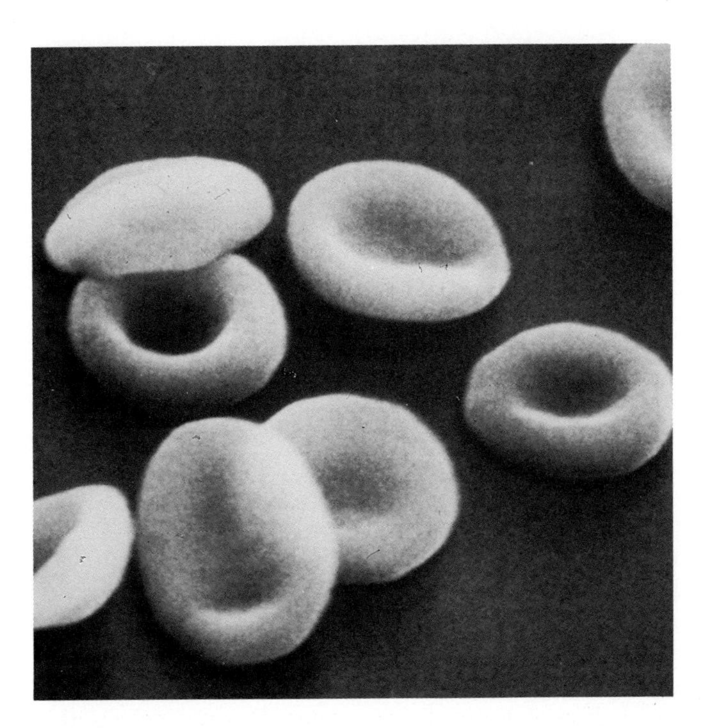

(a) Normal.

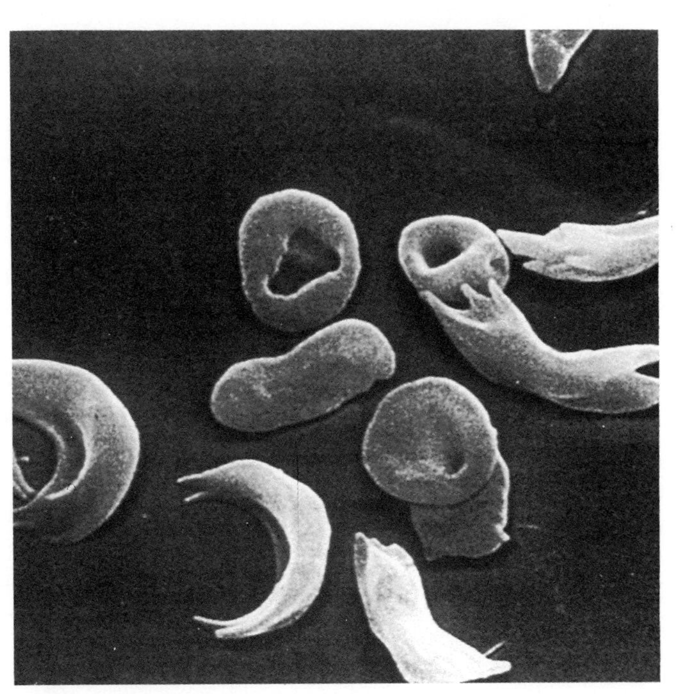

(b) Sickled red blood cells.

Fig 2.5 Electronmicrograph of red blood cells.

Sickle-cell anaemia

This genetic disease of humans provides an interesting example of dominance. The alleles involved affect haemoglobin. The three genotypes have three different phenotypes as shown below.

Hb^AHb^A	:	Normal. Red blood cells never sickled.
Hb^SHb^S	:	Severe, often fatal anaemia. Red blood cells sickle-shaped.
Hb^AHb^S	:	No anaemia. Red blood cells sickle only under abnormally low oxygen concentrations.

An example of sickle cells is shown in Fig 2.5. Sickle-cell anaemia demonstrates the arbitrary nature of the distinction between complete dominance, incomplete dominance and co-dominance.

- With regard to anaemia the Hb^A allele is dominant to the Hb^S allele.
- With regard to blood cell shape the Hb^A allele is incompletely dominant to the Hb^S allele.
- With regard to haemoglobin (Fig 2.6) there is co-dominance of the Hb^A and Hb^S alleles.

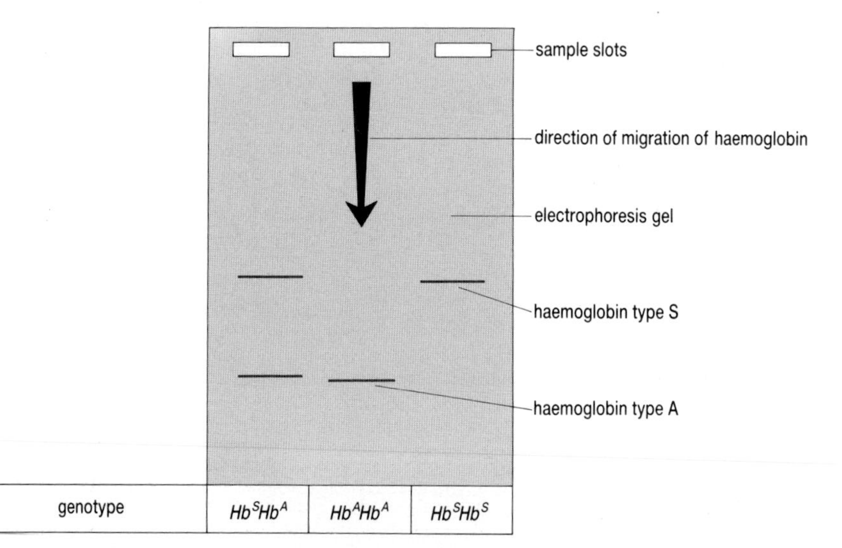

Fig 2.6 An electrophoresis gel stained to show haemoglobin. The position of the different bands reflects the genotypes of individuals who vary with respect to the sickle-cell trait.

In this case the type of dominance effect that we observe in the phenotype depends on the phenotypic level at which the observation is made – organismal (anaemia), cellular (sickle-cell shape) or molecular (haemoglobin).

Conclusion

Dominance is clearly a complex topic but the following two points should be remembered.

1. Complete dominance, incomplete dominance and co-dominance are phenotypic effects which result from the interaction between two alleles at a single locus and are not properties of individual alleles.

2. Dominance is not inherited. Remember, dominance is a property of a phenotype produced as a result of the interaction between two alleles located on a pair of homologous chromosomes. During meiosis this pair of homologous chromosomes will separate so that each gamete contains only one copy of each gene, thus destroying the interaction which produced the observed dominance effect.

2.3 INTERACTIONS BETWEEN GENES AT DIFFERENT LOCI

In the dihybrid cross shown in Fig 2.2 each of the gene loci controls a separate character – seed shape and seed colour. However, the development of a single character may often be due to the interaction between alleles at several loci. In the examples discussed below we will only consider situations where two different loci are involved in the development of a single character. Under these circumstances the exact nature of the interaction is often revealed by modified Mendelian ratios in the F_2 generation. However, the development of many characters is influenced by the interaction of alleles at many different loci. For example, coat colour in mice is determined by the interaction of at least five different genes.

Complementary gene action

It is possible to obtain two different pure-breeding varieties of sweet pea both of which have white flowers. Since flower colour in sweet peas is controlled by two loci you might expect a standard 9:3:3:1 dihybrid ratio of flower colours in the F_2 generation obtained from crossing these two white varieties. However, if we cross these two varieties, the F_1 generation all have purple flowers. Selfing the F_1s produces F_2 progeny in the ratio of 9 purple to 7 white.

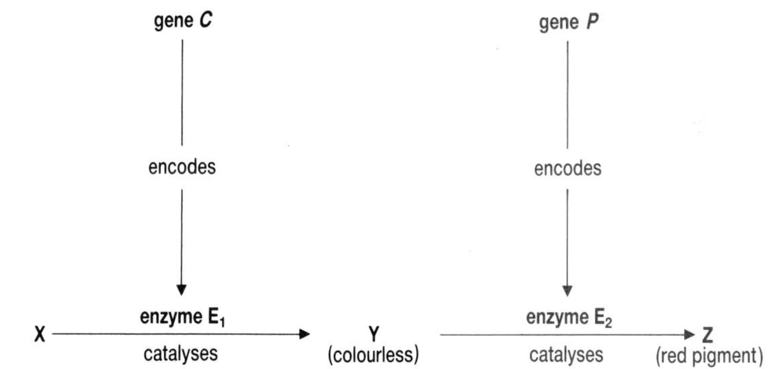

Fig 2.7 The interaction of gene products, showing complementary gene action, in the metabolic pathway producing a flower pigment in sweet peas. Functional enzymes encoded by both gene *C* and gene *P* are essential for the development of red flowers.

Again, we need to consider gene expression at a biochemical level to understand these results. Fig 2.7 shows the biochemical pathway leading to the development of the purple flower colour. The compounds X and Y are both colourless. Conversion of X into Y is catalysed by an enzyme designated E_1 whilst conversion of Y into the purple pigment Z is dependent on the enzyme E_2. The synthesis of functional E_1 and E_2 is directed by the alleles *C* and *P* respectively, whilst the alleles *c* and *p* code for the production of nonfunctional E_1 and E_2. So a plant which has the genotype *CCpp* will be white since it cannot produce functional E_2 and cannot, therefore, convert Y into the purple pigment. A sweet pea which is *ccPP* will also be white since it cannot produce functional E_1 and hence cannot produce Y which is essential for the synthesis of the purple pigment.

QUESTIONS

2.4 Given that the genotypes of the two varieties of pure-breeding white sweet peas described above are *CCpp* and *ccPP*, explain
- **(a)** why the F_1 generation produced by crossing the two white varieties are all purple;
- **(b)** the 9:7 ratio of purple to white flowers observed in the F_2 generation.

2.5 If the initial cross had been between parental sweet pea plants with the genotypes *CCPP* and *ccpp*, what phenotypic ratios of flower colour would have been observed in the F_1 and F_2 generations?

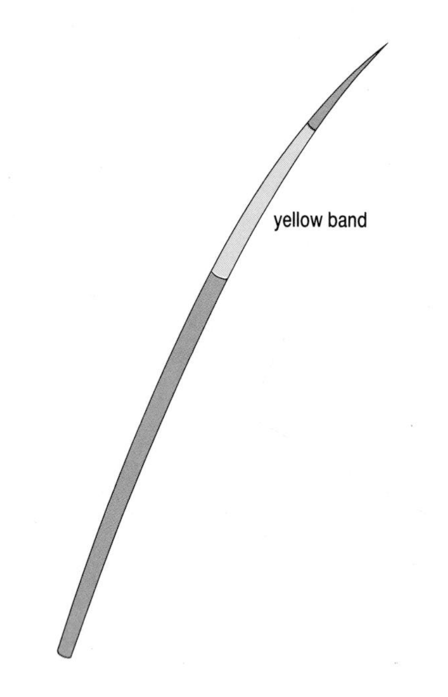

Fig 2.8 A single guard hair from an agouti mouse. The interaction between the yellow band and the black background of the remainder of the hair is responsible for the 'mousey' colour of wild mice.

Epistasis

This is defined as an interaction between two different genes so that an allele of one of them (the **epistatic** gene) interferes with, or even inhibits, the phenotypic expression of the other (the **hypostatic** gene). For example, in the sweet pea example described above, *cc* is epistatic to the *P* locus because *cc* prevents the phenotypic expression of genes at the *P* locus. Similarly, *pp* is epistatic to the *C* locus.

Recessive epistasis. Coat colour in mice provides another example of epistasis. The fur colour of wild mice is described as agouti and is the result of a yellow band around an otherwise black hair (Fig 2.8). Individuals with the genotype *AA* or *Aa* are agouti whilst individuals with the genotype *aa* have black fur. So agouti is dominant to black.

Another, independently inherited, gene is required for the synthesis of hair pigment. Again the gene has two alleles, with colour expression (*CC* and *Cc*) being dominant to non-expression (*cc*). So an individual with the genotype *AAcc* will be albino because *cc* is epistatic to the colour gene.

This recessive epistasis produces a modified F_2 ratio of 9:3:4 from a cross between true-breeding agouti and albino mice, as shown in Fig 2.9.

QUESTION

2.6 What are the phenotypes of mice with the following genotypes:
(a) *AaCc* **(b)** *AACc* **(c)** *aaCC* **(d)** *Aacc* ?

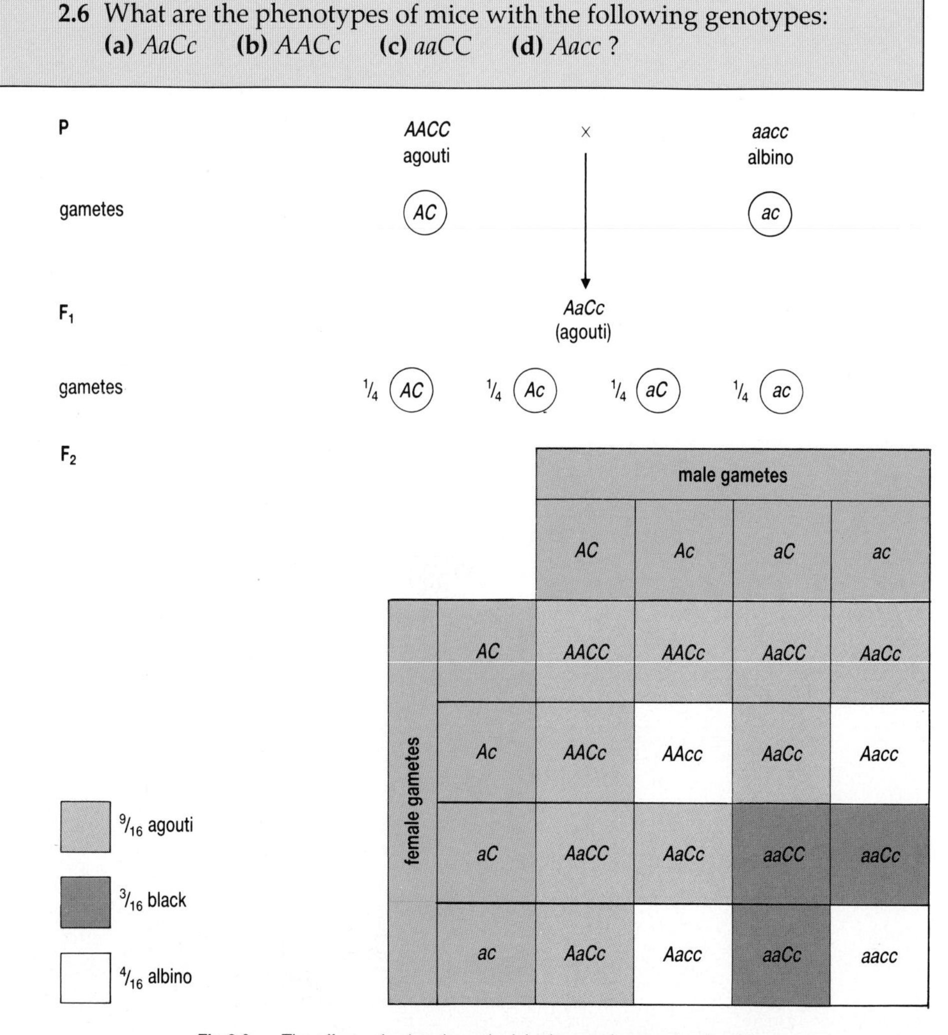

female gametes \ male gametes	AC	Ac	aC	ac
AC	AACC	AACc	AaCC	AaCc
Ac	AACc	AAcc	AaCc	Aacc
aC	AaCC	AaCc	aaCC	aaCc
ac	AaCc	Aacc	aaCc	aacc

Fig 2.9 The effects of epistasis on the inheritance of coat colour in mice.

Dominant epistasis. This phenomenon is illustrated by the inheritance of feather colour in chickens. Here there is a gene, *I*, that is epistatic to the colour gene, *C*. Individuals that carry the dominant allele *I* will have white plumage even if they are carrying the dominant allele *C* for colour. For example, an individual with the genotype *IiCC* will be white. Notice that individuals which are homozygous recessive for the colour gene (*cc*) will also be white.

QUESTION

2.7 White Leghorn chickens have white feathers because their genotype is *IICC*. White Wyandotte chickens also have white feathers but their genotype is *iicc*. Work out the expected phenotypes of the F_1 and F_2 generations produced by crossing white Leghorns with white Wyandottes. Explain how you have reached your conclusion.

Conclusion

- A summary of the modified dihybrid Mendelian ratios produced by the gene interactions described above are shown in Table 2.2.
- Epistatic interactions, like dominance interactions, are not inherited.
- The effects of both dominance and epistasis are to reduce the number of different phenotypes appearing in the F_2. In other words, dominance and epistasis reduce phenotypic variation.

In Table 2.2 the – means that the second locus can be occupied by any allele of the gene. For example *A–* could be *AA* or *Aa*, *b–* could be *bb* or *Bb*.

Table 2.2 A summary of the modified F_2 dihybrid Mendelian ratios produced by the various types of gene interaction

Type of gene interaction	Genotypes *A–/B–*	*A–/bb*	*aa/b–*	*aa/bb*
dominance	9	3	3	1
complementary gene action	9	7		
recessive epistasis by *aa* of *B/b* genes	9	3	4	
dominant epistasis by *A* of *B/b* genes	12		3	1
	Phenotypic ratios			

QUESTIONS

2.8 When pure-bred brown cats are mated with pure-bred white cats, all the F_1 kittens are white. When $F_1 \times F_1$ crosses are made the results are 118 white, 32 black and 10 brown kittens. What is the genetic basis for these results?

2.9 In poultry the genes for rose comb, *R*, and pea comb, *P*, when present together, produce walnut comb. Individuals homozygous for the recessive alleles at both loci (*rrpp*) have single combs. The two pairs of genes are inherited independently.

(a) Determine the kinds and expected proportions of the phenotypes expected in the progeny of the following crosses:
(i) *RRPp* (walnut) × *rr Pp* (pea)
(ii) *RrPp* (walnut) × *RrPp* (walnut)
(iii) *RrPp* (walnut) × *Rrpp* (rose)
(iv) *Rrpp* (rose) × *rrPp* (pea)

(b) What are the genotypes of two parents, one walnut and the other single, that produced $\frac{1}{4}$ single, $\frac{1}{4}$ rose, $\frac{1}{4}$ pea and $\frac{1}{4}$ walnut in the F_1 progeny?

(continued)

2.10 In sweet peas, alleles *C* or *P* alone produce white flowers, the purple colour being due to the presence of both these genes. What will the expected ratio of flower colour in the offspring from the following crosses be?

(a) *CcPp* × *ccPp*
(b) *ccPp* × *CCpp*
(c) *CcPp* × *CcPP*
(d) *Ccpp* × *ccPp*

2.11 The compound malvidin is found in certain species of plant of the genus *Primula*. If two pure-breeding strains, one lacking malvidin and the other having it, are crossed, the F_1 plants are found to have no malvidin. Crossing the F_1 plants produces plants that lack malvidin and plants that have it in a ratio of 13:3. Suggest an explanation for these results.

SUMMARY ASSIGNMENT

1. Identify and give examples of different kinds of gene interaction. What are the differences and similarities between epistasis and dominance?
2. Give a biochemical explanation, along the lines of that given for complementary gene action, for recessive and dominant epistasis.

Chapter 3

QUANTITATIVE GENETICS

So far we have only considered phenotypic characteristics which fall into clearly distinguishable categories. Sweet peas are white or purple, you are blood group O or blood group A, male or female. Such traits are said to show **discrete, discontinuous** or **qualitative variation**. However, such characteristics only form a small fraction of naturally occurring variation. Think about the heights of your friends. Do they fall into distinct categories? Can you classify one group as short or tall? Variation of this sort, without obvious categories, is called **continuous variation** and characters that exhibit it, for example height, are called **quantitative** or **metric characters** because their study depends on **measurement** rather than on **counting**. The branch of genetics concerned with analysing continuous variation is called **quantitative genetics**. The importance of this branch of genetics cannot be overemphasised; most of the characters of interest to animal and plant breeders, such as milk production, growth rate, yield and so on, are quantitative characters.

LEARNING OBJECTIVES

After completing the work in this chapter you will be able to:

1. recognise and give examples of characters showing continuous and discontinuous variation;
2. explain the inheritance of quantitative characters in terms of a polygenic model of heredity;
3. solve problems involving quantitative inheritance.

3.1 INVESTIGATING THE INHERITANCE OF QUANTITATIVE CHARACTERS

The main differences between quantitative and qualitative characters are summarised in Table 3.1. An example of a quantitative character is shown in Fig 3.1. Notice that the frequency diagram is roughly bell shaped with the frequencies being highest in the middle and falling off towards both ends. If the class intervals were made smaller and we were to measure an infinite number of individuals, the histogram would look like a normal distribution curve. So the frequency diagram for most quantitative characters appears to approximate to a normal distribution curve.

We can explain the inheritance of characteristics showing discrete variation using the rules outlined in section 2.1. Does the inheritance of characteristics which show continuous variation also obey these rules? The answer is yes, but quantitative characters are under the influence of many **polygenes** rather than just one or two **major genes**. Polygenes are not a separate type of gene. Rather polygenes are simply genes which produce small phenotypic differences whilst the major genes we looked at in the last chapter produce large phenotypic differences. To understand how polygenes can produce a characteristic showing continuous variation we need to develop a model of polygenic inheritance which will generate a frequency distribution like the one in Fig 3.1.

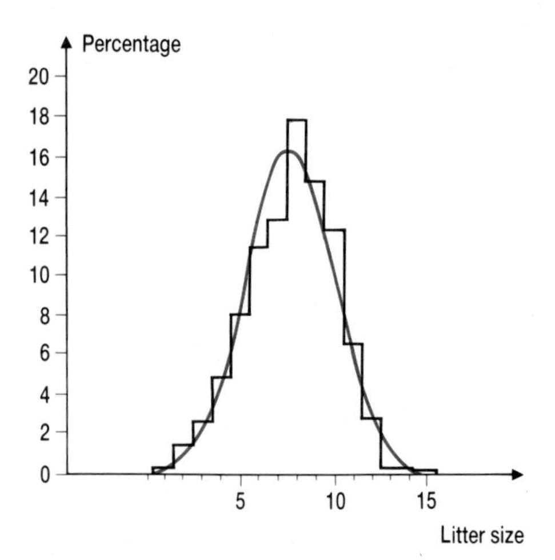

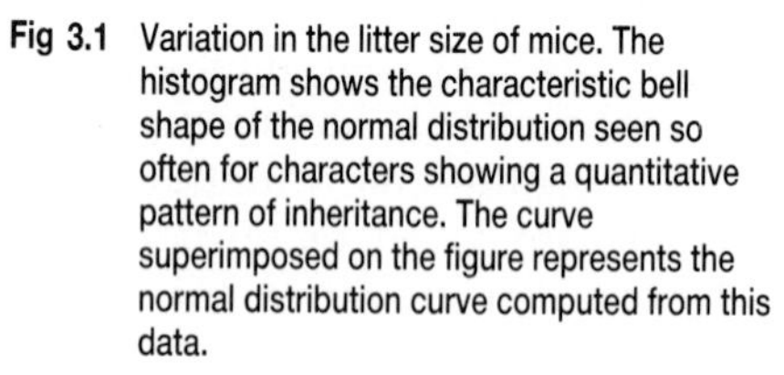

Fig 3.1 Variation in the litter size of mice. The histogram shows the characteristic bell shape of the normal distribution seen so often for characters showing a quantitative pattern of inheritance. The curve superimposed on the figure represents the normal distribution curve computed from this data.

Table 3.1 A comparison of qualitative and quantitative characters

Qualitative	Quantitative
discontinuous variation giving discrete phenotypic classes	continuous variation giving a range of phenotypes
effect of individual genes can be observed	effect of individual polygenes cannot be observed. Effect of polygenes is additive
the environment has a small effect on the appearence of the phenotype	the environment has a large effect on the phenotype
mechanisms of inheritance investigated by counting and comparing ratios in the offspring	mechanisms of inheritance investigated using statistical methods

Getting started – Andalusian fowl

Andalusian fowl are a variety of chicken which come in three different colours: black, white and 'blue' (actually grey!). If a black bird is crossed with a white bird all the offspring are blue. If we now cross lots of blue birds and look at their offspring we get the ratio 1 black : 2 blue : 1 white, as shown in Fig 3.2.

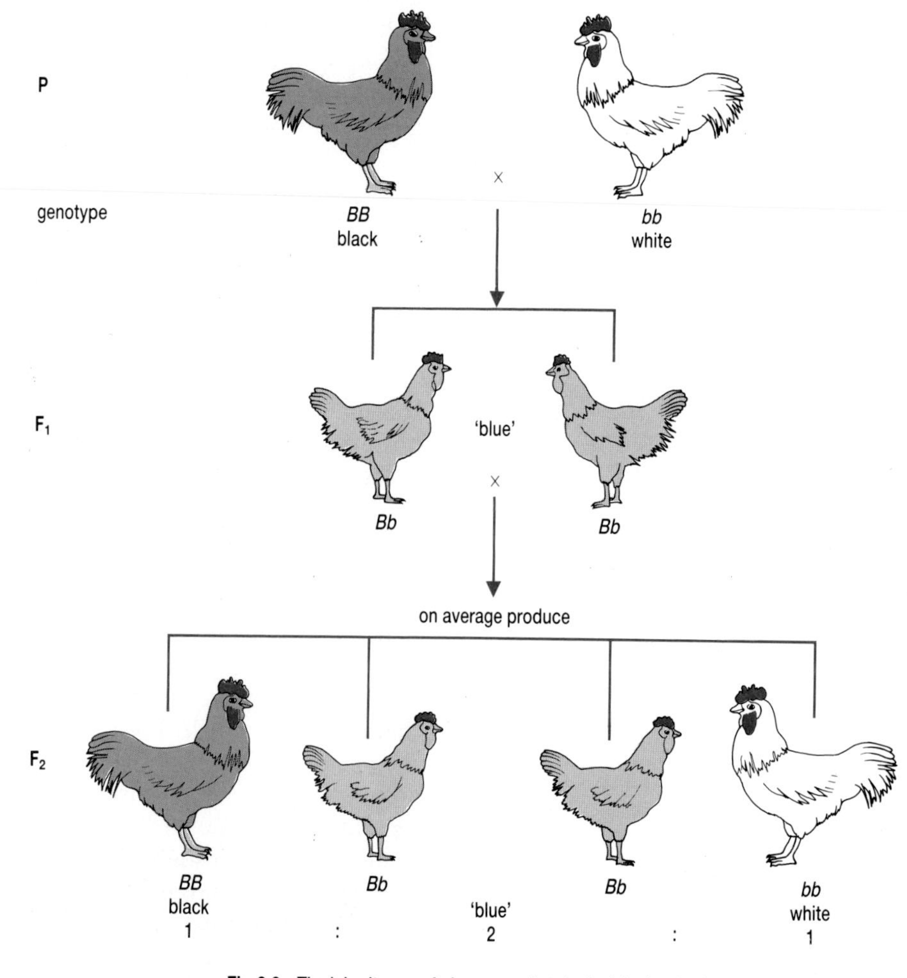

Fig 3.2 The inheritance of plumage colour in Andalusian fowl.

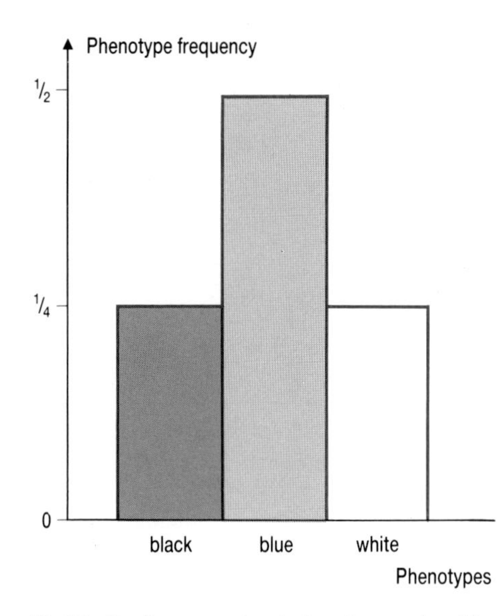

Fig 3.3 The frequency of each phenotype produced in the F_2 of a cross between a black and white Andalusian fowl.

Let the gene for feather colour = *F*. This gene has two alleles, F^B and F^W. So a black Andalusian fowl is F^BF^B whilst a white one is F^WF^W and a blue one is $F^B F^W$. Assume that the F^B allele codes for the production of 1 unit of feather pigment. Crossing the hybrids we get:

Parents	F^BF^W	×	F^BF^W
Progeny	1 F^BF^B :	2 F^BF^W :	1 F^WF^W
Units of feather pigment	2	1	0

In other words, the effects of the alleles are **additive**. Each allele makes a specific measurable contribution to the phenotype. If the bird has two F^B alleles then it has 2 units of feather pigment and is therefore black. If it has no F^B alleles it has no units of feather pigment and so it is white. 'Blue' birds receive 1 unit of feather pigment and so they are grey. Plotting the frequency of each phenotype produced from our hybrid cross we get Fig 3.3, which is a sort of 'stepped' bell shape. We have produced this distribution using only one locus and two alleles. What will happen to the distribution of phenotypes if we increase the number of loci affecting the character?

3.2 POLYGENES AND QUANTITATIVE INHERITANCE

Towards a model of quantitative inheritance

Let us now investigate an *imaginary* genetic system which controls the length of corn cobs. For now we will assume that the environment has no effect on the size of cobs so that all the phenotypic variation we see is due to genetic variation. You should of course realise this is a ridiculous assumption, but nonetheless it will help us to understand how quantitative inheritance works. Scientists, particularly physicists, often make simplifying assumptions if it will help them get to the root of a problem. Indeed, this is a fundamental step in building scientific models.

Assume that the length of corn cobs is controlled by a pair of genes, *A* and *B*. Since maize is diploid there must be a pair of alleles for gene *A* and a pair for gene *B*. Let us call the alternative alleles at each locus *L* and *S*. The contribution made by each allele to the length of the cob is shown in Table 3.2. What is the length of cobs produced by a plant which has the genotype $A^SA^SB^SB^S$? Since each *S* allele will contribute 1.5 cm to the overall length of the maize cob the length of the cobs will be 1.5 cm + 1.5 cm + 1.5 cm + 1.5 cm = 6 cm. Again, the alleles are acting additively. The phenotype is determined by adding the effects of each allele at the two loci. If we add an *S* allele then the phenotype changes by a fixed amount, in this case 1.5 cm, regardless of the other alleles present at the same or different loci.

Table 3.2 The additive effects of alleles on cob length

Locus	Allele	Contribution to cob length (cm)
A	*S*	1.5
	L	6.0
B	*S*	1.5
	L	6.0

QUESTION

3.1 (a) Explain why a maize plant with the genotype $A^LA^LB^LB^L$ will have cobs 24 cm long.

(b) What will the average cob lengths be of maize plants with the genotypes (i) $A^LA^SB^LB^S$ (ii) $A^LA^SB^LB^L$ (iii) $A^SA^SB^LB^L$?

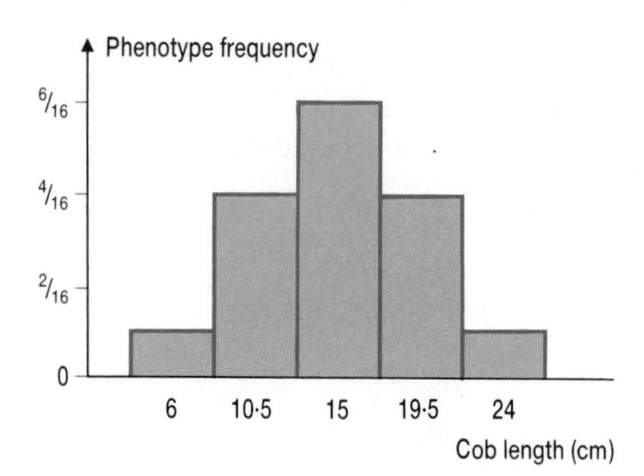

Fig 3.4 The frequency of each F_2 phenotype shown in Table 3.3.

Let us do some genetics with these imaginary plants. We will start by crossing the two extreme parental genotypes – $A^LA^LB^LB^L$ and $A^SA^SB^SB^S$. The F_1 generation all have the genotype $A^LA^SB^LB^S$ and will produce cobs 15 cm long. The F_1s are now selfed to produce the F_2 generation, details of which are summarised in Table 3.3 and Fig 3.4. Again, the distribution of phenotypes in the F_2 generation is a stepped bell, but now the number of phenotypic classes has increased from three to five.

Table 3.3 Variation in cob length between parents and F_2 generation

Generation	Genotype	Relative frequency	Additive effect	Mean cob length (cm)
P_1	$A^LA^LB^LB^L$	all	6+6+6+6	24
P_2	$A^SA^SB^SB^S$	all	1.5+1.5+1.5+1.5	6
F_1	$A^LA^SB^LB^S$	all	6+1.5+6+1.5	15
F_2	$A^LA^LB^LB^L$	$\frac{1}{16}$	6+6+6+6	24
	$A^LA^LB^LB^S$	$\frac{2}{16}$	6+6+6+1.5	19.5
	$A^LA^SB^LB^L$	$\frac{2}{16}$	6+1.5+6+6	19.5
	$A^LA^LB^SB^S$	$\frac{1}{16}$	6+6+1.5+1.5	15
	$A^LA^SB^LB^S$	$\frac{4}{16}$	6+1.5+6+1.5	15
	$A^SA^SB^LB^L$	$\frac{1}{16}$	1.5+1.5+6+6	15
	$A^LA^SB^SB^S$	$\frac{2}{16}$	6+1.5+1.5+1.5	10.5
	$A^SA^SB^LB^S$	$\frac{2}{16}$	1.5+1.5+6+1.5	10.5
	$A^SA^SB^SB^S$	$\frac{1}{16}$	1.5+1.5+1.5+1.5	6

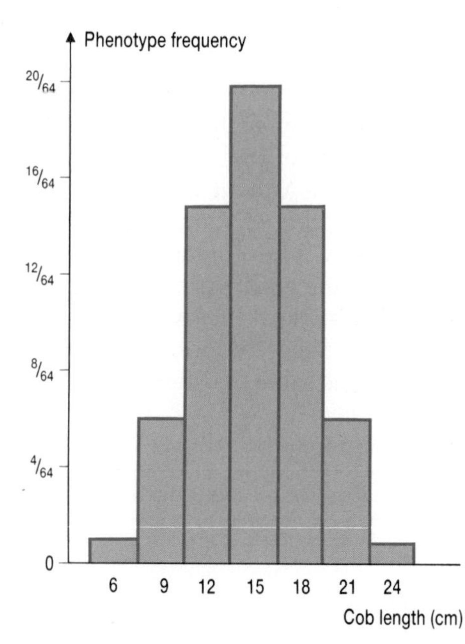

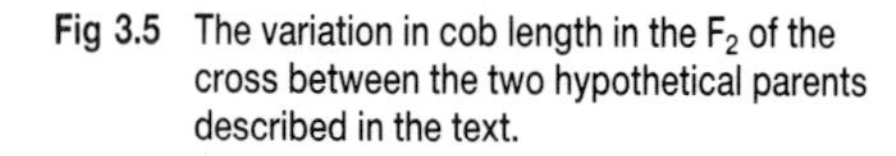
Fig 3.5 The variation in cob length in the F_2 of the cross between the two hypothetical parents described in the text.

Let us add another locus into our imaginary genetic system so that now three gene pairs are controlling cob length. Now a 6 cm plant has the genotype $A^SA^SB^SB^SC^SC^S$ and a 24 cm plant is $A^LA^LB^LB^LC^LC^L$. So this time each L allele contributes 4 cm to the length of the corn cob whilst each S allele contributes 1 cm. If we carry out the same crosses as before we will get the distribution of phenotypes in the F_2 shown in Fig 3.5. Notice that the distribution is even more bell shaped and that the number of phenotypic classes has increased even further.

This bell shaped curve we are producing from our theoretical model is very similar to the distributions that are obtained from investigations of real quantitative characters (Fig 3.1). So the simple model developed above seems to provide a reasonable explanation for the inheritance of quantitative traits.

To summarise the main points:

- The inheritance of quantitative traits is controlled by a large number of gene pairs rather than just a few.
- As the number of gene pairs increases, the contribution of each locus becomes smaller and the expected number of phenotypes in the F_2 increases.
- Environmental variation will tend to smooth out the differences between phenotypes so producing continuous variation.

In short, quantitative variation can be explained by assuming that it results from the interaction of a large number of genes (polygenes) each of which has a small but additive effect on the character in question.

QUESTION

3.2 Table 3.4 summarises the results for our hypothetical corn plants assuming the differences in cob length are determined by one, two, three and six gene pairs.

(a) Using the information in Table 3.4, plot histograms to show the proportion of individuals in each class for the F_2 generation from each cross.

(b) What do you notice about the shape of your histograms as the number of gene pairs controlling cob length is increased? Explain why this observation is consistent with a polygenic model of quantitative inheritance.

(c) In constructing your histograms you have assumed that the environment does not affect cob length. Predict what would happen to the shape of your histograms if the environment also contributed to the phenotypic variation.

Table 3.4 Frequency of cob sizes in the F_2 generation of the cross described above.

	Frequency of cob lengths												
No. of gene pairs	Cob length (cm) 6	7	8	9	10	11	12	13	14	15	16	17	18
1	$\frac{1}{4}$						$\frac{2}{4}$						$\frac{1}{4}$
2	$\frac{1}{16}$			$\frac{4}{16}$			$\frac{6}{16}$			$\frac{4}{16}$			$\frac{1}{16}$
3	$\frac{1}{64}$		$\frac{6}{64}$		$\frac{15}{64}$		$\frac{20}{64}$		$\frac{15}{64}$		$\frac{6}{64}$		$\frac{1}{64}$
6	$\frac{1}{4096}$	$\frac{12}{4096}$	$\frac{66}{4096}$	$\frac{220}{4096}$	$\frac{495}{4096}$	$\frac{792}{4096}$	$\frac{924}{4096}$	$\frac{792}{4096}$	$\frac{495}{4096}$	$\frac{220}{4096}$	$\frac{66}{4096}$	$\frac{12}{4096}$	$\frac{1}{4096}$

3.3 TESTING THE MODEL

The experiments of Ralph Emerson and Edward East on corn

Our polygenic model of inheritance seems to work fine in theory but does it work in practice? We need to test our model against experimental data. Look at Fig 3.6 which shows the inheritance of cob length in two varieties of maize. Both of the parental varieties, Tom Thumb and Black Mexican, are pure-bred lines. Notice that the F_1 hybrids are, on average, about halfway between their parents. Similarly, the F_2 (produced by selfing the F_1) fall about midway between the original parents (P_1 and P_2). Notice also that whilst the F_1 are more or less uniform, the F_2 plants showed great variation, ranging continuously in length almost from one parental extreme to the other. These data are consistent with our model of polygenic inheritance.

How many loci affect the expression of a trait?

The effects of environmental variation will mean that quantitative traits will rarely fall into classes which directly reflect the genotype. How, then, can we estimate the number of genes which affect a trait when we cannot see the effects of individual loci in terms of distinct classes which we can count? This is actually very difficult when a large number of loci are involved. For example, 43 million different genotypes can be produced from a cross between two individuals which are heterozygous at just 16 loci. Economic traits like yield or growth rate are usually controlled by far more than 16 loci.

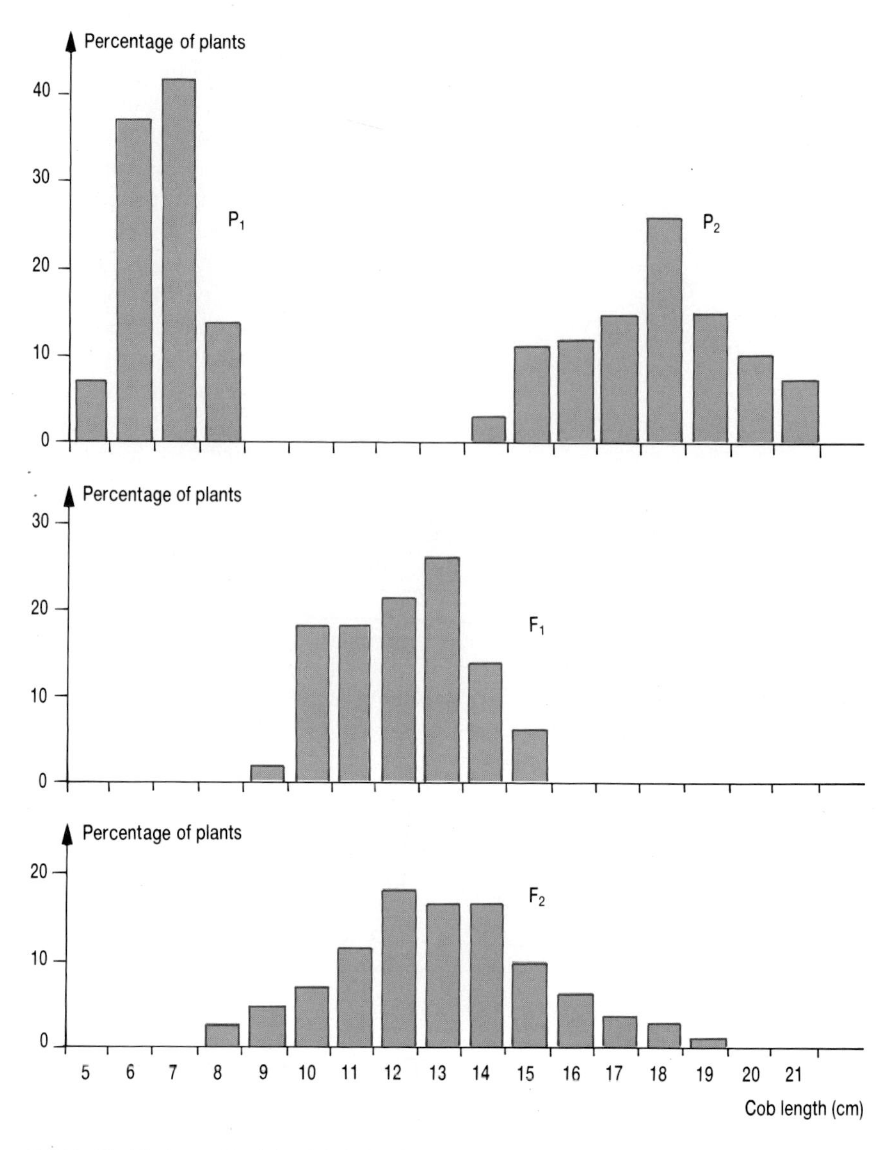

Fig 3.6 The inheritance of cob length in maize. P_1 = Tom Thumb; P_2 = Black Mexican.

However, Emerson and East were able to conclude from their experiments on maize that at least four loci were involved in determining the differences between the original parental varieties Tom Thumb and Mexican Black. How did they reach this conclusion? There are two pieces of information we can use to predict the minimum number of loci affecting a particular character.

1. The number of phenotypic classes appearing in the F_2 generation.
2. The proportion of the F_2 phenotypes which resemble one of the original homozygous parents.

INVESTIGATION 3.1

1 Complete Table 3.5 for n = 4, 5 and 6.

2 Using the results in your table deduce two simple formulae which relate the number of loci affecting a character to:

(a) the maximum number of phenotypic classes which appear in the F_2 generation;

(b) the proportion of F_2 individuals which resemble one of the original homozygous parents.

(continued)

Table 3.5

Number of loci (n)	Number of phenotypic classes in the F_2 generation	Proportion of F_2 individuals similar to one original homozygous parent
1	3	$\frac{1}{4}$
2	5	$\frac{1}{16}$
3	7	$\frac{1}{64}$

QUESTION

3.3 (a) Why did Emerson and East conclude that at **least** four loci were involved in determining the differences in cob length between Tom Thumb and Black Mexican?

(b) If you backcrossed the F_1 to the Tom Thumb parent, what might the average ear length of the progeny (the B_1) be?

A worked example – the experiments of Nilsson-Ehle on wheat

Nilsson-Ehle developed several true-breeding lines of wheat with kernels (part of the seed) ranging from white through various shades of red to dark red. He then carried out a number of crosses using these true-breeding lines.

Cross 1: white varieties crossed with light red varieties.

All the F_1 generation are intermediate in colour. F_1 plants are now crossed and the F_2 generation has white, intermediate, light red kernels in the ratio of 1:2:1.

Cross 2: white varieties crossed with red varieties.

All the F_1 generation are light red. F_1 plants are now crossed and the F_2 has five phenotypic classes ranging from white to red.The proportion of white kernels is approximately $\frac{1}{16}$.

Cross 3: white varieties crossed with dark red varieties.

All the F_1 generation are red. F_1 plants are crossed and the F_2 has seven phenotypic classes ranging from white to dark red with the intermediate forms being most common. The proportion of white kernels is approximately $\frac{1}{64}$.

What clues do we have which will enable us to construct a genetic model to explain these results?

The first step is to work out how many loci are involved in determining kernel colour. Cross 3, between the two extreme parental phenotypes, white and dark red, produced seven phenotypic classes in the F_2 generation. Furthermore, we are told that approximately $\frac{1}{64}$ of the F_2s had white kernels, that is, the same colour as one of the original homozygous parents. From the table you produced in Investigation 3.1, we can conclude that three loci are involved in determining the colour of wheat kernels.

Now let us assume that the alleles at each locus are additive in their effect. Since both the white kernelled and dark red kernelled varieties are pure breeding they must both be homozygous at all three loci affecting kernel colour.

This suggests that the plants with white kernels have the genotype $r_1r_1r_2r_2r_3r_3$ and the plants with dark red kernels have the genotype $R_1R_1R_2R_2R_3R_3$, where each R allele adds 1 unit of red pigment. We can now build a model of cross 3 and compare the model's predictions with the actual results of the experiment. The details of the model, and its predictions, are shown in Fig 3.7.

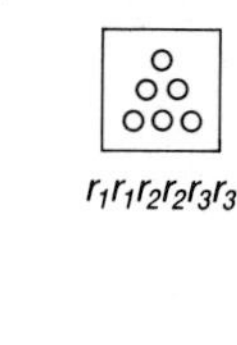

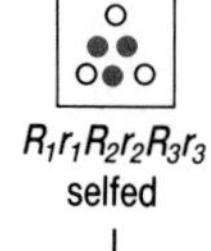

Fig 3.7 The inheritance of kernel colour in wheat. The presence of 'red' alleles is indicated by •; the presence of 'white' alleles is indicated by ○. The effect of all the alleles is additive. Thus dark red varieties are assumed to have six alleles each encoding for the production of one unit of red pigment.

Clearly our model provides an adequate explanation of cross 3. We now need to modify the model to explain the results of cross 1 and cross 2. For cross 1, the genotype of the white parent will be the same as before, $r_1r_1r_2r_2r_3r_3$. The light red variety is, again, pure breeding so it must be homozygous at the three loci affecting kernel colour. Its colour means that it only contains a small number of R alleles which suggests the following genotypes are possible: $R_1R_1r_2r_2r_3r_3$, $r_1r_1R_2R_2r_3r_3$ or $r_1r_1r_2r_2R_3R_3$, each of which produces only 2 units of red pigment.

QUESTION

3.4 (a) Confirm that the genotypes given for the light red parent are consistent with the observed results of cross 1.

(b) Develop a genetic model to explain the results of cross 2. Explain the reasoning behind each assumption that you make. Don't forget to test your model against the actual experimental results.

The effects of dominance and epistasis

So far we have assumed that the only interaction between alleles in our polygenic model of heredity is additive. However, the alleles of polygenes could also interact to produce dominance and epistatic effects. For example, Fig 3.8 shows the expected distribution of phenotypes in the F_2 of the cross between dark red and white kernelled varieties of wheat when the *R* alleles are dominant to the *r* alleles at each locus. Notice that the number of phenotypic classes segregating in the F_2 has been reduced from seven to four which is consistent with the conclusion reached in Chapter 2 that dominance reduces phenotypic variation. Notice also that the distribution is no longer bell shaped but is skewed to one side. Epistatic interactions would have a similar effect on the expected distribution of the F_2 phenotypes.

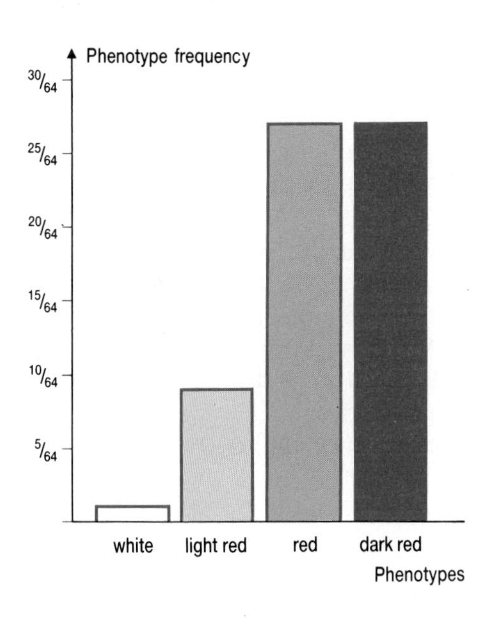

Fig 3.8 The effects of dominance on the number and frequency of phenotypic classes produced in the F_2 of a cross between a dark red and a white kernelled variety of wheat. Compare this figure with Fig 3.7. Notice that the effect of dominance is to decrease the number of phenotypic classes and to increase the number of individuals in one of the extreme classes, here dark red.

QUESTIONS

3.5 Assume that the difference between a pure-breeding line of oats yielding about 4 g per plant and one yielding 10 g is due to three equal and additive gene pairs. Cross one pure-breeding line with the other. What will the phenotypes of the F_1 and F_2 be?

3.6 Fruit length in a particular plant is determined by three unlinked loci. In long fruited plants (fruit length 12 cm) all alleles are 'long' (*L*), in short fruited plants (fruit length 6 cm) all alleles are 'short' (*S*). Each *L* gene contributes 1 cm to the length of the fruit.

(a) Give the genotypes of the parents and the F_1 of a cross between a long fruited plant and a short fruited plant. State the phenotype of the F_1.

(b) Give a possible genotype for a plant with a fruit length of 8 cm.

3.7 Assume that the difference between a corn plant with 6 cm cobs and one with 18 cm long cobs is due to

(a) two gene pairs

(b) three gene pairs

(c) four gene pairs

Assume that in each case the genes have equal and additive effects on cob length. The two plants are crossed and the F_1 is backcrossed to the parent with long cobs.What proportion of the progeny is expected to produce 18 cm long cobs in each case?

3.8 Nilsson-Ehle crossed two types of wild oat, one with white seeds, the other with black seeds. The F_1 had black seeds. The F_2 consisted of 560 plants as follows: 418 black, 106 grey and 36 white. How can the inheritance of seed colour be explained in this case?

3.9 A cross was made between two maize (*Zea mays*) varieties which differed markedly in cob length. The cob lengths of both parents and the F_1 and F_2 generations were measured to the nearest centimetre and the number of cobs in each category was counted as shown in Table 3.6. For example, 14 of the F_1 cobs were 12 cm in length.

(continued)

Table 3.6 Numbers of cobs of each length

	Ear length (cm)																
	5	6	7	8	9	10	11	12	13	14	15	16	17	18	19	20	21
Black Mexican parents									3	11	12	15	26	15	10	7	2
Tom Thumb parents	4	21	24	8													
F_1					1	12	12	14	17	9	4						
F_2			1	10	19	26	47	73	68	68	39	25	15	9	1		

(a) Both parents showed variation in cob length. The variation in the F_1 is comparable to the average of the parental variations while the variation in the F_2 is greater than that found within the parental lines or the F_1. How would you explain this?

(b) How could you explain the whole spectrum of different degrees of expression of this particular characteristic?

(c) What is the term used to describe the range of phenotypes in the above example?

SUMMARY ASSIGNMENT

1. Using specific examples, explain what is meant by the terms
   (a) continuous variation
   (b) discontinuous variation.
   How are these two forms of variation produced?
2. Make a copy of Table 3.1 to remind yourself of the differences between qualitative and quantitative characters.
3. What are polygenes?
4. What is the difference between additive and non-additive (dominance and epistasis) gene interaction?
5. How would you determine the minimum number of gene pairs affecting a particular quantitative trait? Could you do this if the number of gene pairs was large?

Theme 2

GENETICS AND AGRICULTURE

The population of the world now exceeds 5 billion. By the end of this century this figure will have passed 6 billion and by AD 2050 will probably exceed 10 billion. The only way of feeding all these people will be by pursuing more and more efficient methods of agricultural production that will produce the increased amounts of food needed to prevent the mass starvation seen in Ethiopia in 1984.

To achieve the aim of increased food production, two processes need to operate. Firstly, plants and animals need to be looked after; the ground must be prepared, nutrients and water supplied, pests and diseases controlled. Secondly, plants and animals need to be altered genetically so that they respond better to cultivation and husbandry. The methods used to produce improved strains and varieties of plants and animals are the main concern of this theme.

PREREQUISITES

To complete the work in this theme successfully you will need to have completed the work in Theme 1.

The advantages of selective breeding are obvious if you compare the two heads of millet shown in the photograph. The larger, improved strain not only yields more grain but is also resistant to fungal attack.

Chapter 4

HERITABILITY AND SELECTION

Agriculture probably started in the Middle East some 10 000 to 30 000 years ago. Since then, farmers have steadily improved their plants and animals by the process of selective breeding. This involves:

1. selecting those organisms which show the desired characteristics, such as resistance to disease;
2. crossing them with organisms which show other desirable characteristics, such as high yield;
3. identifying and selecting the offspring (progeny) that combine the qualities of both parents, in this case disease resistance and high yield;
4. multiplying these offspring so that there are enough to supply farmers and growers.

Initially this process was highly erratic. Farmers crossed the best with the best and hoped for the best. The laws of heredity discovered by Mendel and subsequent research in genetics has laid a stronger theoretical framework for selective breeding programmes. However, the practice of scientific breeding is still in its infancy and there is much still to be discovered. Given the basic necessity for more food in the face of a rapidly increasing world population, investigating the genetics of plant and animal breeding must rate as one of the most important undertakings of the human species.

For any selective breeding programme to be successful, a number of conditions must be met. Firstly, the character in question, say milk yield, must show variation. Secondly, at least part of that variation must be inherited so that superior genes can be selected and combined to produce new varieties of animals and plants. Variation was considered in Theme 1. In this chapter you will investigate how geneticists decide what proportion of the variation in a character is inherited and how this information can then be used to predict the success of a breeding programme.

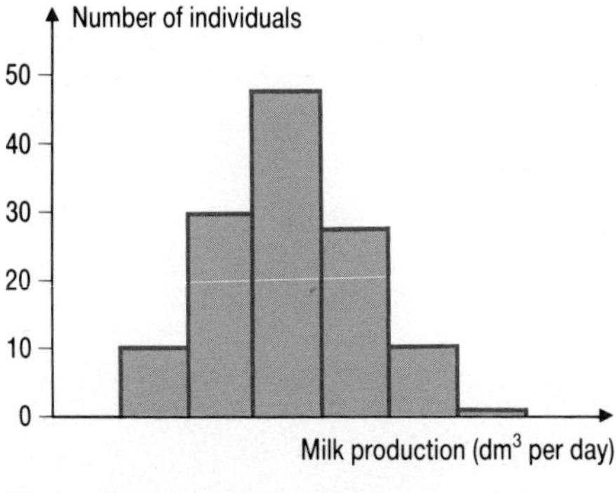

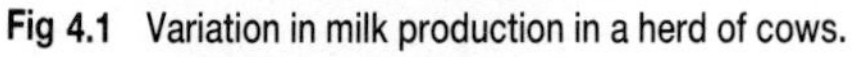
Fig 4.1 Variation in milk production in a herd of cows.

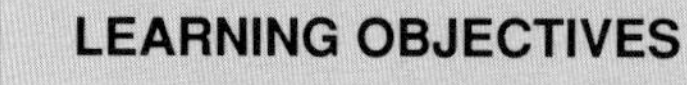

LEARNING OBJECTIVES

After completing the work in this chapter you will be able to:

1. define, calculate and interpret broad- and narrow-sense heritability;
2. use heritability data to evaluate response to selection;
3. define and explain the use of progeny testing.

4.1 DESCRIBING BIOLOGICAL VARIATION

Consider a herd of cows. The milk production for every cow is measured daily for a period of, say, one month. At the end of that time the average daily milk production for each cow is calculated and the results are plotted either as a frequency diagram or a bar chart, (Fig 4.1). Box 4.1 gives details of how frequency diagrams are constructed.

BOX 4.1

Producing frequency diagrams

Frequency diagrams or **histograms** are simply a way of showing variation graphically. Assume you are investigating variation in the mass of trout in a fish farm and you want to produce a frequency diagram to show you how much variation there is.

1. Weigh a large sample of trout – say 1000. This gives you a mass range from 10.24 g to 147.6 g.
2. Divide the mass range into equally spaced classes, e.g. 10.00 – 11.99 g, 12.00 – 13.99 g and so on. The spacing of the size classes will depend on the data you have available.
3. Now count the number of trout in each class
   4 trout have a mass of between 10.00 and 11.99 g,
   6 trout have a mass of between 12.00 and 13.99 g.
4. Now calculate the proportion of trout in each class as

$$\frac{\text{number in class}}{\text{total number}}$$

   proportion of trout in 10.00 – 11.99 g class = 4/1000 = 0.004 = 0.4%
   proportion of trout in 12.00 – 13.99 g class = 6/1000 = 0.006 = 0.6%.
5. Finally, plot the proportion in each class on the vertical axis and the class sizes on the horizontal axis.

INVESTIGATION 4.1

Produce histograms to show the variation in some quantitative characters. Use characters which are easy to measure and where 'individuals' are available in large numbers, for example petal length in daisies, masses of seeds.

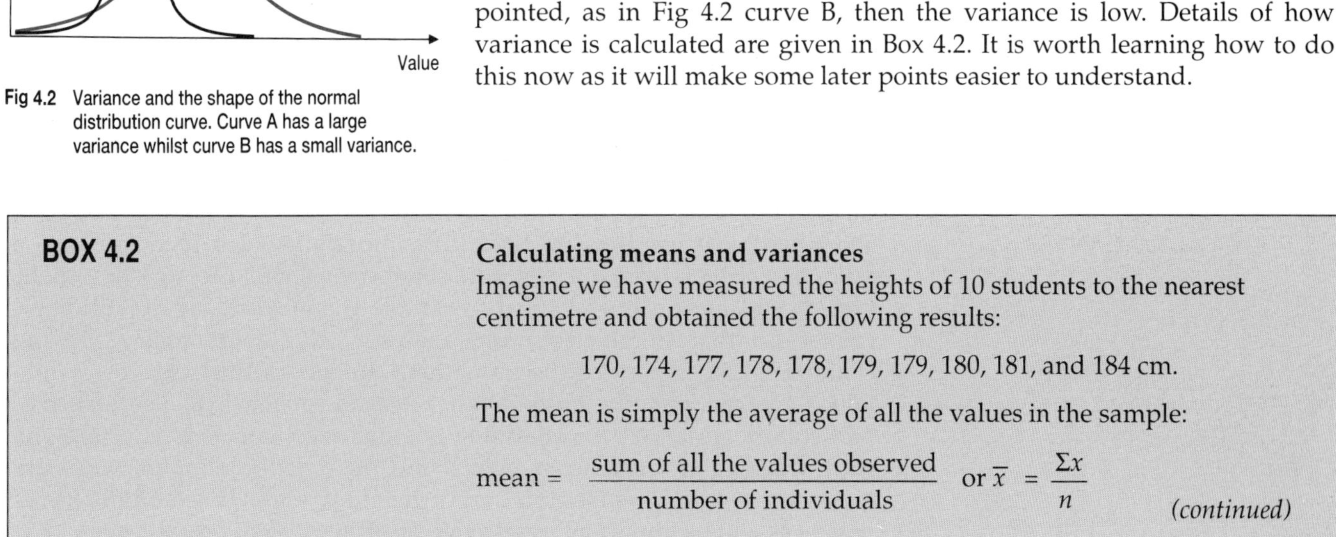

Fig 4.2 Variance and the shape of the normal distribution curve. Curve A has a large variance whilst curve B has a small variance.

At the beginning of chapter 3 we saw that the frequency diagram for most quantitative characters approximates to a normal distribution. This important observation means that we can use the properties of the normal distribution to investigate quantitative characters. In particular, we are interested in the amount of variation shown by a quantitative character. This can be measured and expressed as the **variance**, which is simply a measure of how 'spread out' a normal distribution curve is. If the curve is flat, as in Fig 4.2 curve A, then the variance is high. If the curve is very pointed, as in Fig 4.2 curve B, then the variance is low. Details of how variance is calculated are given in Box 4.2. It is worth learning how to do this now as it will make some later points easier to understand.

BOX 4.2

Calculating means and variances

Imagine we have measured the heights of 10 students to the nearest centimetre and obtained the following results:

170, 174, 177, 178, 178, 179, 179, 180, 181, and 184 cm.

The mean is simply the average of all the values in the sample:

$$\text{mean} = \frac{\text{sum of all the values observed}}{\text{number of individuals}} \quad \text{or } \bar{x} = \frac{\Sigma x}{n}$$

(continued)

where $\overline{x}$ = the mean; x = an observed value, e.g. 178 cm; n = the number of observed values or observations, in this case 10. Σ simply means add up all the observed values.
For our sample, $\overline{x}$ = 178.0 cm

The variance is defined as 'the sum of the squares of the differences between each value and the mean divided by the number of observations minus one' or,

$$s^2 = \frac{\Sigma (x - \overline{x})^2}{n - 1}$$

where s^2 is the variance and all the other values are the same as before. This may look and sound complicated but providing you lay the calculation out properly it is reasonably straightforward:

x	$x - \overline{x}$	$(x - \overline{x})^2$
170	–8	64
174	–4	16
177	–1	1
178	0	0
178	0	0
179	1	1
179	1	1
180	2	4
181	3	9
184	6	36
	$\Sigma (x - \overline{x})^2 =$	132

$$s^2 = \frac{\Sigma (x - \overline{x})^2}{n - 1} = \frac{132}{9} = 14.67 \text{ cm}^2$$

QUESTION

4.1 (a) A farmer has 10 cows which yield 24, 23, 18, 26, 32, 34, 41, 12, 63 and 14 dm^3 of milk per day. Calculate the mean and the variance of this data.

(b) The farmer buys two new cows which produce 73 and 12 dm^3 of milk per day. What is the effect of these two new cows on the variance of the daily milk yield for the herd?

(c) The farmer is being encouraged to reduce the variance in the herd's daily milk yield. What advice would you give?

4.2 USING VARIANCES

You should have found question 4.1 relatively straightforward, but you might be wondering, quite rightly, what calculating variances has got to do with genetics. In analysing characters showing discrete variation you count the number of individuals which occur in each class – 92 tall pea plants, 18 short pea plants and so on. From this information you then calculate a ratio, say 9:3:3:1, which gives you some idea about the sort of genetic system you are dealing with, for example monohybrid or dihybrid. We cannot do this simple 'counting analysis' with continuously varying characteristics, like mass, because there are no distinct categories into which we can put our observations. Instead we have to use statistics, particularly variances, for analysing quantitative characters. Furthermore, since most of the characters which are of agricultural importance are quantitative, it follows that this sort of genetic analysis based on the use of variances will be important in selective breeding.

The variance in milk production which you calculated earlier is called the **total phenotypic variance – V_P.** We can recognise two major components of phenotypic variance: the genetic component (V_G) and the non-genetic or environmental component (V_E). These are related by the following equation:

$$V_P = V_G + V_E \qquad (1)$$

The worked example in Box 4.3 will give you a better idea of how this equation can be used.

BOX 4.3

Partitioning phenotypic variance using tobacco plants
The principles of partitioning variance can be demonstrated using the data on corolla length in the tobacco plant, *Nicotiana longiflora*, shown in

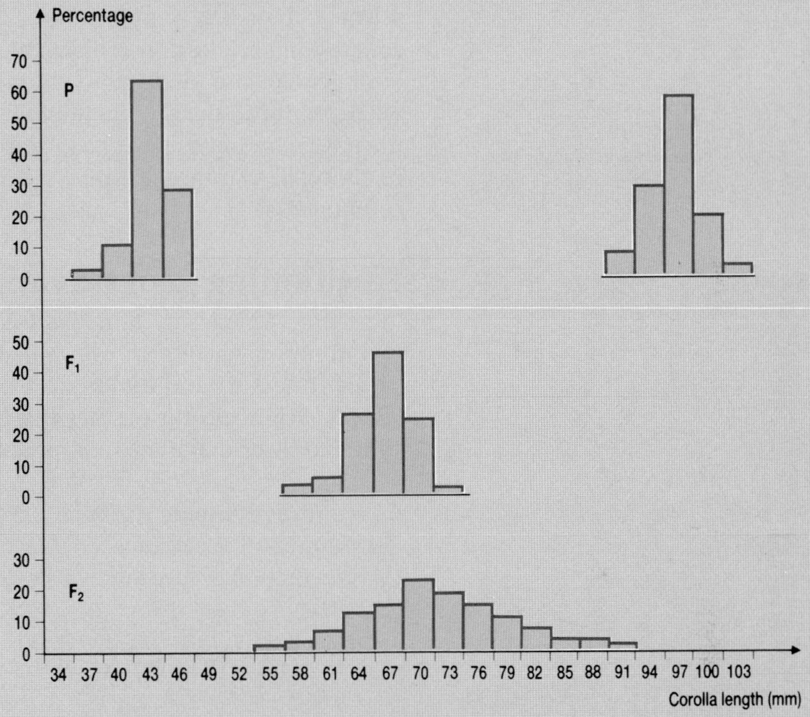

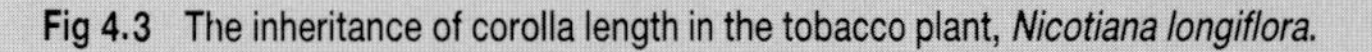

Fig 4.3 The inheritance of corolla length in the tobacco plant, *Nicotiana longiflora.*

Fig 4.3. The varieties of *N. longiflora* used as parents were pure breeding and would therefore have been homozygous at most loci, that is V_G is practically zero. So the variance within each parental group which we can see in Fig 4.3 is all due to the environment. The variance among the F_1 offspring is all environmental as well; all F_1 individuals are genetically identical to each other (although not homozygous) because the gametes produced by each parental strain are identical.

The average variance within each of the two parental varieties and the F_1 offspring is 8.76. This is an estimate of environmental variance (V_E) in the environment where the experiments were conducted. So

$$V_E = 8.76$$

The genes inherited from the two parents segregate in the F_2. So the phenotypic variance of the F_2 (calculated as shown in Box 4.2) contains both genetic and environmental variance:

$$V_P \text{ (of } F_2) = V_G + V_E = 40.96$$

So rearranging equation (1):

$$V_G = V_P - V_E = 40.96 - 8.76 = 32.20$$

4.3 HERITABILITY

Now let us return to the farmer who wants to increase milk yield. There are a number of ways of doing this. For example, the cows could be fed a special diet or kept in special sheds which keep them dry and warm so that they spend less energy keeping warm and more making milk. Alternatively, the farmer could try breeding cows which produce more milk. Only those cows which produce the most milk are allowed to reproduce in the hope that their offspring will inherit the superior milk-producing ability. The problem is that the farmer does not know how much of the variation in milk yield in the herd is due to environmental factors and how much is inherited. If all the variation is caused by subtle differences in the environment, then no amount of selection will produce cows which produce more milk. Under these circumstances the only way to increase milk yield is to change the environment. Alternatively, all the differences between the cows could be due to inheritance. Under these circumstances a selective breeding programme should be highly successful. In reality of course the variation in milk yield probably results from a combination of both genetic and environmental factors. What the farmer needs to know is what proportion of the phenotypic variance in milk yield is due to environmental factors and what proportion is inherited. In other words, the total phenotypic variance in milk yield needs to be divided into its different components.

Partitioning genetic variance

In section 4.2 we saw how the total phenotypic variance can be split into two components which we called V_G and V_E. The farmer needs to go further than this and divide up the genetic component (V_G). The geneticist Sewall Wright proposed that the genetic component of variance could be split into three major sub-components:

V_A = additive genetic variance
V_D = dominance variance } non-additive components
V_I = interaction or epistatic variance } of genetic variance
so that

$$V_G = V_A + V_D + V_I \qquad (2)$$

From the point of view of an animal or plant breeder, it is the additive genetic variance (V_A) which is critical since this is the chief cause of resemblance between parents and their offspring. It is this component of the overall phenotypic variance which is important in determining if the daughters of cows which produce a lot of milk will also be good milkers.

Table 4.1 Essential equations used by plant and animal breeders

(a) Definitions

V_P = total phenotypic variance	V_G = genetic variance	V_E = environmental variance
V_A = additive genetic variance	V_D = dominance variance	V_I = interaction or epistatic variance

(b) Equations

$V_P = V_G + V_E$
$V_G = V_A + V_D + V_I$
$V_D + V_I$ = non-additive genetic variance

broad-sense heritability $= \dfrac{V_G}{V_P} = \dfrac{V_G}{V_E + V_G}$

narrow-sense heritability (h^2) $= \dfrac{V_A}{V_A + V_D + V_I + V_E}$

Put crudely, the additive genetic component of variance is inherited whilst the dominance and interaction variances are not. All these relationships are summarised in Table 4.1.

So, as far as the farmer is concerned, the critical component of equation (2) is V_A since this will determine how successful a programme of selective breeding will be. Estimating the proportion of the total phenotypic variance which is due to V_A will help us decide if a selective breeding programme will be successful. This is achieved by calculating the heritability of a trait.

Defining heritability

To fully understand and apply the concept of heritability it is necessary to define the term mathematically. From equation (1) we know that total phenotypic variance (V_P) can be divided into two components, V_G and V_E. The ratio

$$\frac{\text{genetic variance}}{\text{phenotypic variance}} = \frac{V_G}{V_P} = \frac{V_G}{V_G + V_E} \qquad (3)$$

is called **broad-sense heritability** or **the degree of genetic determination**. This ratio is a measure of the proportion of the total phenotypic variance which is due to total genetic variance. This information is of little use to an animal or plant breeder since it does not tell us the extent to which phenotypes are determined by inheritance. To obtain an insight into the extent to which phenotypic variance is due to inherited differences, we have to use equation (2). The ratio

$$\frac{\text{additive genetic variance}}{\text{phenotypic variance}} = \frac{V_A}{V_P} = \frac{V_A}{V_A + V_D + V_I + V_E} \qquad (4)$$

is called **narrow-sense heritability** which is usually abbreviated to h^2. Narrow-sense heritability is a measure of the proportion of the total phenotypic variance that can be attributed to additive genetic variance. It is this heritability that is useful in determining whether a programme of selective breeding will succeed in changing the population.

QUESTIONS

4.2 Calculate the narrow- and broad-sense heritability for each of the four characters shown in Table 4.2.

Table 4.2

Component of variance	Number of bristles	Length of thorax	Size of ovary	Number of eggs laid
V_A	52	43	30	18
$V_D + V_I$	9	6	35	44
V_E	39	51	35	38

4.3 Estimation of variances in body mass (kg) in a population of 190 day-old pigs gives the following results:

total genetic variance = 117
variance due to dominance effects = 27
variance due to epistatic effects = 13
total environmental variance = 211

Calculate the narrow-sense and broad-sense heritabilities from these variance estimates.

Estimating heritability

In the examples given above you have been presented with the results of experiments designed to estimate the various components of total phenotypic variance. Measuring the total phenotypic variance of a trait in a group of individuals is not difficult (see Box 4.1 for details). However, partitioning the total phenotypic variance into environmental and genetic components is not so straightforward. Box 4.3 gives details of how V_G and V_E can be estimated for corolla length in *Nicotiana longiflora*. This allows an estimate of the broad-sense heritability to be made, but this is of little use to a plant breeder. Furthermore, whilst it is possible to devise experiments which estimate V_A, V_D and V_I, this approach is not a particularly efficient way to estimate h^2, and a number of different methods have been developed by geneticists. One of these is described below.

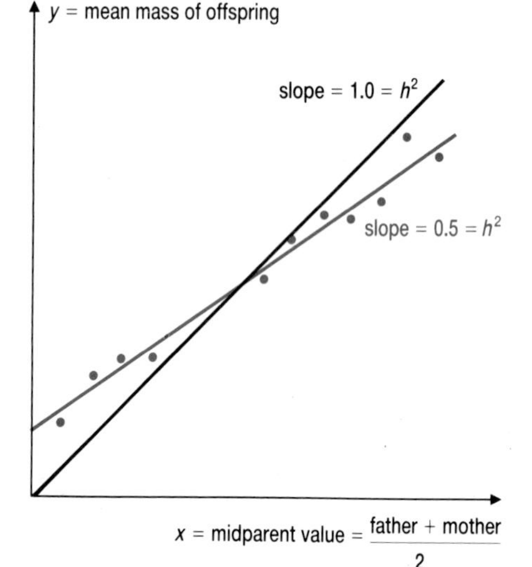

Fig 4.4 The regression (red line) of offspring mass (y) on midparent values (x). The slope of this line, which is equivalent to narrow-sense heritability (h^2), is 0.5. The black line shows the regression slope if the trait were perfectly heritable.

Parent-offspring regression. Imagine we are trying to estimate the heritability of mass in pigs. We carry out a number of crosses using different sets of parents. For each cross we record the mass of the parents and the mass of the offspring produced. Using this information we now plot the mean mass of the offspring from each cross against the average masses of their parents, the so called **midparent value**, and obtain the relationship shown in Fig 4.4.

Notice the following:

- The slope of the regression line of offspring on mid parent value is an estimate of h^2. The steeper the line the higher the heritability.
- The slope of the regression line is positive. This means that heavier parents tend to have heavier piglets.
- The slope of the regression line is less than one. This means that very light parents tend to have offspring slightly heavier than they are whilst heavy parents have slightly lighter offspring. This slope of less than one for the regression line arises because heritability is less than perfect.

QUESTION

4.4 A geneticist has been asked to estimate the heritability of leg length in 6-week-old turkeys. The geneticist carries out a series of crosses and obtains the results shown in Table 4.3. What is the heritability of this trait?

Table 4.3

Midparent value (cm)	Mean offspring value (cm)	Midparent value (cm)	Mean offspring value (cm)
5.0	6.5	22.0	17.5
7.5	5.5	24.0	18.0
9.0	8.5	24.5	20.0
11.0	5.5	25.0	35.0
11.0	11.5	27.0	19.0
12.5	14.0	27.5	14.0
15.0	10.5	28.0	24.5
15.5	14.0	31.5	21.5
19.0	16.5	21.0	15.0
22.0	12.0	35.0	25.0

What does heritability mean?

The concept of heritability is complex. Since it is often used to justify statements about people, for example the inheritance of IQ, it is worth spending some time making sure you fully understand the concept. To do this we will use a series of imaginary experiments using *Potentilla glandulosa* (see Fig 1.2) devised by Francisco Ayala and John Kiger. This plant can be reproduced by cuttings which makes it possible to obtain a group of individuals genetically identical to one another because they have all been produced from the same plant.

Experiment 1

One plant is divided into several pieces and the individual plantlets so produced are planted out on a hillside. Here they are exposed to considerable environmental differences in the quality of the soil, the amount of sunlight, moisture and so on. At the end of the growing season each plant is collected and the total mass (biomass) of each plant is determined. Since all the plants are genetically identical any variation in biomass is all environmental. So the heritability of biomass as determined from this group of plants is zero.

Experiment 2

Plants are collected from different areas so that they are genetically different. A small cutting from each plant is placed in an experimental garden and all the plants are provided with exactly the same optimal conditions. Again, the heritability of biomass is determined for this group of plants but this time a high value, say 0.96, is obtained because the plants are genetically different (V_G is high) but the environment is quite uniform (V_E is low).

Experiment 3

Cuttings from the plants used in experiment 2 are planted in another experimental garden. Again, all the plants receive the same treatment, but this time the conditions are sub-optimal – poor light, no fertiliser and marginal amounts of water. At the end of the growing season the plants are all small but the estimate of heritability will still be high because the plants are genetically different but the environment is uniform.

Experiment 4

Again, cuttings from the plants used in experiments 2 and 3 are taken, but this time they are planted out all over the hillsides where the plants grow naturally. So here there is variation both in the genotype and in the environment. This time the estimate of heritability is 0.6.

These experiments illustrate a number of fundamental points about heritability. Firstly, heritability is a property of a population of organisms not of an individual. Remember, heritability measures **not the degree to which a trait is determined by genes, but rather the proportion of the phenotypic variance that is due to the additive genetic variance**. So the value of 0.96 obtained in experiments 2 and 3 for the heritability of biomass does not mean that 96 per cent of the biomass of an individual plant is due to the genes it contains and 4 per cent is due to its environment. Rather, the value indicates that 96 per cent of the phenotypic variance is due to the additive genetic variance.

Secondly, heritability is a population specific measurement, valid only for a given population in a given environment. In all four experiments we estimated the heritability of the same trait, the total biomass of the plant. Yet the heritability of the trait is 0 in experiment 1, 0.96 in experiments 2 and 3 and 0.6 in experiment 4. The heritability of a trait is different when estimated in different groups of individuals (as in experiment 1 versus 2) or in different environments (as in experiment 3 versus 4).

Thirdly, just because the heritability of a trait is high in two populations does not mean that the differences between the means of the populations are due to genetic differences. To illustrate this point imagine someone told you that the differences in size between the plants in experiments 2 and 3 above are largely genetic, on the grounds that heritability is very high in both cases. This claim is obviously nonsense since cuttings from the same set of plants were used in both experiments. The differences between the two populations are the result of the different environments to which the plants were exposed during the experiments.

QUESTIONS

4.5 Describe a method using parents and their children that could estimate the heritability of cranial capacity (the volume of the skull). You will need to consider and give details of how you would measure cranial capacity.

4.6 Egg weight in chickens is said to have a heritability of 60 per cent. What does this mean?

4.7 Numerous studies have shown that the average IQ score is higher in American whites than in American blacks. Other studies have also shown that the heritability of IQ is high in both populations. (Note that there is considerable debate among psychologists as to the extent to which IQ scores actually reflect intelligence.) Some people have argued that because IQ is so highly heritable, the difference in average IQ between white Americans and black Americans is largely genetic. Explain why this is an invalid conclusion.

4.4 USING HERITABILITY

Even though h^2 is a number which applies only to a particular population and a given environment, it is still of great practical significance to breeders. Using a variety of techniques, heritability estimates for a number of traits of agricultural importance have been made. Some of these are shown in Table 4.4. In particular, heritability is useful in predicting the effects of artificial selection. Basically, the breeder wants to know, given a certain programme of selective breeding, how much the average phenotype can be changed and how fast the change will occur. Heritability can be used to calculate this response to selection.

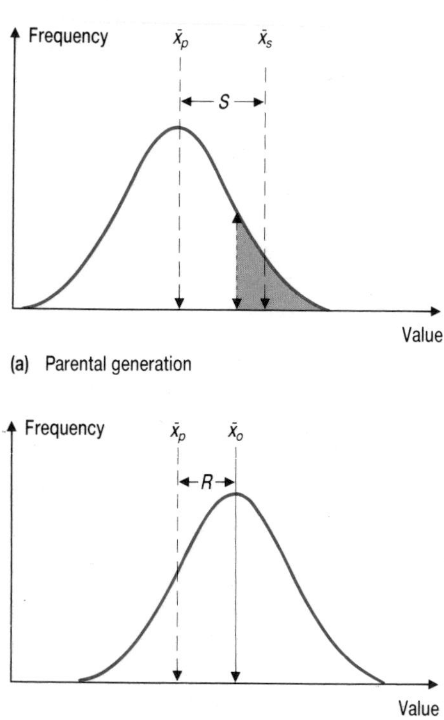

Fig 4.5 The relationship between selection differential (S) and selection response (R). Other symbols are defined in the text.

Response to selection

Look carefully at Fig 4.5(a) which shows the variation of a trait we are interested in, say mass in chickens, in a parental population. The mean mass of the parental population is shown as $\bar{X}_p$. The aim of the breeding programme is to increase the mean mass of the chickens. To achieve this we select the individuals who fall in the shaded section of the parental distribution curve to be the parents of the next generation. The mean mass of this selected group is $\bar{X}_s$. These individuals are allowed to reproduce and the mass of the offspring is recorded giving the distribution shown in Fig 4.5(b). The mean mass of the offspring is shown as $\bar{X}_o$.

The difference between the mean mass of the original population and the mean mass of the selected parents ($\bar{X}_p - \bar{X}_s$) is called the **selection differential (S)**. The difference between the mean of the original population and the mean of offspring ($\bar{X}_o - \bar{X}_p$) is called the **selection response (R)**.

$$\text{selection response } (R) = h^2 \times \text{selection differential } (S) \qquad (5)$$

Table 4.4 Heritability (h^2) estimates for some characters of agricultural importance

Trait	Heritability (h^2)
slaughter weight in cattle	0.95
plant height in maize	0.70
egg weight in poultry	0.60
fleece weight in sheep	0.40
milk production in cattle	0.30
yield in maize	0.25
ear length in maize	0.17
conception rate in cattle	0.05

or

$$h^2 = \frac{\text{selection response } (R)}{\text{selection differential } (S)} \quad (6)$$

This last equation provides yet another way of estimating h^2: by selecting for one generation and comparing the selection response with the selection differential. Usually this is carried out over several generations and the average response is used.

More importantly, equation (5) provides a means for assessing how successful a selective breeding programme will be. The larger the heritability the larger the selection response for a given selection differential. To show this is so, try the following investigation.

INVESTIGATION 4.2

The relationship between heritability and selection response

Using a selection differential of 3.52 calculate the selection response when heritability increases from 0 to 1 in steps of 0.1 (0.1, 0.2, 0.3 etc.). Plot a graph of your results with heritability on the x- (horizontal) axis and selection response on the y- (vertical) axis. Describe the relationship between heritability and selection response.

QUESTION

4.8 The mean mass of 8-week-old pigs on an experimental farm is 5.7 kg. Two sets of parents are used to produce the following generation:
(a) heavy pigs with a mean mass at 8 weeks of 6.9 kg
(b) light pigs with a mean mass at 8 weeks of 4.4 kg.
The mean mass of the offspring, produced by these two groups of pigs, at 8 weeks are: **(a)** 6.2 kg and **(b)** 5.1 kg respectively. Calculate the heritability of this characteristic in each of the two sets of progeny.

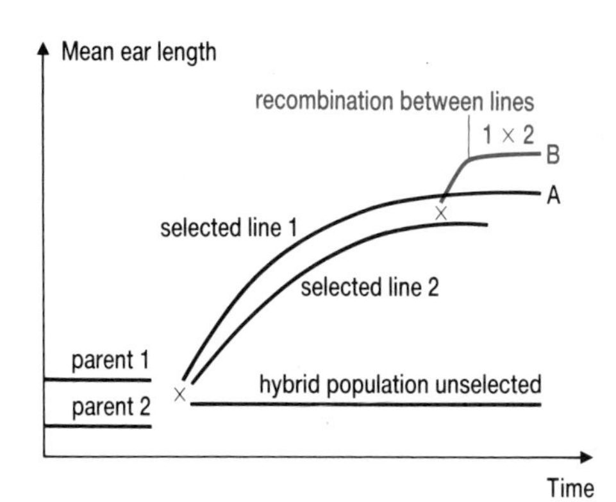

Fig 4.6 A diagram to show response to selection over time. Two original plants (parent 1 and parent 2) were crossed to produce a hybrid. Two plants from this hybrid population were selected to produce two lines, 1 and 2. Within each of these lines selection was practised for increased ear length. Case A represents the situation often seen in practice. Case B represents a renewed response to selection resulting from an introduction of new genetic variation as a result of crossing line 1 and line 2.

The limits to selection

A large heritability enables us to predict that the response to selection will be rapid whilst a small heritability implies a slow response. But how long can we go on selecting for? Will the improvement carry on for ever with the pigs or sheep or cows getting bigger and bigger? Unfortunately the answer is no. To understand why selection will not produce infinitely large increases in the trait under consideration, imagine we are trying to breed wheat plants which have large ears (see Fig 4.7). We produce a hybrid by crossing together two different parents. We sow the seed from this cross, allow the plants to grow and select a group of plants with the largest ears to be the new parents of the next generation. These plants are self-fertilised, the seeds are sown and new parents selected. If mean ear length is plotted against time we get the graph shown in Fig 4.6. Initially mean ear length increases rapidly, but eventually a plateau is reached where, despite continued selection, we get no further increase in ear length. We have produced a pure-bred line where all the individuals in our population are genetically identical. Any variation that we see is due to the environment. The response to selection shown in Fig 4.6 suggests, therefore, that two factors are critical if a selective breeding programme is going to be successful. Firstly, the character in question should have a high heritability. Secondly, there must be genetic variation within the population for the selected character.

Fig 4.7 Ears of bread wheat, the most widely cultivated cereal crop in the world. Each of the four ears contains many grains of wheat.

QUESTIONS

4.9 Sketch a graph to show the change in heritability of ear length in line 1 of Fig 4.6 against time. Explain the shape of your graph.

4.10 Why will changing the environment or adding in extra genetic variation possibly give a renewed response to selection in inbred lines?

4.11 Why will a cattle breeder choose to work with the herd which contains the greatest genetic variance?

4.12 In question 4.6 the heritability of egg weight was estimated as 0.60. What does the heritability value tell you about the prospects for selecting for an increase in egg weight?

4.5 PROGENY TESTING

In many selective breeding programmes, what the breeder actually wants to know is which parents to use to produce the best results. For example, suppose an animal breeder was trying to produce cows which gave a lot of milk. The breeder can look at all the cows which could be used as parents and choose those which are the best milkers. But bulls don't give milk. How do you choose the male parent when you have no idea about the quality of the milk-producing genes it contains? If you remember, V_A is a measure of how closely offspring resemble their parents. So examining the performance of the offspring produced from a cross will tell us something about the value of the parents. This is the basis of progeny testing.

A **progeny test** is defined as **a test of the value of a genotype on the performance of the offspring produced in some definite system of mating**. Progeny testing is used in both plant and animal breeding but is especially important in the latter because of the limited number of crosses that can be made and the problems of sex limitation of characters like milk yield. A typical progeny test would involve crossing, say, a bull with a number of cows of known quality. The offspring of this cross are then assessed for the character in question, say milk yield. The average milk yield of the offspring is then taken as a measure of the breeding value of the bull. The higher the milk yield the greater the breeding value for this particular character. Comparing the breeding values of different bulls determined by progeny testing will allow the breeder to select the best bull for a particular breeding programme.

4.6 PULLING THE THREADS TOGETHER

This has been a complex chapter, so a summary is worthwhile. By subdividing total phenotypic variance into environmental and genetic components, valuable information about plant and animal breeding can be obtained. In particular, narrow-sense heritability (h^2) can be estimated. Whilst this value only applies to a particular population in a particular environment, it is of great practical importance to breeders. A cattle breeder interested in increasing, say, growth rate is not concerned with genetic variance over all possible herds and all possible environments. The question is, given a particular herd (or a choice between a few herds) under the environmental conditions approximating present husbandry practice, can a selection scheme be used to increase growth rate and, if so, how fast? If one herd has a lot of genetic variation and another only a little, the breeder will choose the former to carry out selection. If the heritability of growth rate in the herd is high then the mean of the population will respond rapidly to selection because most of the superiority in the parents will appear in the offspring. The higher the value of h^2 the higher is the parent-offspring correlation (see Fig 4.4). If, on the other hand, h^2 is low, then only a small fraction of the increased growth rate of the selected parents will be reflected in the next generation.

SUMMARY ASSIGNMENT

1. Make sure you have definitions and explanations of the following terms: broad- and narrow-sense heritability, progeny testing.
2. Make a copy of the equations given in Table 4.1.
3. Write an essay to answer the following:
   'Describe three ways of estimating narrow-sense heritability. What are the limitations and uses of such estimates?'

Chapter 5

SELECTIVE BREEDING – GENERAL PRINCIPLES

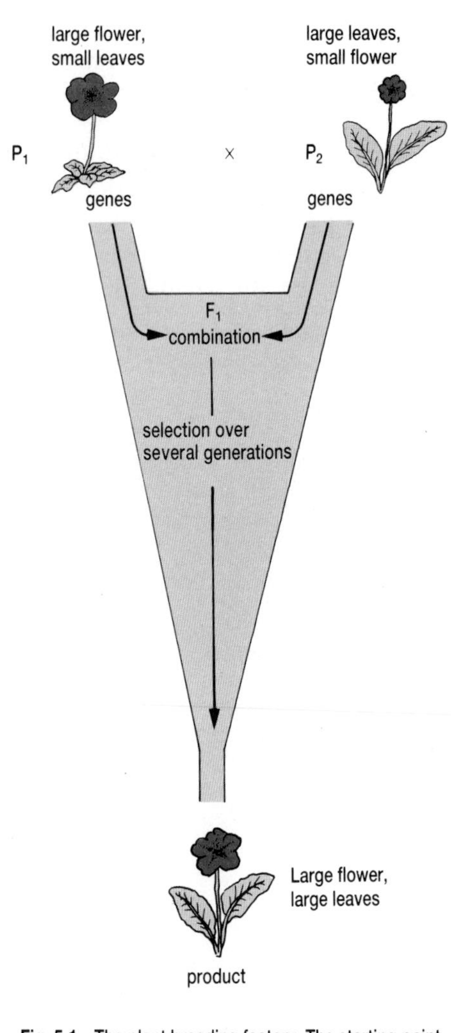

Fig 5.1 The plant breeding factory. The starting point for selection is a genetically variable population. This could be a natural population of a plant, a land race or it could be, as shown here, produced by crossing plants from two or more different inbred lines. The plant breeder then refines down this variable population, using one or more selection schemes, to produce the final product, a new inbred line with the desired characteristics.

In terms of meeting the food requirements of people, plants are fundamentally more important than animals. Consequently this chapter and the next concentrate on plant breeding. Nonetheless, many of the ideas developed are directly applicable to animal breeding.

The idea of a plant breeding programme is to combine superior genes in one parent with superior genes in another parent so producing superior offspring. This means that there must be a supply of superior genes available, a means of combining the genes by crossing plants and then a means of selecting those offspring with the desired characteristics. This process can be likened to a factory, as shown in Fig 5.1. The raw materials, genes, are fed in at the top. In the factory they are joined together in new ways and the best forms are selected. Finally, an end product is produced which can be sold to farmers as an improved variety of plant.

This process is both expensive and time consuming, so plant breeders need to define their objectives carefully before starting a programme of selective breeding. Such a definition has to take into account not only the biological aspects of the selective breeding process but also the economic and social aspects. One example will serve to highlight this. Scientists at the International Maize and Wheat Improvement Centre (CIMMYT) in Mexico developed an improved strain of yellow maize for use in Africa. The local people, however, rejected the improved variety since they had traditionally used white maize rather than yellow. Selective breeding is not simply a question of juggling with genes. You must be certain that there is a need and a commercial market for the end product.

LEARNING OBJECTIVES

After completing the work in this chapter you will be able to:

1. explain the genetic consequences of inbreeding and outbreeding;
2. define the terms inbreeding depression and heterosis;
3. describe the roles of selection, mutation, introgression and scientific breeding in the development of modern crop plants;
4. explain the importance and mechanics of backcrossing;
5. explain artificial selection in terms of allele frequencies and gene pools.

5.1 BREEDING SYSTEMS

Table 5.1 provides a glossary of terms which you will find useful in this and subsequent chapters. In the previous chapter we saw that two factors affect the response to selection; heritability of the trait in question and the amount of genetic variability (measured as the genetic variance) in the population. We also need to consider a further factor if a programme of selective breeding is going to be successful: how the organisms involved reproduce; their **breeding system**. There are two reasons for looking at this aspect of the biology of agriculturally important plants and animals. Firstly, the breeding system affects the genetic structure of the population – the extent to which individuals are homozygous or heterozygous. As you will see, this directly affects the selective breeding strategies that can be used. Secondly, genes can only be moved from one organism to another if the breeder can find a way of introducing them into the plants or animals. This requires a good understanding of the process of sexual reproduction in the organism the breeder is interested in.

Table 5.1 A glossary of terms used in animal and plant breeding

Term	Definition
Gene pool	the sum total of all the alleles in a breeding population
Species	groups of interbreeding natural populations that are reproductively isolated from other such groups, i.e. they do not usually interbreed and produce fertile offspring
Breed	as applied to animals, a genetically distinct population within a species. For example, the different breeds of domesticated dogs
Variety	the plant equivalent of a breed. For example, Maris Piper and Pentland Javelin are varieties of potato
Cultivar	a cultivated variety
Pure line	in plants, a population produced by selfing, the members of which are genetically identical and homozygous at all (or nearly all) gene loci. In animals, a population where individuals are homozygous for some genes but heterozygous for others
Inbred line	a population where a large number of genes are homozygous as a result of inbreeding

Some basic concepts

The basic reproductive strategies of plants and animals are summarised in Fig 5.2. Notice that an organism can reproduce both asexually and sexually. For example, aphids reproduce asexually during the summer but sexually in the autumn. Asexual reproduction in plants is an important means of increasing the numbers of superior varieties. The remainder of this section is only concerned with sexual reproduction.

Inbreeding involves the fusion of gametes from close relatives. The ultimate inbreeders are those plants which are naturally self-fertile and where self-fertilisation (selfing) is the usual method of sexual reproduction, for example wheat, tomatoes. **Outbreeding** involves the mating of unrelated organisms. There are many mechanisms in plants which prevent self-fertilisation and so promote outbreeding. **Self-incompatibility systems**, where pollen will only germinate on certain stigmas, are of particular importance. Incompatibility systems are discussed further in Chapter 7.

asexual reproduction

Offspring identical and form a **clone**. Members of a clone may be homozygous or heterozygous.

sexual reproduction

Inbreeding. Offspring are highly homozygous and phenotypically uniform, e.g. wheat, tomatoes.

Outbreeding. Offspring are highly heterozygous and phenotypically variable, e.g. wild populations of maize.

Fig 5.2 The basic reproductive strategies of plants and animals.

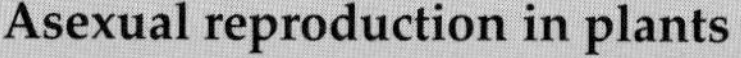

INVESTIGATION 5.1

Asexual reproduction in plants

Asexual methods of propagating plants are an important means of increasing the number of plants in a superior variety. Prepare a report, using the following headings, which summarises the main methods of asexual reproduction in plants: tubers, bulbs, stolons, runners, suckers, cuttings, bud grafting and whip or tongue grafting. You should include details of the techniques used and, where possible, give examples of the crop plants which are propagated in this way.

You may like to divide up the tasks between individuals and then present your findings to the rest of your group or class.

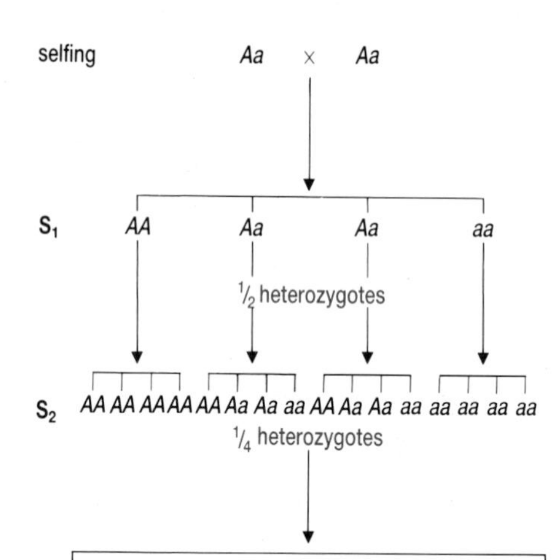

Fig 5.3 The effects of inbreeding. Starting with one heterozygous locus, repeated self-fertilisation over just a few generations rapidly breeds out the heterozygotes, producing two homozygous pure-breeding lines (*aa* and *AA*). We can go through exactly the same process with a population of plants which are heterozygous at many loci, e.g. *AaBbCcDd*, though it will take longer and a large number of different pure-breeding lines will be produced, e.g. *AAbbCCdd, aaBBCCDD, aabbccDD* and so on.

The genetic effects of inbreeding and outbreeding

Inbreeding will tend to promote homozygosity in a population whilst outbreeding tends to promote heterozygosity. The effects of inbreeding are shown in Fig 5.3. By crossing together close relatives or by selfing, we can rapidly produce pure-breeding or inbred lines which contain little genetic variability because they are homozygous at practically all loci. This distinction between the genetic structure of populations of outbreeding and inbreeding organisms is of fundamental importance. The main features are summarised in Table 5.2.

Table 5.2 A comparison of inbreeders and outbreeders

Outbreeder	Inbreeder
has mechanisms which promote cross-fertilisation	has mechanisms which promote self-fertilisation
individuals heterozygous at many loci	individuals approach homozygosity
carries deleterious recessives	deleterious recessives tend to be eliminated
intolerant of inbreeding	tolerant of inbreeding
much heterozygote advantage	less heterozygote advantage?

QUESTION

5.1 Consider two pure-breeding plants with the following genotypes:
(i) *AAbbccDD* and (ii) *aaBBccdd*.

(a) Give the genotype of the F_1 produced by crossing these two plants.

(b) How many pure-bred lines could be produced by selfing the F_1 and subsequent generations until all loci are homozygous?

Inbreeding depression and heterosis

The major point for you to grasp from Table 5.2 is that outbreeders are adapted to being heterozygous and inbreeders are adapted to being homozygous. Interference with this natural state of affairs through a programme of selective breeding can have surprising consequences. This can be seen by forcibly inbreeding a natural outbreeder like maize.

Consider a field containing a wild population of maize plants. There will be considerable variation in the population for a whole host of characteristics. Imagine we find a particular plant in this field which has got extremely large cobs and therefore produces a high yield. It would make sense to breed from this plant. You plant the seeds and when the offspring are ready to reproduce you self them. You want to inbreed the plants because you want to conserve the superior genes that these plants contain. Unfortunately, trying to inbreed any normally outbred organism can have fairly drastic effects upon the plants, as shown in Fig 5.4. There is always a decline in general vigour, size and fertility. Often the inbred lines simply die out. This deterioration in the quality of an outbreeder which is forcibly inbred is called **inbreeding depression**.

Fig 5.4 The effects of inbreeding depression in maize over eight generations. The plant on the left is a very vigorous hybrid whilst the seven plants on the right show the steady decline in vigour as a result of inbreeding.

Fig 5.5 An example of hybrid vigour in maize. The plant in the middle is the hybrid produced by crossing the two parent plants on either side. Notice that the hybrid is not only taller but has much bigger cobs – the structures half way up the stalk.

An interesting phenomenon occurs if we now cross these highly inbred lines of maize to produce hybrids. These tend to show increased vigour relative to the parents, a phenomenon called **hybrid vigour** or **heterosis** (Fig 5.5). As we shall see in the next chapter, the production of better varieties of maize by selective breeding makes use of this phenomenon.

Another excellent example of inbreeding depression is seen in the production of pedigree dogs. Here, dogs from the same breed, for example labradors, are always mated together. These highly inbred strains suffer from a number of distressing conditions which in the wild would seriously reduce their fitness. For example, labradors are particularly prone to arthritis whilst bulldogs and pekinese suffer from problems with their breathing.

The genetics of inbreeding depression can be explained in terms of the accumulation of recessive alleles which, when present in the homozygous state, reduce an organism's fitness. In natural inbreeders, such alleles will tend to be eliminated, by natural selection, from the gene pool since they rapidly become homozygous (see Fig 5.3; assume a is a deleterious allele). However, in natural outbreeders, where most gene loci will be heterozygous, such recessive alleles will not affect the phenotype. Consequently, deleterious alleles can accumulate and reach quite high frequencies in the gene pools of outbreeders. However, inbreeding a natural inbreeder will increase the possibility that an individual will be homozygous for these deleterious alleles. We can demonstrate this effect using the Hardy-Weinberg equilibrium.

Consider a deleterious recessive allele which has an equilibrium frequency $q = 0.0033$. Imagine we are developing an inbred line using a population which contains this recessive allele. Each generation we mate brothers and sisters. The genetic consequences of inbreeding are measured by the coefficient of inbreeding, F. This is discussed in more detail in Chapter 15, but accept for now that when brothers and sisters are mated, $F = 0.25$.

In a randomly mating population the frequency of the homozygote for the recessive allele will be

$$q^2 \approx 10^{-5}$$

In our inbred population, the frequency of the homozygotes for the allele will be given by

$$\begin{aligned} q^2 + pqF &= 10^{-5} + (0.9667 \times 0.0033 \times 0.25) \\ &= 10^{-5} + 8 \times 10^{-4} \\ &\approx 8 \times 10^{-4} \end{aligned}$$

The frequency of the homozygous recessives is thus about 80 times greater in the inbred population compared with the outbred population. A similar increase in the frequency of homozygotes will also occur with respect to other deleterious recessive alleles in the population. Continued inbreeding will amplify this effect further.

QUESTION

5.2 A recessive allele has an equilibrium frequency of 0.0027. Compare the frequency of homozygotes for the recessive allele in a randomly mated population and in an inbred line produced by
(a) selfing ($F = 0.5$) **(b)** mating first cousins ($F = 0.0625$).

pure-breeding line 1 | pure-breeding line 2

P *AAbbccdd* × *aaBBCCDD*

F_1 *AaBbCcDd*

Fig 5.6 A possible explanation of hybrid vigour or heterosis. Both the parental lines are suffering from inbreeding depression because they are homozygous for deleterious recessive alleles. By contrast, the F_1 hybrid is heterozygous at all loci.

One possible explanation of heterosis is that different inbred lines will become homozygous for different deleterious recessive alleles. Crossing two different inbred lines will make gene loci heterozygous with respect to these recessive alleles, as shown in Fig 5.6. An alternative explanation involves the phenomenon of overdominance, where the heterozygote is fitter than either homozygote. A simple example of this is provided by sickle-cell anaemia. Here, individuals who are heterozygous for the sickle-cell gene, Hb^AHb^S, do not suffer from the effects of anaemia and are less likely to catch malaria than individuals with the Hb^AHb^A genotype. However neither of the above explanations is likely to be the whole story and it is probable that epistatic and linkage effects are also involved.

Even though inbreeding outbreeders often results in inbreeding depression, the technique does have advantages because it will increase the phenotypic uniformity of the population. This may be advantageous in terms of, for example, producing plants which grow to the same height, enabling them to be harvested mechanically; or developing strains of animals which grow at an even rate and can thus be slaughtered at the same time. Such advantages may outweigh the problems presented by inbreeding depression.

QUESTION

5.3 A friend is planning to buy a dog as a pet. Explain the advantages of buying a mongrel (that is, a hybrid) compared with a pure-bred animal.

5.2 GENES – THE BASIC RAW MATERIAL

The main aims of the selective breeder are to:

1. refine the existing gene pool of the crop by removing deleterious or sub-standard alleles;
2. recombine the existing alleles in a gene pool in new and interesting ways; or
3. introduce new alleles into the gene pool.

Both 2 and 3 are achieved by making carefully planned **crosses**, that is, mating two plants, between different varieties or even different species. The process of crossing is described in the next chapter. Here we shall look at the possible sources of new genes which a breeder could use.

In the beginning

All modern crop plants ultimately started their existence as wild plants. The early farmers would have made a selection from the wild plants which grew in the neighbourhood and planted them in their fields or gardens. For example, suppose some farmers were trying to grow a plant which produces grain, say wheat. They notice that some wild wheat plants seem to produce more grain (seeds) than other plants. They collect the seed from these plants, plant it in their fields and so produce a new generation of wheat plants. Providing this is an inherited characteristic, the farmers have made a selection of alleles from the total gene pool of the wheat population. They have not got all the possible wheat alleles in the plants growing in their fields, only some of them.

After the initial selection involved in domesticating the crop plant, the farmer would not have stopped there. Every generation seeds would have been collected and planted from those plants which had produced most seed in the previous generation. Indeed, only seed from one plant may be used to raise the next generation of plants. So the gene pool of the now domesticated wheat plants becomes smaller and smaller as the farmer selects only certain plants to breed from. Eventually a situation may be reached when all the alleles of a particular gene are lost except for one and all the plants are homozygous for that gene. At this point the gene is said to be **fixed** within the population. As selection proceeds, more and more genes become fixed.

The farmer may now hit a problem. If the plant being used is a natural inbreeder, like wheat or barley, then the increase in homozygosity brought about by selection will present no problem. Indeed, selection would probably have brought about rapid changes in the quality of the crop in just a few generations. However if the plant is a natural outbreeder, like maize, then the increase in homozygosity caused by selection will start to cause inbreeding depression and the quality of the crop will fall rapidly. So with crops like maize the farmer would need to maintain a variety of different individuals and avoid selecting too rigorously, so maintaining at least some heterozygosity in the population.

Adding new genes

Whilst the gene pool of the domesticated plants was being reduced by the process of selection, two other factors, **mutation** and **introgression**, would have been adding new alleles to the gene pool. Mutation was discussed in Chapter 2, but an important point to notice is that mutations which may be harmful to a wild plant may actually be very useful to the farmer and are therefore preserved by cultivation. For example, at some time in the past a mutation must have occurred which locked maize seeds into a tight cob. This is, of course, extremely convenient for the farmer and would have been selected as a desirable characteristic, but for a wild plant it would have been disastrous since seeds which are tightly bound to a cob cannot be scattered. Nowadays, breeders may induce mutation deliberately in order to provide new variation. This technique is discussed further in Chapter 7.

Introgression is the addition of genes from wild plants to the cultivated plants' gene pool. Whilst natural outbreeders will be more prone to introgression, even natural inbreeders like wheat are cross-pollinated to some extent. The effect of introgression is to add new genes to the gene pool and therefore increase genetic variation. Sometimes this added variation may prove valuable but it may also destroy years of selective breeding.

Land races

By about 150 years ago farmers had made considerable progress towards developing the modern varieties of crops. By conscious selection they had refined the wild gene pool until it contained few deleterious alleles but many desirable ones. They had taken advantage of mutations and the beneficial introduction of new genes by introgression. In the middle of the last century farmers were using available locally adapted populations of inbred plants called **land races**. Whilst these land races would have been considerably less variable than their wild counterparts, they were still quite variable and far from being the simple mixture of pure-bred lines which are available today. Land races are still used today in some parts of the world. For example, many local tropical rices (Fig.. 5.7), sorghums (Fig 5.8) and pulses (Fig 5.9) are still of this character, whilst in Peru farmers may plant a different land race of potatoes in each field since each field is different.

Fig 5.7 A high yielding variety of rice. The rice grains are in the little pod-shaped structures. White rice has all the outer part of the pod removed.

Fig 5.8 A geneticist and a field assistant cross-pollinating varieties of sorghum in Uruguay.

Fig 5.9 Soya beans – an example of a pulse. Each pod will contain many beans.

The use of land races is now being superseded in all parts of the world by varieties which are far less variable. It is these modern varieties which have been produced by the process of scientific breeding based on genetic theory.

The advent of scientific breeding

Modern scientific breeding differs in five important ways from the early attempts of farmers and growers.

1. Modern breeders can add genes to the gene pool of their crops from all over the world. For example, African sorghum is not resistant to mildew (a fungal disease) and the local sorghum gene pool does not contain mildew-resistance genes. But Indian sorghums are resistant to mildew. In the past these genes would have been unavailable to the African farmer but the advent of international organisations like ICRISAT (International Crops Research Institute for the Semi-Arid Tropics) allows the appropriate gene to be brought from India and added to the local gene pool. The only limitation is that the breeder has, of course, got to find a way of introducing the gene into the African sorghums. This is where crossing comes in.
2. Modern breeders have a much better understanding of the genetic processes operating in crop plants. Consequently they can arrange the appropriate crosses to produce the desired offspring. Such forward planning is necessary if breeding programmes are going to achieve their objectives in a reasonable time and at a reasonable cost. However, you should not get the impression that everything is cut and dried. There is still an awful lot that is unknown about plant genetics and you still have to raise thousands of plants to obtain the desired result.
3. Modern breeders understand the reproductive systems of their plants better. For example, the realisation that the sort of breeding programme which is successful with inbreeders will not work with outbreeders was a major step forward in plant breeding.
4. It is now easier to monitor the progress of a breeding programme using modern analytical techniques to assess the quality of, say, seed proteins or specific resistance to a disease.
5. The modern breeder has a complete battery of special techniques which enables specific problems to be overcome. These include induced polyploidy, tissue culture, chromosome manipulation and most recently genetic engineering. These special techniques deserve a section to themselves so they are dealt with in Chapter 7.

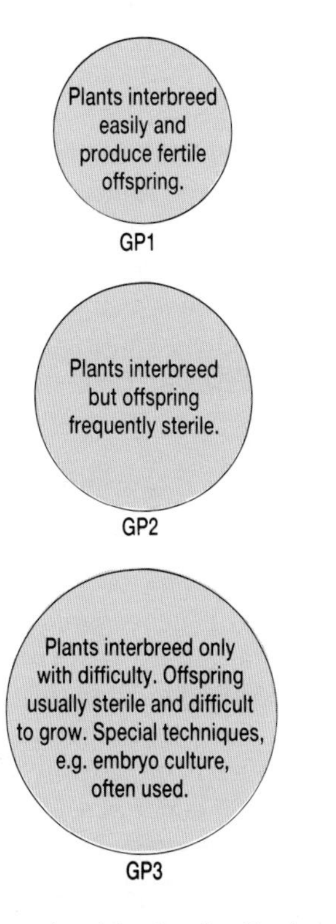

Fig 5.10 Gene pools and plant breeding. The diameter of the circles is proportional to the number of species of plants in each gene pool.

Finding new genes

The starting point for any selective breeding programme must be the identification of a particular characteristic that a breeder wants to incorporate in a new variety. The next problem is to locate the appropriate genes and combine them to produce the desired progeny. Here the breeder might hit a problem. Look at Fig 5.10. This shows a useful way of thinking about the relationship between plants.

Plants of the same or closely related species which interbreed easily producing fertile offspring are said to belong to the primary gene pool, GP1. Exchanging genes between the members of GP1 is relatively straightforward. Plants which belong to related species where crossing is possible but where only a few of the offspring are fertile belong to the secondary gene pool, GP2. Plants in the third gene pool, GP3, can only be crossed using special techniques and the offspring are usually infertile or are difficult to raise to fertility.

As an example, consider the primary, secondary and tertiary gene pools for the domestic tomato, *Lycopersicon esculentum*. GP1 contains all the cultivated varieties of *L. esculentum* plus the wild *L. cerasiforme*, *L. pimpinellifolium*, *L. cheesmanii* and two other wild species of *Lycopersicon*. All of these will interbreed freely with *L. esculentum*. In GP2 there are three more wild members of the genus *Lycopersicon* plus a potato, *Solanum pennelli*. GP3 of *L. esculentum* contains the two remaining wild *Lycopersicons* and two more potatoes, *S. lycopersicoides* and *S. tuberosum*, the potato you eat.

The significance of these different gene pools to the plant breeder lies in the ease with which crosses between plants can be arranged. In the ideal situation the breeder will be able to find the required genes in GP1. Only if GP1 cannot supply the necessary genes will the search be extended to GP2 and GP3. In addition to the difficulty of arranging crosses with members of GP2 or GP3 and the fact that the offspring of such crosses are likely to be sterile, there is another reason for using closely related individuals as a source of new genes.

Background genes

The Plant Breeding Institute (PBI) at Cambridge has achieved notable successes in improving wheat grown in the United Kingdom. This steady progress has been accomplished by crossing varieties of wheat which are already well adapted to growing in the United Kingdom. These varieties contain the correct balance of genes which adapt the plants to life in this country. Indeed, since they have been bred in the United Kingdom they are likely to have many genes in common. Consequently, when two different varieties are crossed they will pass on the vast majority of these essential **background genes** to their offspring, in addition to the special genes that the breeder is trying to combine.

Now consider a variety of wheat from, say, India. This has some excellent genes, perhaps for disease resistance, that the breeder would like to incorporate into a new variety for growing in Britain. The problem that the breeder has is that in addition to these excellent disease-resistance genes, the Indian wheat also contains a whole range of genes which, while adapting it to life in India, would be a disaster in, say, Scotland. Consequently a cross between a British variety and an Indian variety, whilst producing offspring that might well be resistant to a particular disease, would not contain the necessary balance of background genes that would enable it to flourish in the United Kingdom. The plant breeder has to consider the *totality* of the genes in a plant, not just a single gene. In the extreme case, a wild species, which again might contain one extremely useful gene, could also contain a number of positively harmful genes which selective breeding in the past had removed from the cultivated varieties' gene pool.

Backcrossing

Imagine a wild variety of potato contained a gene which conferred resistance to a particular virus. Cultivated potatoes are particularly prone to viral disease so it would be useful if we could transfer this resistance gene in the wild plants to the cultivated ones. The problem is that the wild potato also contains a lot of genes that we don't want. Ideally we would like to simply transfer the resistance gene, but that is not possible.

Look at Fig 5.11. The offspring (F_1) of this **wide cross** (a cross between different species) will get half of their genes from the wild parent and half from the cultivated one. What we now need to do is to dilute the genetic contribution of the wild parent whilst retaining the resistance gene. This is achieved by backcrossing.

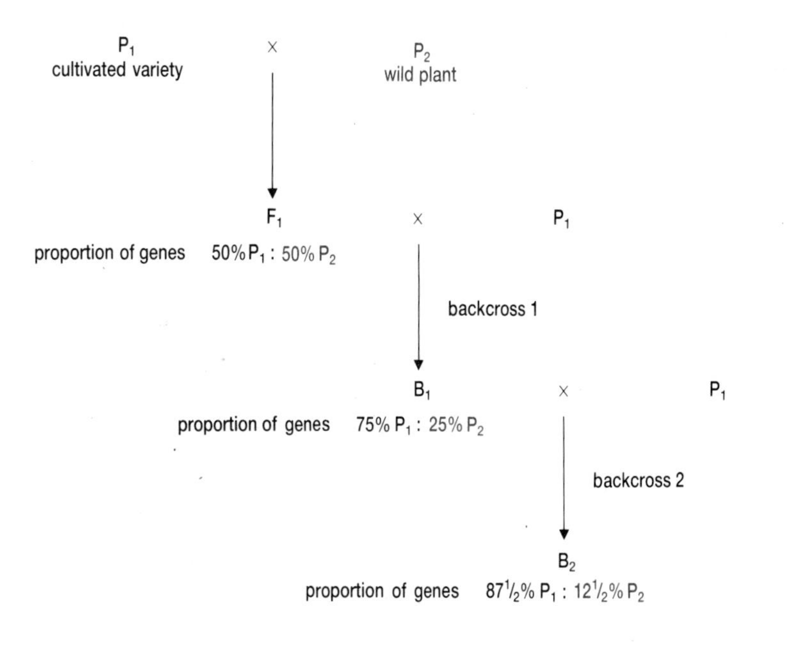

Fig 5.11 Backcrossing. After a wide cross ($P_1 \times P_2$) the number of undesirable genes in the progeny contributed by one parent (P_2) can be diluted by repeatedly crossing the offspring to the other parent (P_1).

The F_1 generation from the cross shown in Fig 5.11 is now crossed with the plants from the same line which provided the domesticated parent (P_1 in Fig 5.11) to produce a new set of offspring. These offspring will be checked for the presence of the resistance gene contributed by the wild parent. Only those plants which contain the resistance gene will be selected to be parents in the next generation. Again, these plants are backcrossed to the domesticated parental line to produce another set of offspring. This process continues for another five or six generations.

This backcrossing procedure achieves two things. Firstly, it preserves the desired gene contributed by the wild parent. Secondly, it dilutes the rest of the genes contributed by the wild parent. The F_1 generation in Fig 5.11 receives 50 per cent of its genes from each parent. After the first backcross the offspring have 25 per cent of their genes from the wild parent and 75 per cent from the domesticated parent. After the second backcross the proportions are 12.5 per cent to 87.5 per cent and so on.

Thousands of genes have been transferred by backcrossing in the last 50 years. For example, 320 of the 596 cultivars of potato grown in Europe contain genes from wild species. Thus *Solanum acaule* has provided resistance genes against potato virus X and potato leaf roll virus whilst *S. spegazzinii* has donated genes conferring resistance to nematodes and the fungus *Fusarium coeruleum*.

Once the foreign gene has been introduced into a domesticated variety then that variety can be used to introduce the gene into other domesticated varieties. For example, genes for resistance to the fungus disease eyespot were transferred from goat grass, *Aegilops ventricosa*, to the French wheat line VPM1 by backcrossing. These genes have now been passed from VPM1 to another variety of wheat called Rendezvous.

QUESTIONS

5.4 Draw a diagram to summarise the relationship between GP1, GP2 and GP3 of the domestic tomato. Why would such a diagram be useful to a selective breeder?

(continued)

5.5 **(a)** What is backcrossing?

(b) Calculate the relative proportions of wild and domesticated genes in the offspring after 0, 1, 2, 3, 4 and 5 backcrosses. Plot a graph to show the relationship between the proportion of genes derived from the wild parent and the number of backcrosses. Describe the shape of your graph.

(c) Why do you think breeders usually stop after five or six backcrosses?

5.6 Why would breeders prefer to use a cultivated variety, like VPM1, to introduce a wild gene into another cultivated variety rather than the wild plant itself?

5.7 The following extract was written by William Cobbett in his *Cottage Economy* 1822:

This very year, I have some Swedish turnips, so called, about 7000 in number, and should, if the seed had been true, have had about twenty tons weight, instead of which I have about three. Indeed they are not Swedish turnips, but a sort of mixture between that plant and rape. I am sure the seedsman did not wilfully deceive me. He was deceived himself. The truth is that seedsmen are compelled to buy their seeds of this plant. Farmers save it: and they but too often pay very little attention to the manner of doing it. The best way is to get a dozen of fine turnip plants, perfect in all respects, and plant them in a situation where the smell of blossoms of nothing of the cabbage or rape or even of the charlock kind, can reach them.

Using the above extract answer the following questions:

(a) What do you think William Cobbett meant by the 'smell of the blossoms'?

(b) What sort of breeding system does the 'Swedish turnip' have? Explain your answer.

(c) What plants would you expect to find in the primary gene pool of the 'Swedish turnip'?

(d) If the 3 tons of plants which grew in Cobbett's garden were all sterile would this change your answer to question (c)?

(e) Imagine you were to meet Cobbett. Try to explain the genetics behind his disastrous Swedish turnip crop bearing in mind the date is 1822.

5.3 A CLOSER LOOK AT SELECTION

Investigation 5.2 is designed to show you how powerful selection can be and how fast it can change the appearance of a population.

INVESTIGATION 5.2

A population of plants is growing in a field. The seeds of the plants germinate in the autumn and the plants overwinter as small seedlings. In the following spring the seedlings continue to develop, they flower in the summer and produce seeds and then die. The seeds germinate in the late summer and the cycle continues.

Two distinct phenotypes can be recognised in the population. Plants of one type are resistant to mildew which attacks the seedlings during the winter. The other type of plant is susceptible to attack by mildew. The two phenotypes are called 'resistant' and 'susceptible' respectively.

Some research on this population of plants uncovers the following points.

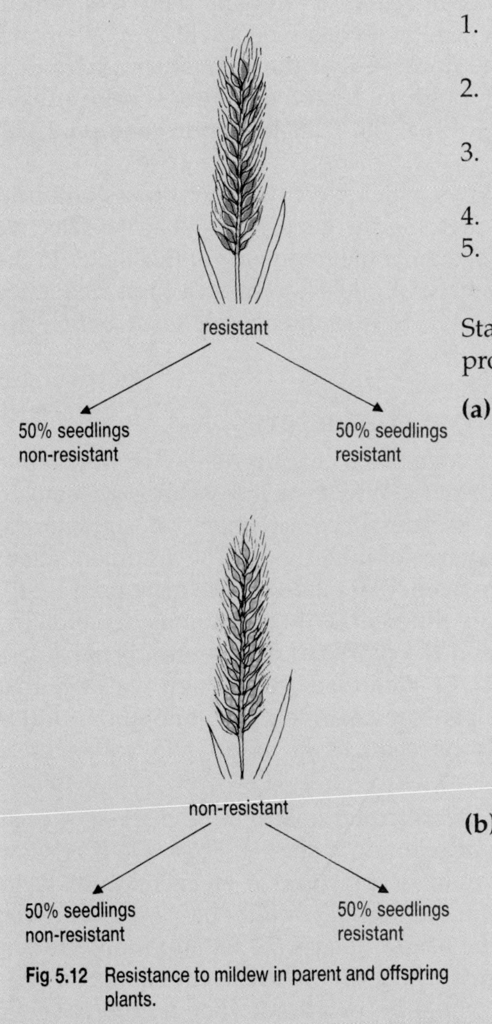

Fig 5.12 Resistance to mildew in parent and offspring plants.

1. 30 per cent of resistant plants survive the winter on average, whereas only 10 per cent of susceptible plants do.
2. At other times of the year, the two phenotypes are equally viable, that is, they are equally liable to survive.
3. The two types of plant produce the same amount of seed in the autumn.
4. All plants that survive the winter reproduce.
5. Each plant that reproduces produces five seedlings which develop in the late autumn.

Starting in the autumn of year zero, you are going to monitor the progress of 100 resistant seedlings and 100 susceptible seedlings.

(a) First assume that half of the seedlings that develop from the seeds of any plant are resistant and half are non-resistant, regardless of whether the parent plants are non-resistant or resistant (see Fig 5.12). That is, assume that offspring do not inherit resistance or susceptibility to mildew from their parents. Using this assumption, work out the numbers of resistant and susceptible seedlings that would be expected in the population each autumn during the three years following the start of the experiment. Remember, in the first autumn there are 100 resistant and 100 susceptible seedlings in the field. Plot the result as a graph of numbers of seedlings present in each autumn against the number of autumns after the start of the project. Your graph will have two lines, one for susceptible and one for resistant seedlings.

(b) Now assume that all of the seedlings that develop from resistant plants are themselves resistant and all of the seedlings that develop from susceptible plants are susceptible. Again, starting with 100 resistant and 100 susceptible seedlings in the field calculate the number of each type of seedling expected in the population each autumn for the three years following the start of the experiment. Plot the numbers of resistant and susceptible seedlings that would be expected each autumn during the three years following the start of the experiment. Again, you will have two lines.

(c) Now compare the two graphs. What conclusions can you draw from them?

(Adapted from an exercise in S102 – a Science Foundation Course, Open University.)

Selection and gene pools

The investigation you have just completed shows that provided a characteristic is inherited then selection can rapidly alter the composition of a population. This is what a selective breeder is trying to do. You can imagine that instead of letting the mildew kill off the susceptible plants, the breeder was the selective agent. However, we are only looking at the change in phenotypes. We also need to understand what is happening at the genetic level and particularly what is happening to the frequency of different alleles in the gene pool. The **allele frequency** is defined as **the relative proportion of the alleles of a gene present in a population.**

Let us assume that resistance to mildew is controlled by one gene which has only a pair of alleles, *R* and *r*, where *R* is dominant to *r*. Individuals which are resistant to mildew are *rr*. Look at Fig 5.13. The plants in the field are outbreeders so mating between the plants occurs at random. Consequently three genotypes can occur in the population. However two of those genotypes, *RR* and *Rr*, are susceptible to mildew and so they do badly. As a result the frequency of the *R* allele in the gene pool decreases

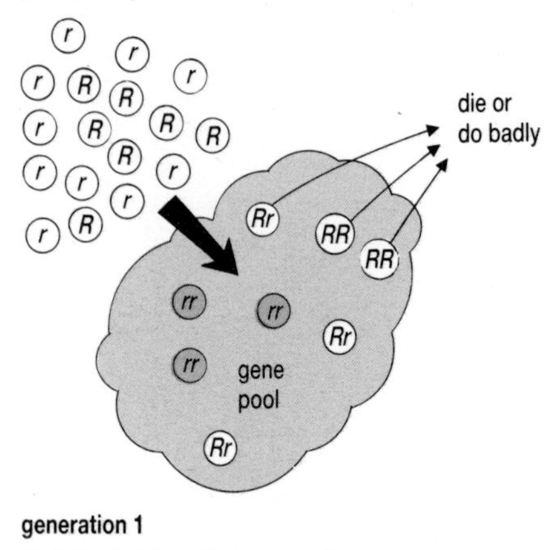

generation 1

R alleles lost from the gene pool

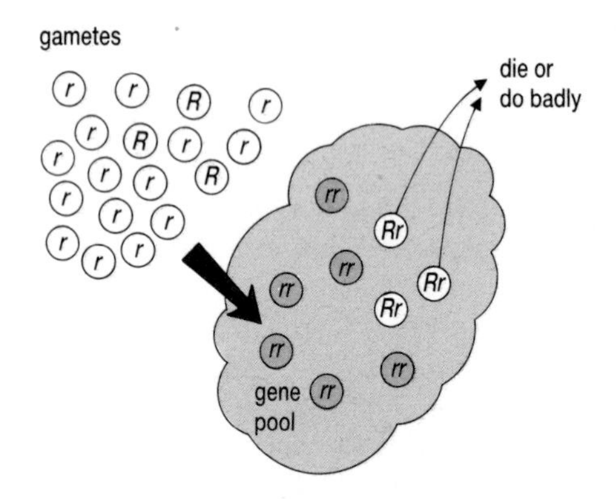

generation 2

Increased frequency of *r* alleles in the gene pool

Fig 5.13 Changes in the frequency of alleles in a gene pool as a result of selection. Allele *r* confers resistance but is recessive to allele *R* which confers susceptibility to mildew infection on the phenotype. Consequently both *RR* and *Rr* genotypes are eliminated from the gene pool.

whilst the frequency of the *r* allele increases. Eventually there will come a time when the only allele of this gene in the gene pool will be *r*. When one allele has spread through a gene pool, so that the only other alleles of it present in the gene pool arise through mutation, the allele is said to have achieved **fixation**. Given enough time, the *r* allele in our gene pool will become fixed.

So selection acting on phenotypes which have different fitness will lead to a change in the frequency of alleles in the gene pool. What the selective breeder is trying to do, then, is to change the frequency of the alleles in the gene pool of the crop so that beneficial alleles occur at a high frequency whilst deleterious alleles occur at a low frequency or, even better, are eliminated altogether.

Selection and quantitative characters

The example we looked at above concerns a character showing qualitative variation controlled by one major gene. Whilst such resistance systems do occur in plants (see Chapter 8) most characteristics of agricultural importance show quantitative patterns of inheritance. The argument about selection changing the relative frequency of alleles in the gene pool is still valid but the phenotypic effects are different. With continuous variation the relationship between the genes and the character they control is much less obvious. Individual genes cannot be identified and though we *know* that the allele frequencies in the gene pool are changing we cannot monitor this directly. Instead we have to rely on changes in means and variances to chart the progress of selection.

Three types of selection can be identified for quantitative characters, each of which could be important in developing a specific crop (Fig 5.14). For example, directional selection would be practised if an increase in yield was required, whilst stabilising selection would be important for maintaining uniformity in, say, the height of a crop – an important characteristic on highly mechanised farms where the crops are cut using combine harvesters. Disruptive selection would be practised when a crop is being developed for two different markets. For example, varieties of barley have been developed which, because they contain low levels of nitrogen in the grain, are suitable for malting and making beer. By contrast, barley used for feeding livestock requires high levels of nitrogen.

QUESTION

5.8 For making biscuits, wheat flour with a low gluten level is needed whilst for making bread, flour with a high gluten content is needed. Gluten content is a quantitative character.

(a) Starting with a wheat with intermediate gluten content, describe the selection procedure you would use to produce the two varieties of wheat needed for these purposes.

(b) How would you estimate the response of the plants to your selection procedure?

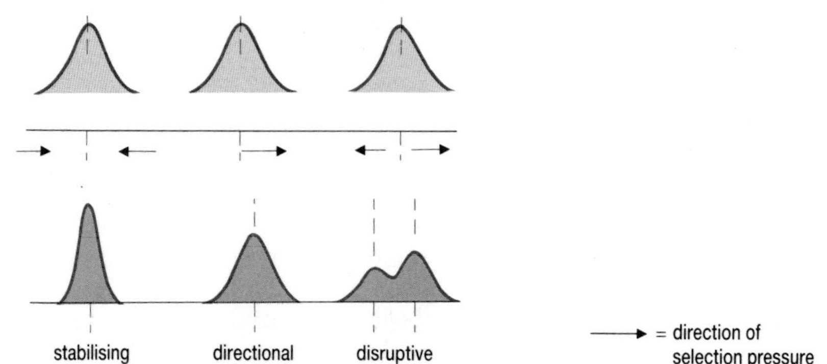

Fig 5.14 Stabilising, directional and disruptive selection. Note how the original frequency distribution of the phenotypes (upper row) may change according to the nature of the selection pressures operating in the environment.

SUMMARY ASSIGNMENT

1. Make sure that you have definitions, explanations and, where appropriate, examples of the following: inbreeder, outbreeder, inbreeding depression, heterosis, gene pool, allele frequency, backcrossing, fixation, land race.
2. What advantages do modern breeders have over plant breeders working in the middle of the last century?
3. **(a)** Why do modern varieties of crop plants and domestic animals contain only a small proportion of the total genetic variation present in the species?
   (b) How can plant breeders introduce new genes into a variety?
   (c) What are the disadvantages of introducing genes from wild species into specialised varieties of domesticated plants?

Chapter 6

SELECTIVE BREEDING IN PRACTICE

In the previous chapter we looked at where plant breeders could obtain the genes they require (GP1, GP2 or GP3) and how a gene from a wild plant could be introduced into a cultivated variety by backcrossing. Ideally, a plant breeder would like to combine genes from plants which both belong to GP1 and which are adapted to the same environment. The actual breeding scheme used will depend upon the breeding system of the plant. In particular, selective breeding is more straightforward with inbreeders than outbreeders.

LEARNING OBJECTIVES

After completing the work in this chapter you will be able to:

1. describe the major breeding methods used with inbreeding and outbreeding plants;
2. assess the role of hybridisation in plant breeding programmes;
3. evaluate the different forms of plant breeding in terms of their usefulness for producing new crop varieties in different parts of the world.

Fig 6.1 Varieties of barley in a breeding trial in Iraq. The genotypes of the plants in row 7 will be different to those in row V.

Fig 6.2 Varieties of wheat in a breeding trial. Notice that the plants in this trial are being grown in blocks.

6.1 THE PLANT BREEDING PROCESS

As you work through this chapter you will find it useful to have a mental picture of the plant breeding process. Generally, plant breeders do not work in the laboratory. Rather they use experimental fields like the one shown in Fig 6.1. Notice how all the plants are in rows. Usually each row will contain a **line** of plants. The term line has a very special meaning in plant breeding. For example, an inbred line is a group of plants which are nearly homozygous and are practically identical to each other. So the plants in each row in the field in Fig 6.1 will have practically identical genotypes, but the plants in different rows will have different genotypes. Sometimes the breeder will grow the plants in experimental plots (Fig 6.2) rather than rows. Each plot will then usually contain plants from the same line. So when you meet the term 'selection between lines' you know that the plant breeder is comparing plants in *different* rows or plots, while 'selection within lines' means that the plant breeder is comparing plants in the *same* row or plot.

6.2 INBREEDERS

The strategy the plant breeder employs with natural inbreeders like wheat will depend upon the material available. If the starting point is a variable land race then the breeder will start by refining the gene pool of this population. If the starting point is the variable population produced as a result of crossing two well-defined inbred lines then the breeder will use a different set of selection techniques. Each of these is reviewed in turn.

Refining the gene pool

Over the last 150 years, plant breeders have refined the variable gene pool of land races to produce the uniform gene pool of modern cultivars. For example, the wheat plants growing in a field in Turkey or Iran 30 years ago would have been genetically quite different from each other. By contrast, the wheat plants that you would probably see growing in the same fields today are, to all intents and purposes, practically identical to each other. In this section we are going to examine the techniques which plant breeders have used to effect this revolutionary change.

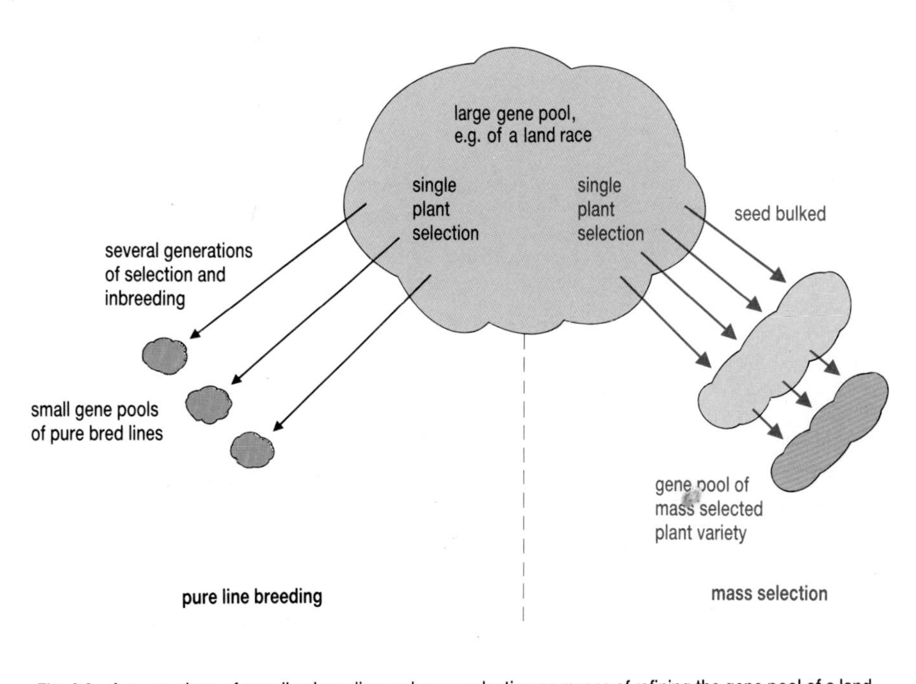

Fig. 6.3 A comparison of pure line breeding and mass selection as means of refining the gene pool of a land race. Both methods depend initially on single plant selection, but pure line breeding produces several varieties (pure-bred lines) with small gene pools whilst mass selection produces one variety with a larger gene pool.

Look at Fig 6.3. This summarises the processes involved in the initial stages of selective breeding. The individual plants forming the land race would each be highly homozygous but they would be different from each other, for example one plant might be *AAbbccDDee* whilst another might be *aaBBccDDEE*. Starting with the land race, the gene pool is fragmented and reduced to a series of inbred lines. Because these lines have been produced by inbreeding, the plants in each line are genetically identical but the plants in separate lines are very different. You can see this in Fig 6.4. Remember, this continual selfing can only be tolerated by natural inbreeders because they are adapted to high levels of homozygosity. This effect can be achieved in two ways.

Fig 6.4 A breeding trial to test different inbred lines of barley. The three rows on the middle left are a short Mexican variety, which is more resistant to wind damage than the taller variety on the right.

Pure line breeding. This approach can be broken down into three stages.

Step *1.* A large number of individual plants (so-called single plant selection) are selected from the genetically variable land race. The quality of the chosen plants is critical since they will be the founders of the inbred lines. The number of plants chosen at this stage will depend upon a range of factors, such as time available, expense, type of plant, but is usually between a hundred and several thousand.

Step *2.* The seeds collected from individual plants are sown in rows so that the seed from one plant occurs in just one row. So one row in the trial plot contains one line, for example row 6 contains line A35 and so on. The progeny are examined as they grow for any weaknesses, for example susceptibility to disease, and those lines which do not make the grade are eliminated, that is, the whole row is destroyed. This evaluation by eye may continue for several years, with the number of lines of plants being gradually reduced.

Step *3.* Eventually a stage is reached when the breeder can no longer choose between the lines by eye alone, and the plants are now subjected to replicated trials in which each line is exposed to a variety of environments and its commercial value, for example in terms of yield, is compared with existing commercial varieties. The length of time required for this stage varies but usually it lasts for about three years.

Examples. The red winter wheat varieties common in the United States of America in the 1960s, for example Kanred, Blackhull, Nebred and Cheyenne, were all produced by this method from the land races known collectively as Turkey or Crimean wheat.

Mass selection. The aim of pure line breeding is to rigorously refine the gene pool so that the end product is a set of highly homozygous inbred lines. This procedure requires time and is expensive. Mass selection is a more relaxed way of refining the gene pool. The idea is that once again individual plants are selected, but the seed from these plants is not kept separate as in pure line breeding but is put together to form a bulk. This bulk of seeds is then sown to produce the next generation. Again, the best plants will be selected but their seed will be combined to produce the next generation. This procedure can be repeated as many times as is necessary.

So mass selection differs from pure line breeding in two ways. Firstly, a number of different plants contribute seed to the new variety rather than just one. Secondly, whilst the varieties produced by mass selection contain fewer genotypes than the original parental population, they do contain more than the single genotype of varieties produced by pure line breeding.

Mass selection has been of less importance than pure line breeding in refining the gene pool of crop plants used in the developed world. It does, however, enjoy two advantages. Firstly, it is quicker and cheaper to implement than pure line breeding because it reduces step 2 and eliminates step 3 mentioned above. This is important in developing countries where land races are still used for food. Mass selection will eliminate rapidly those plants of low agricultural value, and therefore the alleles they contain, so the land race can be rapidly improved and released to farmers without the need for the extensive and time-consuming testing required to develop pure line varieties. Secondly, mass selection, whilst eliminating poor alleles, will still preserve the local adaptation of the land race. This second aspect is important because farmers in poor countries will not have access to the fertilisers and pesticides needed to ensure that highly inbred, pure line crops stay healthy. By contrast, the land race will already be well adapted to local growing conditions and local pests because the gene pool contains the necessary background alleles. So a land race improved by mass selection will not produce as high a yield as a pure bred variety, but you do not get the high yields from the pure bred varieties unless you pamper them with nitrogen fertiliser and a battery of chemical pesticides. Remember, genetic improvement of crops and improvements in farming techniques go hand in hand.

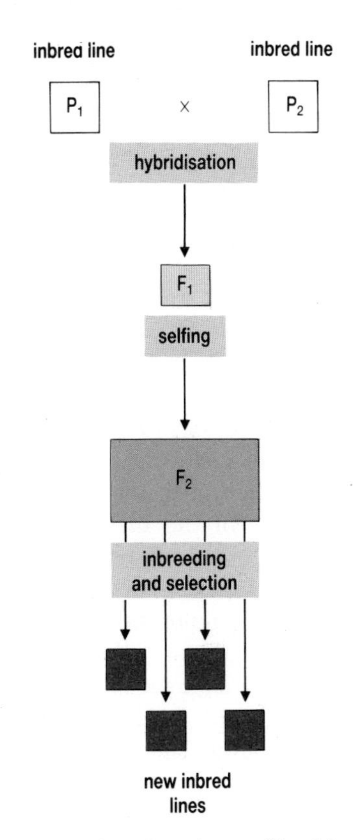

Fig 6.5 A breeding scheme for use with an inbreeder like wheat. Note, the width of the boxes reflects, approximately, the amount of genetic variation present at each stage of the breeding procedure.

Pedigree methods

Having produced a series of highly refined, pure bred lines, what does the breeder do next? Each line will not contain all the desirable genes present in the original gene pool. For example, variety 1 may be an excellent yielder and exactly the right height but it may be susceptible to frost damage. Variety 2, on the other hand, does not produce quite as much grain as variety 1 but it also is the correct height and it is frost hardy. The obvious thing for the breeder to do is to try and combine the attributes of the two varieties. For example, at the Plant Breeding Institute in Cambridge a selective breeding programme is under way to combine the high yields of a variety of wheat called Norman with the superior grain quality of the wheat variety called Avalon. The procedure for doing this is shown in Fig 6.5.

The aim is to cross two (or more) plants from different inbred lines, so producing a variable population. This variable population (the F_2 in Fig 6.5) produced by hybridisation of two inbred lines is the starting point for the programme of selection, based on inbreeding, which will hopefully produce the new variety with the desired characteristics. We will examine each stage of the process in turn.

Choosing parents. The selection of the parents is a critical step in the breeding process since they ultimately provide the genetic variability for the programme of selection. The choice of parents is largely a matter of experience. The number of parents to use is also a problem. In Fig 6.5 only one cross is shown. In reality, many crosses between different parents will be made to produce the variable population which is the starting point for the selection process. For example, the F_1 of the cross shown in Fig 6.5 could then be crossed with another pure-breeding line to produce another F_1 generation which is again crossed with another pure-breeding line. The combinations are seemingly endless. The most common method used to generate the variable population is called **diallel crossing**. Here, all the chosen genotypes are crossed in all possible combinations to produce the variability needed for selection.

Fig 6.6 The challenge faced by maize breeders. To ensure that pollination was controlled rows of plants used to be detasselled by hand. Using the phenomenon of male sterility has now greatly reduced labour costs and so the cost of F_1 hybrid seeds to the farmer.

Hybridisation. The actual process of crossing the parents represents an enormous practical problem. Remember, the plants we are dealing with are natural inbreeders so the breeder has got to stop them from selfing. This could be achieved by removing the anthers from all of the plants of one inbred line, which then becomes the female parent, by hand. The enormity of this task can be judged from Fig 6.6. This is obviously a time-consuming business, so breeders often resort to introducing genes for male sterility into one of the inbred lines. Male sterility will be dealt with in more detail in the next chapter. The effect is equivalent to the hand emasculation described above. The line containing the male sterility genes does not produce any pollen and so it cannot self-fertilise. During selection the male sterility genes are removed so that the final product is once again self-fertile. This technique has been used widely, for example, in barley.

If one of the original parents in Fig 6.5 had been a wild variety which was being used as a source of just one or a few favourable genes, the gene pool of the variable F_2 generation would now be refined using the backcrossing method discussed earlier. However, if the hybridisation programme involved well-adapted, pure-bred lines, two other selection procedures could be used: pedigree selection or bulk selection.

Pedigree selection. A pedigree is a record of the ancestry of an individual. In pedigree selection individual plants are chosen to form new lines, not only on their appearance but also on the basis of their ancestry. So pedigree breeding involves keeping very detailed records of the history of each experimental line. Remember, after the initial cross-fertilisation used to produce the F_1 all reproduction involves self-fertilisation.

Individual plants are selected from the variable F_2 generation to produce the F_3. One obvious basis of selection would be to eliminate all those plants which were carrying undesirable major genes. The seeds from each selected F_2 plant are sown in a row to produce the F_3. So the F_3 plants in each row will be the descendants of just one F_2 plant. This relationship is recorded, for example row 17 contains the progeny of plant A27 and so on. Because the plants in each row are descended from a common ancestor they are said to form a **family**.

In the F_3 generation we will be able to see not only differences between plants within a family (that is, within a row) but also differences between families (that is, between rows). For example, if you were to walk about the experimental field you might notice that plant 6 in family row 1 seemed to be growing better than plant 12 in family row 1. So plant 6 is one of the plants selected to produce the next generation rather than plant 12. This is selection *within* a family. But you might also notice that on average all the plants in family row 1 were growing better than those in family row 8 so the family of plants growing in row 8 might be eliminated from the breeding programme. This is selection *between* families. From the F_3 plants the breeder makes a further selection, but a careful record is made of which family the chosen plants came from.

Using the selected F_3 plants, the F_4 is produced. By now selection in the F_2 and F_3 generations will have eliminated many families with obvious shortcomings. Furthermore, by the F_4 genetic differences between the individuals in the same family will be small because two generations of inbreeding has reduced the level of heterozygosity in the plants. So whilst some single plant selection within a family is still possible, the main emphasis of selection will switch to selection between families and this is where those carefully kept pedigree records start to become useful.

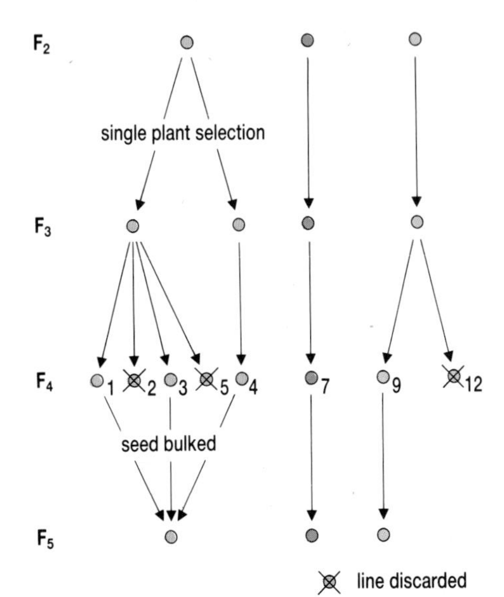

Fig 6.7 Using pedigree records. Three lines of F_2 plants have been produced and these have been used to produce four and eight lines of F_3 and F_4 plants respectively. Pedigree records show that lines 1, 3 and 4 are related. Since all these lines are performing equally well in field trials, the seed from them can be bulked to produce just one line for producing the F_5 Lines 2, 5 and 12 are not performing well so they are eliminated from the breeding programme.

Look at Fig 6.7. Here there are eight lines (families) of F_4 plants in a field, five of which (1, 3, 4, 7 and 9) are growing superbly. The breeder eliminates 2, 5 and 12. However, by looking at the pedigree record of each line, the breeder is able to see that families 1, 3 and 4 all have the same ancestor, that is, they are all part of some bigger family. The same is true of you. You have your immediate family, brothers, sisters, mother and father. But you also have a wider family to whom you are related through your grandparents, aunts, uncles and cousins. Lines 1 and 3 in Fig 6.7 are 'sisters' whilst line 4 is their 'cousin'. One obvious difference is that your family has not been produced by inbreeding so you and all your relations are genetically different from each other. The same is not true of the plants in lines 1, 3 and 4. Since there is so little to choose between these related families, both phenotypically and genotypically, only one line is needed to perpetuate the lineage. So the seeds produced by plants in lines 1, 3 and 4 can be bulked together to produce one line. This is the advantage of pedigree breeding. It allows the number of lines which have to be grown to be reduced whilst still maintaining the best combinations of genes from the original variable F_2 population. Without the pedigree record the breeder would not have been able to bulk the seed from lines 1, 3 and 4 in Fig 6.7 because the relationship between the plants would not have been obvious.

In the F_5 generation the surviving families are usually planted not in rows but in blocks which allow better comparisons among families than do long single rows. By now the differences within a family will be so small that selection is almost entirely between families. Again, using the pedigree record the breeder can further reduce the number of families that have to be tested in the F_6. The selection procedure adopted in the F_5 generation is adopted in the F_6 and F_7 generations. So by the end of the F_7 there will be a few outstanding but only distantly related families left to enter the next stage of the selective breeding programme – trialling.

Trialling. During final testing and trialling the remaining families are examined for any serious weaknesses or defects, their produce is analysed for quality, for example the protein content, and precise yield tests are made. All of this may take five or more years. So from the start of the breeding programme to the end takes at least 12 years. However, if two or more generations can be grown in a year this process can be speeded up. One way of doing this is to use a modification of the pedigree method called single seed descent.

Single seed descent. The pedigree breeding method has two aims: to obtain homozygous or pure-breeding lines and to select among these lines for agriculturally desirable characteristics. If both aims are pursued at the same time then this requires large numbers from the F_2 onwards and only one generation a year may be grown because the plants must be examined under field conditions. The single seed descent modification seeks to separate these two aims.

Fig 6.8 A wheat breeder checking plants in a controlled environment growth room. Using these artificial conditions enables three generations of plants to be raised in one calendar year, so speeding up the breeding process.

Large numbers of F_2 plants are grown, but rather than a whole row of plants being produced from a single F_2 parent, only a few seeds are allowed to grow. In the extreme case only one seed from each F_2 plant selected will be used, hence single seed descent. This same pattern, of only allowing a small number of progeny to be produced by each parent, is continued to the F_6. The advantage is that the F_6 can be reached in only two years because two generations can be grown in a greenhouse during the winter (Fig 6.8) and one generation can be grown in the field during the summer. So each year three generations can be grown. After just two years the breeder has available a large number of inbred lines, but each inbred line is represented by only a few individuals. The number of individuals in each line can now be increased by planting a lot of seed from each individual and selection can now start. This procedure effectively cuts four years off a breeding programme.

Bulk selection

The alternative to pedigree selection is bulk selection. This involves planting the F_2 generation in a plot large enough to accommodate several hundreds or thousands of plants. When the crop is mature the plants are harvested in bulk and the seeds used to plant a similar plot next year. This process is repeated for several years. During this period natural selection will be shifting the gene frequencies by weeding out weak or disease-susceptible plants. Artificial selection can also be practised. Its mode of operation is made clearer by using an example.

Nilsson-Ehle, a Swedish plant breeder, was interested in combining the winter hardiness of the Squarehead variety with the high yield of the Stand-up variety of winter wheat. He crossed the two varieties and then carried out bulk selection from the F_2 onwards. He assisted natural selection by throwing away plants that had suffered winter frost damage but had not actually been killed. In this way he increased the rate of shift towards hardy types, in other words he was practising directional selection. He recognised that growing large populations would increase the chance of high yielding types appearing among the winter hardy selections and that the bulk population method was ideally suited to the large numbers involved because he would not have to keep pedigree records. He also realised that homozygosity would increase during the period of bulk breeding.

After a few generations of bulk breeding, which may or may not involve mass selection, single plant selections are made. In Nilsson-Ehle's case this would have involved choosing the high yielding plants from the winter hardy population produced by the bulk breeding procedure. Pure line breeding or even pedigree selection can then take over to produce the final variety.

Conclusion

This has been a long section, but it was necessary for you to get a feel of what the mechanics of plant breeding are actually like. Each year a plant breeder will be carrying out new crosses as well as continuing to assess the plants from crosses made perhaps 10 or more years ago. Keeping records is essential and the computer has greatly eased this aspect of the plant breede' s work. To give you an example of the scale of the task, read the following extract about the breeding of wheat at the Plant Breeding Institute located near Cambridge.

'At PBI, John Bingham and his colleagues make about 1100 different crosses in a given year. When I visited PBI early in 1985, 2000 – 3000 varieties were being 'actively considered' for inclusion in breeding programmes. In the end, in that particular year, the PBI breeders made about 8000 crosses, between pairs of varieties drawn from 87 different types. Between them, the different crosses generate around two million progeny by the second generation – about 2000 grandchildren from each of the original 1000 crosses.

Inspection of the F_2 invariably shows that many of the original crosses are of no value, and those particular breeding lines are simply discontinued. All the F_2 progeny from that particular cross are rejected. From other crosses, perhaps only half a dozen F_2 individuals are retained and in other cases, several hundred F_2 progeny are kept. In total around 50 000 of the two million F_2 individuals are kept: the seeds from one of the heads of each one are sown in a row, to produce 50 000 seed rows of F_3s.'

Colin Tudge, *Food Crops for the Future* (Basil Blackwell, 1988)

QUESTIONS

6.1 (a) Starting with two loci, each with two alleles, for example *A* or *a*, *B* or *b*, how many different pure-breeding lines could you make? Repeat your calculation using one, three and four loci. Can you detect any mathematical relationship between the number of loci and the number of pure-bred lines that you can produce?

(b) What would happen to the number of pure-bred lines if you had three alleles at each locus rather than just two? Try to make your answer as quantitative as you can.

(c) What are the implications of your findings for a plant breeder faced with the task of refining the gene pool of a land race?

6.2 Briefly summarise the different breeding programmes described above in the form of a table under the following headings:

Parents	Hybridisation	Source of variation	Selection method	End product
land race	none	Differences between homozygous individuals	Single plant selection followed by pure line breeding	Highly homozygous inbred lines

The first entry has been completed to assist you.

6.3 OUTBREEDERS

Any breeding programme involving an outbreeder is essentially a compromise between two conflicting requirements. On the one hand, the breeder wants to produce as uniform a crop as possible. Ideally this would take the form of an inbred line with a high degree of homozygosity. On the other hand, outbreeders in general will not tolerate inbreeding so the breeder has to maintain a certain level of heterozygosity in the plants. The problem here is that these plants will not breed true. The aim of the breeder is therefore to produce a population of plants which is not too variable, otherwise it will not be easy to manage in the field and it will not give the required performance. For example, if we were to produce maize which was too genetically variable we would end up with a field of maize plants showing great variation in height. Since maize is harvested mechanically, this would present real problems for the farmer and increase the cost of crop production. But the plants that we do produce must be sufficiently heterozygous to avoid inbreeding depression. Bearing these constraints in mind, let us now turn to the three breeding methods used for outbreeders.

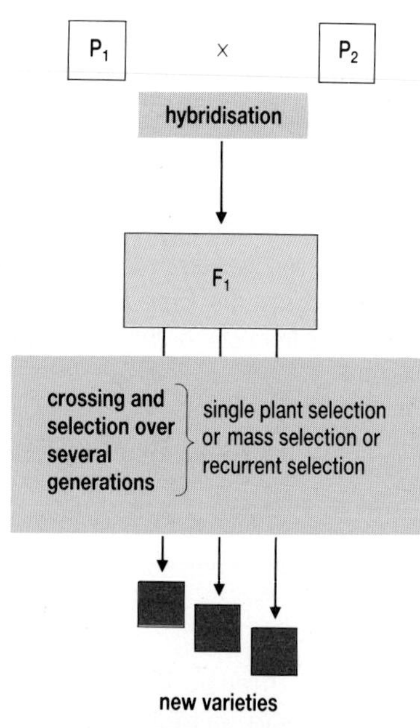

Fig 6.9 A breeding scheme for use with an open-pollinated population, e.g. rye. Note, the width of the boxes reflects, approximately, the amount of genetic variation present at each stage of the breeding procedure.

Open-pollinated populations

With crops like rye, many maizes and sugar beets the strategy adopted by the plant breeder is to change the frequency of alleles in the crop's gene pool so that some favourable alleles approach fixation, that is, are present in the homozygous condition, while maintaining high levels of heterozygosity at other gene loci. This may sound rather daunting, but basically what the breeder is trying to ensure is that all the plants in a variety are homozygous for certain important alleles but remain heterozygous for other background alleles. The breeding scheme is shown in Fig 6.9. Notice that the variation in the F_1 is much higher than in the F_1 of the cross between two inbred lines. The F_1 is examined for the appropriate characteristics and a selection made. These are then used as the parents in the next round of reproduction. This procedure continues until eventually the breeder will end up with the population of plants that is required.

An example: millet at ICRISAT. Millet is an important crop in much of the semi-arid tropics. It is an outbreeder, so a natural population would be highly heterozygous. Such a population would look highly variable, and the gene pool will contain a large number of alleles which may be extremely useful to the farmer but also a large number of alleles which the farmer will not want. Furthermore, if the useful alleles are recessive then they will only be seen in those rare plants which are homozygous for the recessive allele. What the breeders at ICRISAT have tried to achieve is to locate and fix the beneficial alleles, for example those which determine the number of seeds, and decrease the overall level of heterozygosity in the plants. This makes the crop more uniform whilst avoiding the effects of inbreeding depression. The programme is described below.

Plots of millet were established in 1975. Within each plot different characters are selected for. For example, in one plot large head size; in another short plants are selected whilst large ones are rejected. The plants in each plot are only allowed to breed with plants in the same plot. This reduces the overall heterozygosity of the population, that is, the plants within each plot become increasingly inbred. Any plants which suffer from inbreeding depression are rejected so that the total number of deleterious alleles in each plot is slowly reduced and the plants become increasingly tolerant of homozygosity. The overall effect is that plants within plots become more similar but the differences between the plots increases. Occasionally, plants from two different plots are crossed (this could be P_1 and P_2 in Fig 6.9) to begin a new plot. The selection procedure outlined above is then applied so that the gene pool of the plants within the plot becomes increasingly refined until the new variety is produced. This can then be sold to the farmer as a new variety. The effect of this selection procedure can be quite impressive. In 10 years, yield in some plots has increased by 15 per cent and the heads of some plants are now 30 centimetres long.

You should recognise this selection procedure as **single plant selection** but modified so as to reduce inbreeding depression. The breeder also has available three other selection procedures.

Mass selection. A large number of plants from the variable F_1 population or even a primitive land race are chosen and their seed bulked. The seed is planted and selection applied against the resulting progeny. The plants left after selection are allowed to open-pollinate. This means the breeder has no idea which plants are the male parents because pollination has not been controlled. The cycle is then repeated. Since the breeder has no control over the male parent, selection is only being applied against the female.

This procedure is quite good at increasing the gene frequencies of readily observable characters. In maize, for example, mass selection has been used to develop varieties differing in grain colour, plant height, size of cob, date of maturity and percentage of oil and protein in the grain. Mass selection has even been effective at modifying yield in maize. For example, in one programme yields were improved for 16 generations until all the genetic potential for improvement had been exploited. However, mass selection suffers from three main drawbacks (notice we are talking here about mass selection as applied to outbreeders).

1. It is difficult to identify superior genotypes simply by looking at one plant. This can be overcome to some extent by progeny testing where a few of the seeds collected are sown and the performance of the progeny is used to judge how good the parent was.
2. Pollination is uncontrolled so plants are pollinated by both superior and inferior pollen parents.
3. The strict regime of selection can lead to inbreeding depression. This can be overcome by line breeding where, following three or four cycles of mass selection, perhaps including progeny testing, the seed from a number of superior plants is combined and then sown in an **isolated plot**. The plants in the plot can then interbreed, so avoiding inbreeding depression, but only pollen from the superior plants in the plot will be available to fertilise the female flowers.

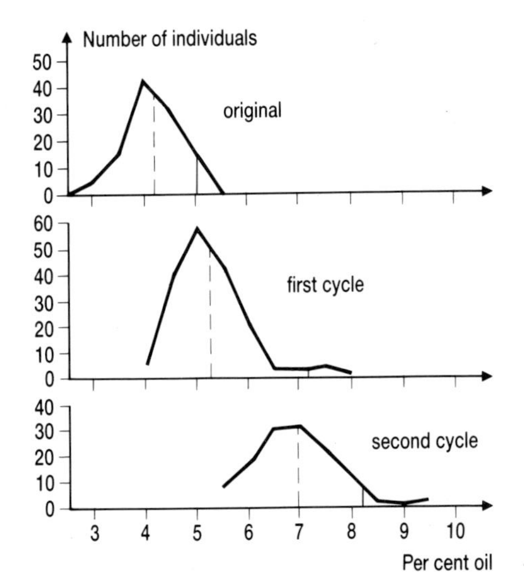

Fig 6.10 Frequency distributions for oil percentage in a maize population undergoing recurrent selection. In each distribution the mean is indicated by a black vertical line and the mean of the 10 ears chosen as parents for the next cycle of selection by a vertical red line.

Recurrent selection. Recurrent selection is a technique used with outbreeders which overcomes the three problems listed above, including control of pollination. The procedure is shown in Fig 6.11. This simple scheme has been refined to provide a wide variety of techniques suitable for a number of different plants. The effectiveness of recurrent selection is illustrated in Fig 6.10.

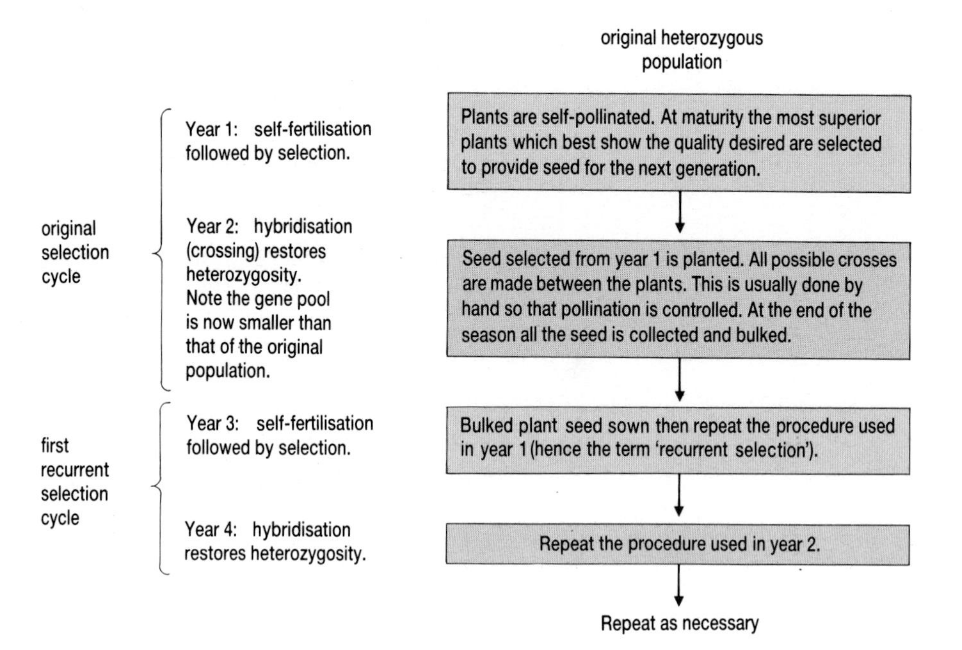

Fig 6.11 A diagrammatic representation of simple recurrent selection.

Backcross breeding. As with inbred plants, beneficial genes from wild plants can be included in outbreeders by backcross breeding. This is proving particularly important with potatoes.

Clones

Perennial plants like potatoes, cassava, sugar cane, strawberries and pineapples are all outbreeders but they can be propagated asexually. This is a great aid to the breeder since, provided the correct combination of genes can be produced in an individual plant, that plant can then be reproduced asexually to produce the new variety. The advantage of this is that the plants in the new variety will be genetically identical, in other words they will form a clone. However, unlike the plants in an inbred line, plants produced by clonal propagation will be heterozygous and will therefore not suffer from inbreeding depression. The breeding scheme is shown in Fig 6.12.

Heterozygous clonal parents (PC_1 and PC_2) are crossed to produce the F_1. Notice once again the variability of the F_1. The appropriate F_1 seedlings are selected and propagated vegetatively with the objective of selecting out those plants which contain the best arrangement of genes. This procedure will be repeated over several generations so that the total number of clonal lines decreases but the number of plants in each clonal line increases. At the end the breeder will have a large number of the few survivors of the original F_1 population, which may have numbered thousands.

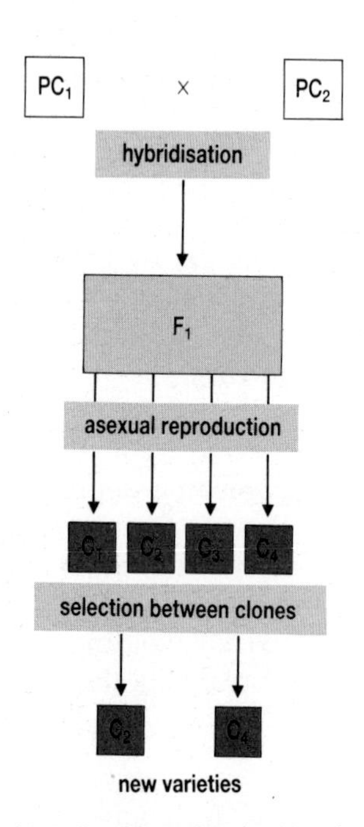

Fig. 6.12 A breeding scheme for use with outbreeders which can be propagated asexually, e.g. potatoes. C = clone; in reality there would be thousands of these. Note, the width of the boxes reflects, approximately, the amount of genetic variation present at each stage of the breeding procedure.

F_1 hybrids

The final method of selective breeding in outbreeders differs substantially from all the other breeding schemes in that a hybrid is the final product rather than the starting point for selection. The breeding scheme is shown in Fig 6.13. The important thing to notice is that the F_1 hybrids are produced from inbred lines, consequently they will be uniform, that is, the plants will all be the same but they will also be heterozygous. What the plant breeder is doing is making use of the phenomenon of heterosis or heterozygous advantage. Whilst fine in theory, F_1 hybrid production poses three problems for the breeder.

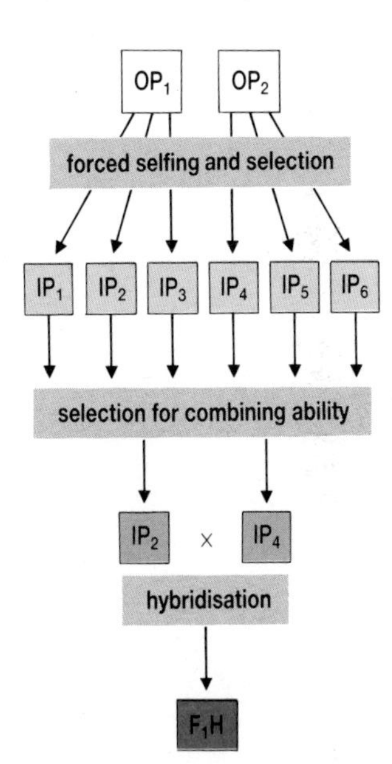

Fig. 6.13 A breeding scheme to produce F_1 hybrids of an outbreeder like maize. OP = open-pollinated population; IP = inbred parental line; F_1H = F_1 hybrid. Selection for combining ability is discussed in the text. Note, the width of the boxes reflects, approximately, the amount of genetic variation present at each stage of the breeding procedure.

1. The F_1 will only be uniform if the parents are homozygous. But producing the homozygous parents (IP_1, IP_2 etc. in Fig 6.13) involves inbreeding, which leads to inbreeding depression in outbreeders. Consequently commercial production of F_1 hybrids is not possible until varieties which can withstand a reasonable amount of homozygosity have been produced using the procedure for open-pollinated populations (Fig 6.9).
2. The commercial production of F_1 hybrids involves crossing plant X with plant Y. But in an open field where the breeding is taking place, what is to stop plant X fertilising plant X? You will see the solution to this problem when we examine F_1 production in maize.
3. F_1 hybrids do not breed true. This means that the farmer cannot simply collect seed at the end of the season to use to produce next year's crop. Instead, the farmer has to go back to the seed merchant each year for new supplies of F_1 seed. This can be expensive, so the costs of production must be kept to a minimum if the crop is going to be economically viable.

Fig 6.14 Farmers and gardeners now have a choice of a large range of F_1 hybrids. These often provide better crops and larger flowers.

A large number of plants are now being produced as F_1 hybrids (Fig 6.14), but the classic example is maize and this is used here to demonstrate the procedure, the problems faced by the breeder and the solutions to those problems.

Breeding maize. Maize is a monoecious (has male and female flowers), wind-pollinated plant where the male flowers or tassels are borne at the top of the plant and the female flowers are carried at the side (Fig 6.15). The pollen is usually shed 2–3 days before the styles (called silks) on the female flowers become receptive, so maize is usually cross-pollinated. A suitable breeding procedure for F_1 hybrids of maize is shown in Fig 6.16. Let us work through this diagram starting at the top.

Fig 6.15 (a) The male flower or tassel of a maize plant. This structure is borne at the top of the plant which assists wind pollination. The individual anthers are just visible dangling beneath the branches.

Fig 6.15 (b) The female flower of a maize plant. Each silky strand is a stigma, the number indicating the number of grains which will eventually develop on the cob.

Selection of parents. The parents are chosen as superior individuals from an open-pollinated population. The breeder looks at a field of open-pollinated maize plants and chooses those which appear to have appropriate qualities.

Selfing. The chosen plants are then selfed for several generations to produce homozygous inbred lines. These will inevitably suffer from inbreeding depression. Remember that the plants in each inbred line will come to resemble each other more and more but plants in different inbred lines will become increasingly different.

First cross. Inbred lines which have good combining ability (more about this later) are chosen as parents and crossed. This is achieved by interplanting two rows of the seed (female) parent to one of the pollen (male) parent. The seed parents are emasculated either by having their tassels removed manually, by chemical treatment or by using genetic phenomena like incompatibility or male sterility (again, these are described in more detail later). The amount of seed produced by this first cross is small since the parents are suffering from inbreeding depression which reduces their vigour.

Second cross. The major breakthrough in the production of F_1 hybrid maize during the 1920s and 1930s was the realisation that whilst the cross of A × B or C × D might not provide economic yields, crossing the F_1 progeny of A and B with the F_1 progeny of C and D would. The progeny of the double cross are reasonably uniform and very vigorous, as shown in Fig 6.17. Note: double crossing is not used in the production of all F_1 hybrids.

Indeed, the maize story has moved on since the early days described above. By the 1950s the maize gene pool had been stripped of many of its deleterious alleles using the sort of breeding scheme described for millet. Now, relatively homozygous lines which can tolerate inbreeding are available and the second cross is often not necessary.

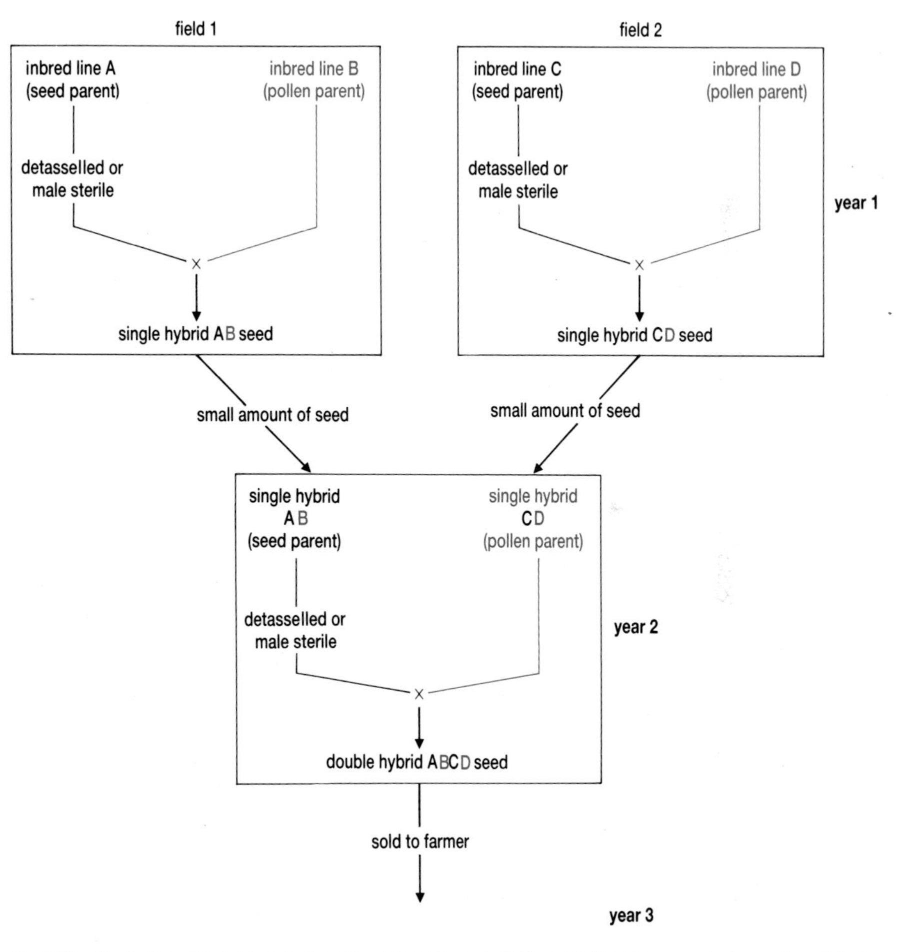

Fig. 6.16 Production of hybrid maize from four inbred lines, A, B, C and D. Paired crosses between the original inbred lines produces two vigorous hybrid plants, AB and CD, which are then crossed to yield the double cross hybrids ABCD.

Combining ability. Whilst some of the improved performance of F_1 hybrids comes from the selection practised in the production of the inbred lines used as parents, by far the major contribution is due to the use of inbred lines which have good combining ability, that is, that complement each other. A major task, then, must be the identification of inbred lines which combine well. This is selection for combining ability and is achieved by a number of test crosses between different inbred lines and looking at the progeny of these crosses. Using these results, the breeder can then predict which crosses will provide the most seed. This experimental work is essential if the production of F_1 hybrid seed is going to be economic.

One of the major costs in producing F_1 hybrids is incurred during the hybridisation stage. Traditionally this has been done by hand. Increasingly the phenomena of male sterility and self-incompatibility are used. These and other breeding 'tricks of the trade' are described in the next chapter.

Fig. 6.17 Maize which has been produced by the double cross method.

QUESTIONS

6.3 In both the breeding schemes used for open-pollinated populations and clones, the genetic variability 'locked up' in the parents is released, after hybridisation, in the F_1 generation. By contrast, the release of variation in inbreeders is not seen until the F_2. Explain the genetics underlying this observation. You may find it useful to use an imaginary plant which has only got three loci with two alleles to construct a simple genetic model.

6.4 Inbreeding depression is thought to result from an accumulation of deleterious recessive alleles. Explain why selfing a heterozygous outbreeder will produce this effect. Can you suggest a reason why inbreeders can tolerate high levels of homozygosity?

6.5 Summarise the breeding programmes used with outbreeders. Explain the advantages and disadvantages of each method. Explain how recurrent selection overcomes the problems of mass selection.

SUMMARY ASSIGNMENT

1. Your answers to questions 6.2 and 6.5 will act as a summary of the different breeding methods.
2. Summarise the role of hybridisation in programmes of selective breeding under the following headings:
   (a) a means of increasing variation;
   (b) a means of introducing new genes;
   (c) a means of exploiting heterosis or hybrid vigour.
3. The Green Revolution introduced many high yielding, inbred cultivars to increase food production in the Third World. Why, with hindsight, might this not have been a particularly wise strategy? Can you suggest alternative solutions to the challenge of increasing food production in the developing world?

Chapter 7

TOOLS, TECHNIQUES AND TRICKS

Plant breeding has come a long way in the last 150 years. International cooperation and gene storage means that, potentially, a breeder can obtain new genes from all over the world to incorporate into plants. Using natural genetic phenomena has made the time-consuming process of hybridisation easier and cheaper, thereby reducing the cost to the farmer and the consumer. Increasingly the plant breeder is moving into the laboratory as the techniques of cell culture and genetic engineering become more powerful. This chapter will look at some of these advances in plant breeding techniques and at some of the technology now used by animal breeders.

LEARNING OBJECTIVES

After completing the work in this chapter you will be able to:

1. explain the need for and methods of storing genes;
2. describe the following techniques and give examples of their application in plant breeding: induced mutation, tissue culture, anther culture, embryo culture, protoplast fusion;
3. describe the biological basis and uses of male sterility and self-incompatibility systems in plant breeding;
4. discuss the use of polyploids in plant breeding;
5. describe the uses of artificial insemination and embryo transfer in animals.

7.1 PRODUCING AND STORING GENES

All selective breeding programmes depend upon a source of genetic variability. If we use up all the genetic variability available, then, as we have seen, selective breeding grinds to a halt. So breeders are always on the lookout for new genes to increase the variability of their crop's gene pool. New genes can be found in other plants and animals or they can be created. We will consider these two alternatives in turn.

New genes

As emphasised in Chapter 5, the plant breeder will always look first to advanced cultivars for new genes. If these are not available then the breeder will turn to other sources. By far the most fruitful sources of new genes are the primitive land races, the origins of which were discussed in Chapter 5. Unfortunately, land races are rapidly being lost as modern cultivars take over. For example, during the Green Revolution, a few new high yielding cultivars of wheat replaced many of the traditional land races in Turkey, Iraq, Iran, Afghanistan and India. As a result it is now difficult to find older wheat varieties that were common only 30 years ago. This represents a major loss of genes which could have been potentially useful in breeding programmes.

A similar situation also applies to wild plants. The International Union for the Conservation of Nature estimates that at least 10 per cent of the world's quarter of a million flowering plants are on the verge of extinction. Again, this would represent a major loss of genes which could be used in plant breeding programmes. Clearly, then, there is a need to conserve plants to preserve genetic variability. Similar considerations also apply to animal breeding.

There are four ways of effecting such conservation.

1. Maintaining plants and animals in wild places or old-fashioned farms. Unfortunately, the demands made by people for living space and raw materials, for example timber, severely limit the number of truly wild places left. Game reserves and parks may provide a refuge especially in countries with developing economies. However, the destruction of natural habitats, particularly rainforests, is a cause for concern since so many potentially valuable plants and animals are being lost.

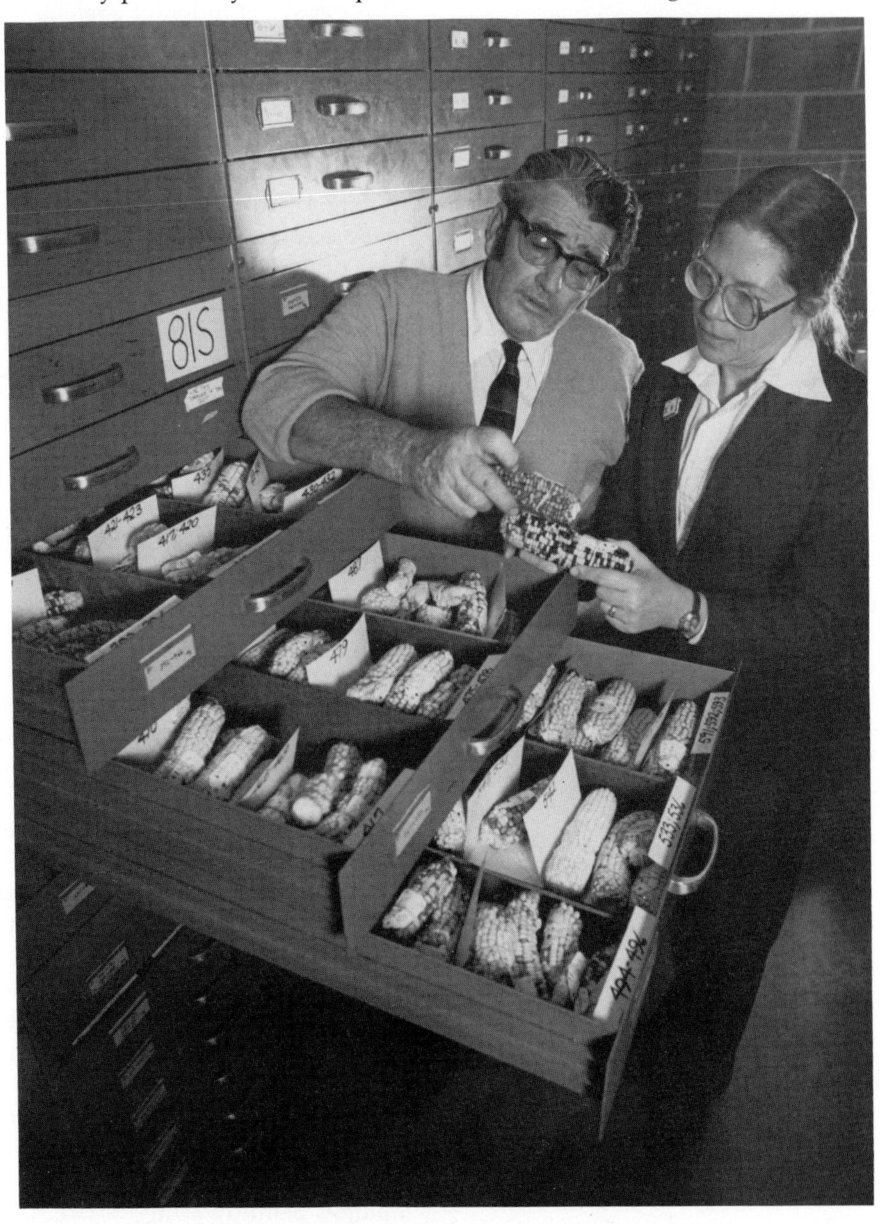

Fig 7.1 The university of Illinois maize gene bank. The seeds are kept both dry and cold. This particular material is used for the studies of plasmids (small pieces of DNA) in maize.

2. Botanical gardens and zoos can act as reservoirs of genetic variability. However, the number of organisms which can be maintained is severely limited and zoo animals in particular suffer from the problems of inbreeding.
3. Rare plants can sometimes prove attractive to the horticultural trade and become established in ordinary gardens. Similarly, rare animals like fish could be maintained in captivity by amateur enthusiasts.
4. Maintaining gene banks – see below.

Gene banks. All the above methods are flawed in the sense that any captive conservation of wild plants and animals will only maintain a limited part of the gene pool. Plant and animal breeders require access to as wide a range of alleles as are available in one species. The only practical way to do this is to store every single variant that is found in the form of germplasm in a gene bank. A gene bank contains a collection of seeds (Fig 7.1) or other suitable tissues, for example meristematic tissue, in tissue culture (see Box 7.1) or frozen sperm or embryos, which between them contain the principle alleles of a particular species or group of species.

A variety of different gene banks now exists. Some contain only one crop plant and its wild relatives. For example, the International Rice Institute in the Philippines holds the world's varieties of rice – all 100-120 000 of them. The central store of maize variants is in Mexico at the International Maize and Wheat Improvement Centre. The International Potato Centre in Peru holds 12 000 land races of potato. Other gene banks may hold a variety of different plants. For example, Kew Garden's gene bank at Wakehurst Place in Sussex contains 5000 species of wild plant, but this only represents 2 per cent of the total number of flowering plants in the world. So whilst most of the world's important food crops are adequately represented in gene banks, only a tiny proportion of the genes of wild plants are in store.

Collection and storage of germplasm of plants like the grasses and legumes of the Mediterranean region is a matter of urgency, and the destruction of the world's rainforests must represent a genetic catastrophe for all people.

Making new alleles – induced mutation breeding

If the plant breeder cannot find the alleles needed in a gene bank then the last resort is to try and create new ones. A variety of agents, for example X-rays, gamma rays, ultraviolet light and chemicals, can induce mutation. Exposing plants to such agents may produce the new allele that the breeder is looking for. This is obviously a hit-and-miss affair. Firstly, mutagenic agents can kill the plant if too much is used. Secondly, beneficial mutations will be rare so a lot of mutagenised plants will have to be screened. One way of doing this rapidly is to use the **anther culture** techniques described later. Thirdly, the mutation must occur in the cells which are going to produce the gametes, or in the gametes themselves if they are going to be incorporated into a breeding programme. Finally, many alleles produced by mutation will be recessive so their phenotypic effect will only be seen when they are in the homozygous condition. So breeding programmes for inbreeders are most likely to benefit from induced mutation as they can tolerate high levels of homozygosity. This leaves a large number of crop plants, outbreeders, for whom induced mutation is unlikely to produce beneficial results.

Despite these limitations, induced mutation has resulted in the production of new varieties of crop plant. For example, barley varieties which contain artificially mutated genes have increased yield and resistance to mildew. Semi-dwarfing in rice has been induced by treatment with X-rays and a gamma irradiated rice variety produced in the 1950s showed improved cold hardiness.

BOX 7.1

Plant tissue culture

Tissue culture is a technique of growing undifferentiated (that is, unspecialised) cells which have the capacity to regenerate new plants. The process is normally started with the production of callus – a piece of undifferentiated, actively growing tissue. Callus can be obtained from many different types of plant tissue including leaves, stems and roots. To be of any use the callus cells must be **totipotent** which means that they have the necessary genetic information and the capacity to regenerate a complete plant. The sequence of photographs shows the stages in the processes of tissue culture.

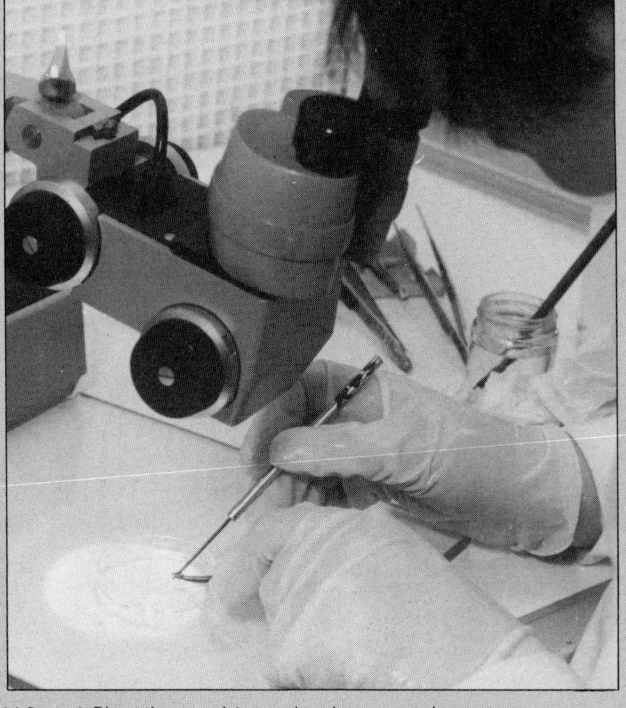

(a) Stage 1. Dissecting a meristem under a low power microscope.

(b) Stage 2. Nodal explant inoculated onto culture medium.

(c) Stage 3. Growth and proliferation of shoots.

(d) Stage 4. Root initiation on shootlet from stage 3 producing plantlet.

INVESTIGATION 7.1

The dangers of nuclear radiation are always being stressed and the impact of Five Mile Island and Chernobyl will remain with us for many years. But how dangerous are nuclear power stations? Imagine the government decides to build a nuclear power station near your home town. The local inhabitants are very concerned about a nuclear accident. They call on you, a geneticist, to inform them about the biological hazards of an accident. What would you say in your speech?

7.2 AIDS TO EASIER CROSSING

Fig 7.2 Hand pollination of the turnip. The anther from another plant, held in the forceps, is being brushed against the stigma – so transferring pollen.

One of the major challenges facing plant breeders is arranging the crosses they want. Pollination by hand is still practised widely (Fig 7.2) but this makes commercial production of F_1 hybrid seed expensive. Breeders have turned increasingly to two natural genetic phenomena, male sterility and self-incompatibility systems, to help them out.

Male sterility

To successfully produce F_1 hybrids all the seed (female) parents must be emasculated (have their anthers removed) so that they can only be pollinated by the desired pollen (male) parent. In the case of maize this used to be achieved by detasselling the seed parents by hand, a major source of holiday work for college students. However, such hand emasculation is very expensive and a great step forward for the breeders (if not the college students) was the discovery of the phenomenon of **male sterility**.

The genes which encode male sterility are in some cases inherited as recessives and in others are carried in the cytoplasm. The phenotypic effects of male sterility genes are shown in Fig 7.3. In maize the male sterility genes were discovered in 1940 in a variety called Texas June. The Texas June male sterility genes are not found in the nucleus but in the mitochondrial DNA, that is, they are cytoplasmic genes. So to produce an inbred male sterile line it is necessary to transfer the Texas June cytoplasm, and mitochondria, into the desired line. The procedure for doing this is shown in Fig 7.4. Notice that all the mitochondria in a plant are derived from the female parent.

The resulting male sterile line can now be used as the seed parent to produce hybrid seed without the need for expensive hand emasculation. However, one problem remains. The hybrid seed sold to the farmer will be sown in monoculture so the plants need to be self-fertile. But the hybrid seed contains Texas June cytoplasm which will make the plants that grow from the seed male sterile. Fortunately, the effects of Texas June cytoplasm can be overcome by so-called restorer (*Rf*) genes. These restorer genes are found in the nucleus and are dominant to cytoplasmic male sterile genes. The details of producing F_1 hybrid maize using male sterile and restorer genes are summarised in Fig 7.5.

Fig 7.3 A male sterile maize plant. If you compare this tassel with the one shown in Figure 6.15 (a), the absence of anthers is obvious.

Self-incompatibility

It has been known for a long time that some plants will simply not self-pollinate. A single plant produces both male and female gametes but no seeds are ever produced. The same plants will cross with certain other plants so they are obviously not sterile. This phenomenon is called self-incompatibility. This phenomenon is now known to have a genetic basis. Two systems of self-incompatibility can be recognised.

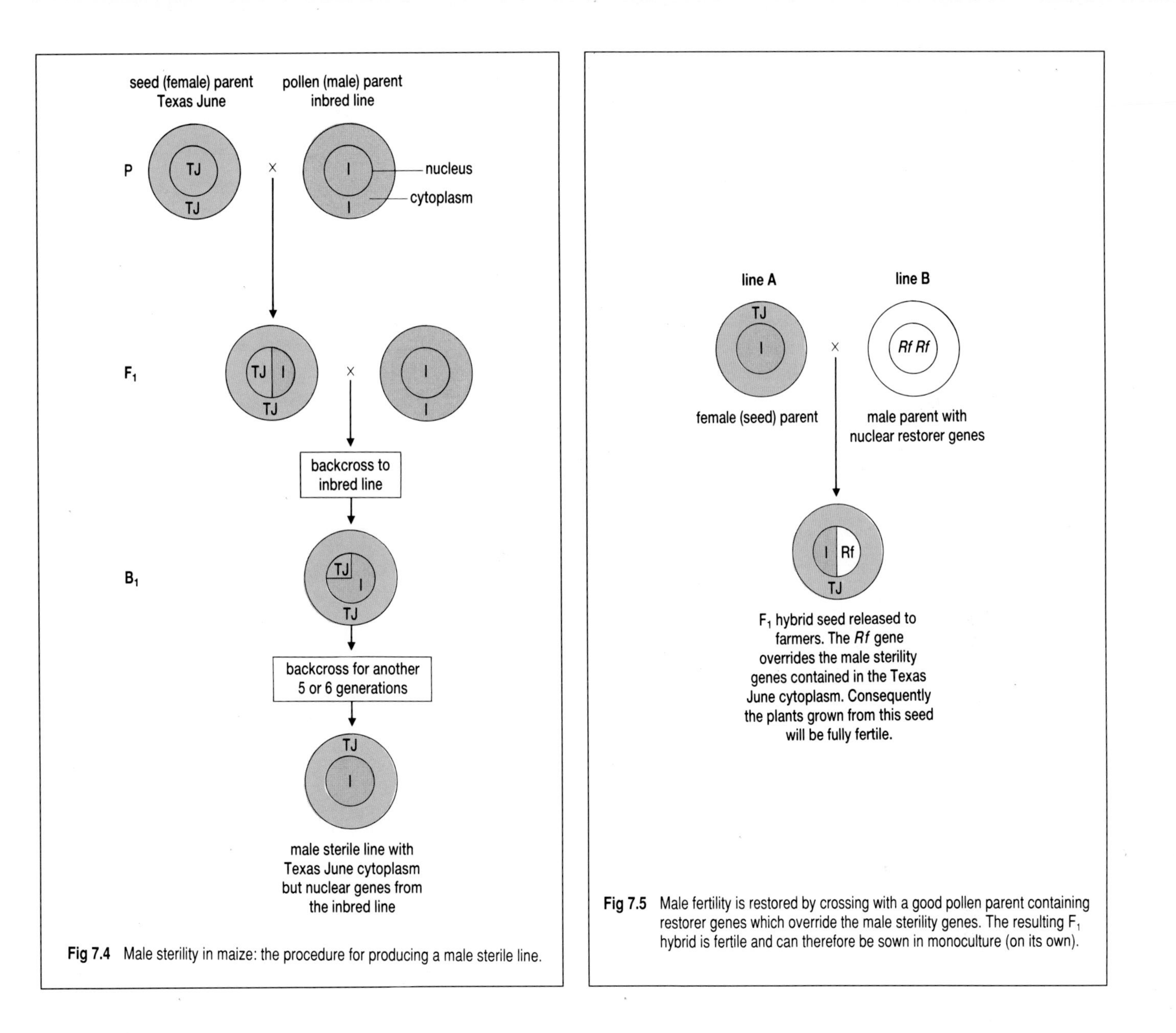

Fig 7.4 Male sterility in maize: the procedure for producing a male sterile line.

Fig 7.5 Male fertility is restored by crossing with a good pollen parent containing restorer genes which override the male sterility genes. The resulting F_1 hybrid is fertile and can therefore be sown in monoculture (on its own).

Gametophytic incompatibility. One of the most common systems is found in many plants including sweet cherry, tobacco, evening primrose and petunia. In each of these species a single gene, called *S*, determines whether a plant is self-compatible or not. The *S* gene exists in many different forms, that is, it is multiallelic. An example of how these multiple alleles control self-compatibility is shown in Fig 7.6.

The basic rule is quite simple. If a pollen grain contains an *S* allele which is also present in the female parent then the pollen grain cannot germinate. If the *S* allele in the pollen grain is not present in the female parent then the pollen grain germinates, producing a pollen tube which grows down the stigma. The male nucleus then passes down the pollen tube and fertilises the female gamete. The number of *S* alleles in a series varies between species and may be very large. For example, evening primrose and clover have over 50 alleles of the *S* gene and some species have in excess of 100.

Sporophytic incompatibility. The generation of F_1 hybrids in brassicas like Brussels sprouts makes use of the phenomenon of sporophytic incompatibility. Here, the incompatibility is determined not only by whether the pollen grain and the stigma have an allele in common but also on a form of dominance. Again, the alleles involved are usually designated as S. The S alleles form a dominance hierarchy in which S_1 is dominant over all other alleles, S_2 is dominant over all alleles except S_1, S_3 is dominant over all other alleles except S_1 and S_2, and so on. During pollen formation all the pollen retains the phenotype associated with the dominant allele in the male diploid tissue which produced the pollen grain, regardless of the genotype of the pollen grain. For example an S_1S_2 male would produce pollen with the S_1 phenotype even though half will have the S_2 genotype. The genetic basis of this rather bizarre behaviour is still uncertain though it could be due to some sort of cytoplasmic genetic system.

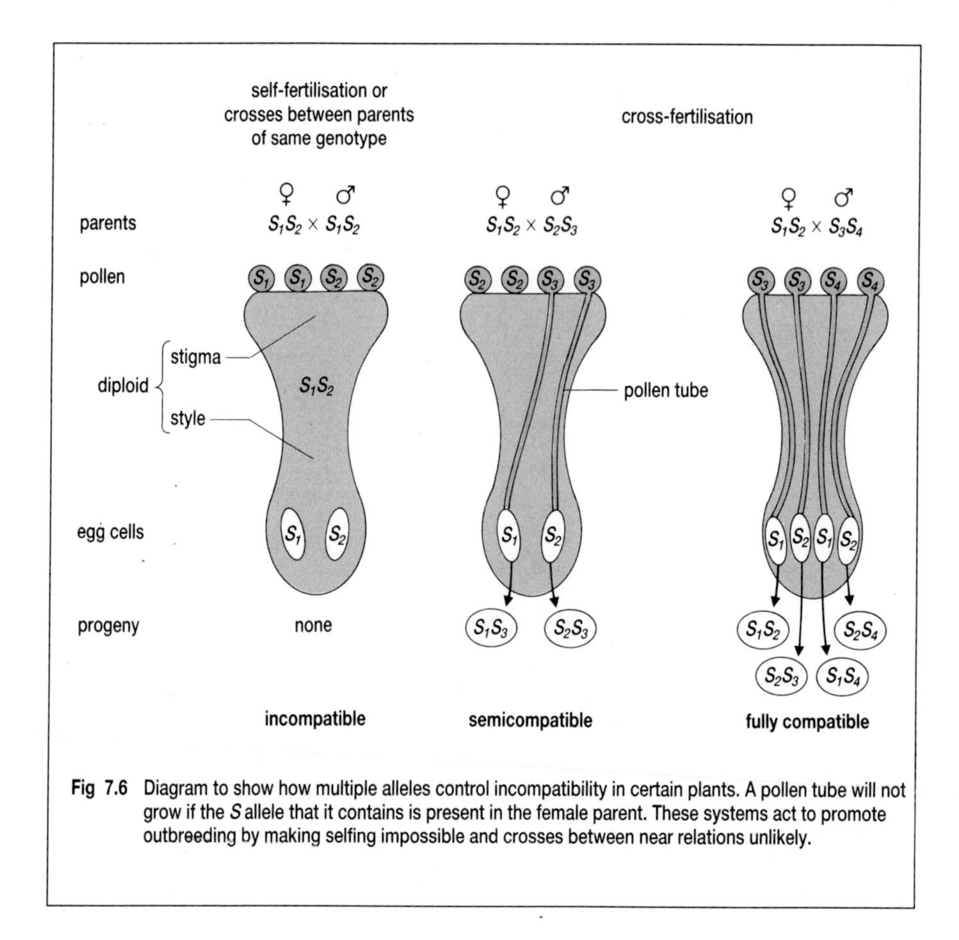

Fig 7.6 Diagram to show how multiple alleles control incompatibility in certain plants. A pollen tube will not grow if the *S* allele that it contains is present in the female parent. These systems act to promote outbreeding by making selfing impossible and crosses between near relations unlikely.

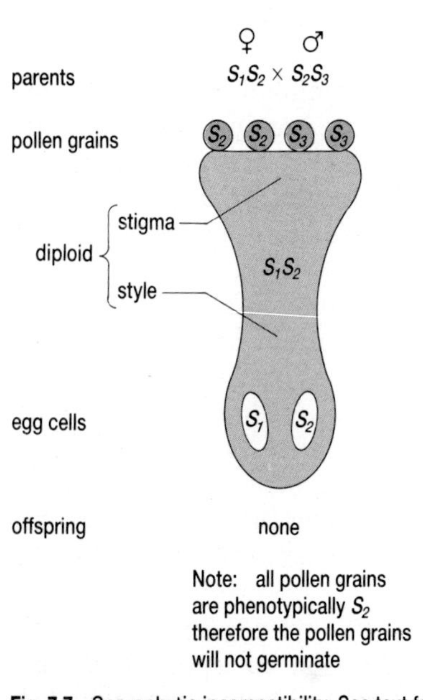

Fig 7.7 Sporophytic incompatibility. See text for explanation.

There is no dominance expressed in the female and the stigmatic tissue functions in exactly the same way as in the gametophytic system. So the cross shown in Fig 7.7 is incompatible in a sporophytic system, though not in a gametophytic system, because all the pollen will be phenotypically S_2.

Cross-incompatibility

In addition to self-incompatibility, some species show cross-incompatibility. For example, in the diploid sweet cherry the varieties fall into groups. Within each group the varieties are incompatible as shown in Fig 7.8. An example of cross-incompatibility is that the varieties Early Black and Bedford Prolific won't cross-pollinate. The fruit grower must plant varieties which are cross-compatible and which flower at the same time to ensure adequate pollination in the orchard.

female \ male	Early Black	Bedford Prolific	Black Tartarian	Early Rivers	Schrecken	Frogmore Big	Waterloo	Napoleon	Emperor Francis	Kentish Big	Elton	Governor Wood
	S_1S_2				S_1S_3			S_3S_5		S_2S_3	S_1S_4	
Early Black	–	–	–	–	+	+	+	+	+	+	+	+
Bedford Prolific	–	–	–	–	+	+	+	+	+	+	+	+
Black Tartarian	–	–	–	–	+	+	+	+	+	+	+	+
Early Rivers	–	–	–	–	+	+	+	+	+	+	+	+
Schrecken	+	+	+	+	–	–	–	+	+	+	+	+
Frogmore Big	+	+	+	+	–	–	–	+	+	+	+	+
Waterloo	+	+	+	+	–	–	–	+	+	+	+	+
Napoleon	+	+	+	+	+	+	+	–	–	+	+	+
Emperor Francis	+	+	+	+	+	+	+	–	–	+	+	+
Kentish Big	+	+	+	+	+	+	+	+	+	–	+	+
Elton	+	+	+	+	+	+	+	+	+	+	–	–
Governor Wood	+	+	+	+	+	+	+	+	+	+	–	–

Fig 7.8 Compatibility relations between 12 types of sweet cherry; there are five incompatibility groups. + indicates successful pollination; – indicates unsuccessful pollination. (After Crane and Lawrence, 1947.)

QUESTIONS

7.1 In a gametophytic incompatibility system, what are the progeny from the following crosses?
(a) $S_1S_2 \times S_3S_4$ **(b)** $S_1S_2 \times S_1S_3$ **(c)** $S_1S_3 \times S_1S_2$

7.2 In a sporophytic incompatibility system, what are the progeny from the following crosses?
(a) $S_1S_2 \times S_3S_4$ **(b)** $S_1S_2 \times S_1S_3$
(c) $S_1S_3 \times S_3S_4$ **(d)** $S_1S_3 \times S_2S_3$.

7.3 A species of plant possesses a gene S with many different alleles (S_1, S_2, S_3, etc.) The cells of each plant contain any two S alleles (one on each homologous chromosome). All S alleles are co-dominant. These alleles control the growth of a germination tube from a pollen grain into the stigmatic tissues of a female flower.

Some typical results from crosses are shown in the table below.

Cross	Male parent	Female parent	F_1 Genotypes
1	S_1/S_2	S_1/S_2	No progeny
2	S_1/S_3	S_1/S_2	S_3/S_1 and S_3/S_2
3	S_1/S_2	S_1/S_3	S_2/S_1 and S_2/S_3

(a) Suggest an explanation for these results, bearing in mind that stigmatic tissue is diploid.
(b) If S_1 was dominant to all other S alleles, what results would have been expected in cross 3?
(c) What important function could these S alleles serve in a plant?
(d) Comment on the significance of genetic systems of the S-allele type for commercial growers.

7.3 CHROMOSOME ENGINEERING

Altering the number of chromosomes in a plant is a very powerful technique in plant breeding. Before looking at the various techniques which plant breeders use to do this, you need some definitions of useful terms.

Basic definitions

The number of chromosomes in a basic set is called the **monoploid** number, x. Do not confuse this with the haploid number of chromosomes, n. The difference will become obvious in a minute. Organisms which contain multiples of the monoploid number of chromosomes are called **euploid**. Various degrees of euploidy can be identified, as shown in Table 7.1. Notice that organisms with more than two sets of chromosomes are called **polyploid**.

Table 7.1 The degrees of euploidy

Number of sets of chromosomes	Term	
$1x$	monoploid	
$2x$	diploid	
$3x$	triploid	polyploid
$4x$	tetraploid	polyploid
$5x$	pentaploid	polyploid
$6x$	hexaploid	polyploid

The haploid number of chromosomes, n, refers to the number of chromosomes contained in a gamete. In most animals and plants that you are familiar with the haploid and monoploid number is the same. For example, n and x in people equals 23. However in many plants, for example modern wheat, n and x are different. Wheat has 42 chromosomes but it is a hexaploid. This means that it contains six sets of chromosomes. These six sets are similar but not identical. So in wheat $6x = 42$ and $x = 7$. However, the gametes of wheat contain 21 chromosomes which means that n, the haploid number, is 21. We will now examine how plant breeders can use monoploids and polyploids.

Monoploids

These are coming to play a major role in plant breeding. For a plant breeder, diploidy is essentially a nuisance since dominant alleles mask the effects of recessive alleles. A monoploid has only got one copy of each allele at each locus, therefore the phenotypic effects of all the genes can be seen. One way of producing monoploid plants is by **anther culture**. The process is shown in Fig 7.9.

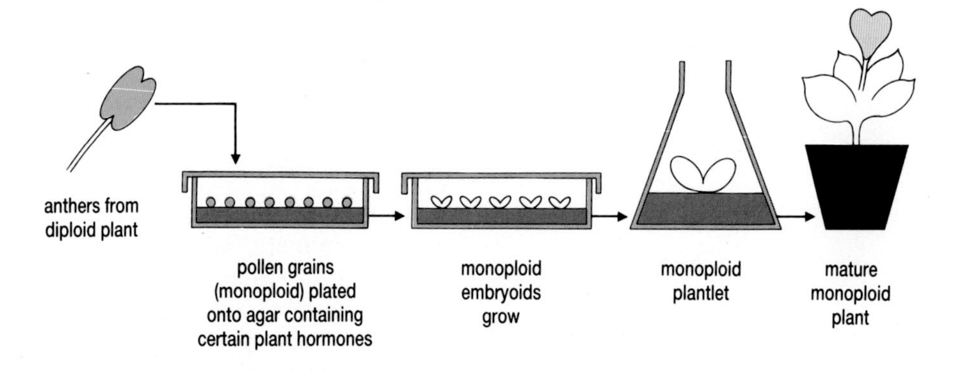

Fig 7.9 Anther culture. Pollen grains are plated onto agar which contains certain plant hormones. Changing the plant hormones induces the monoploid embryoids to grow into monoploid plants.

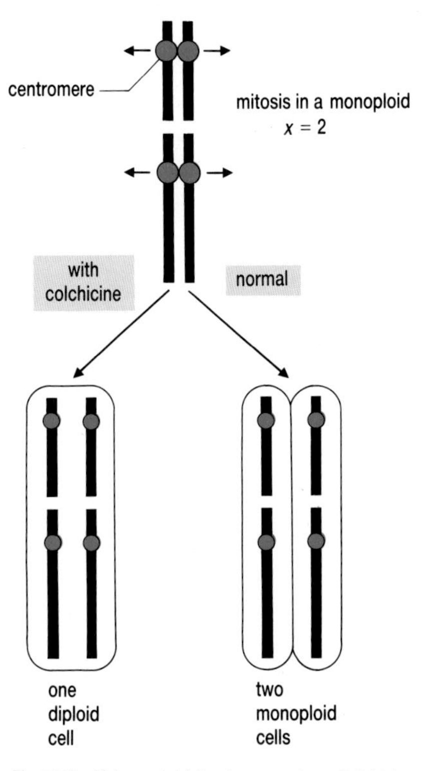

Fig 7.10 Using colchicine to generate a diploid from a monoploid. Colchicine disrupts the formation of the mitotic spindle and so prevents separation of the chromatids after the centromeres split. A single cell is created containing pairs of identical chromosomes which will be homozygous at all gene loci.

Monoploids produced in this way have a number of uses. For example, they can be assessed rapidly for favourable traits, for example resistance to a particular chemical or parasite, because they can be grown in large numbers on petri dishes. Such favourable combinations of genes may arise by the recombination of the genes already present in the parent during meiosis, or they may have been produced by induced mutation. Having identified a useful monoploid, the breeder now needs to reproduce the plant. However, monoploids are sterile. Meiosis and therefore gamete formation cannot occur because the chromosomes have no partners to pair with. You could propagate the monoploid vegetatively but this is time consuming and will not work for many species. This problem is overcome by doubling the number of chromosomes using colchicine. This is shown in Fig 7.10. Cells which contain the double set of chromosomes can be identified under the microscope, isolated and grown in tissue culture. This produces homozygous diploid plants capable of producing seed.

A second way of using monoploids is to treat them rather like bacteria, as shown in Fig 7.11. Here, the plants carrying the characteristic required by the breeder are selected using **selective media**. These powerful plant breeding techniques based on anther culture have been applied successfully in several crop plants, for example soya beans.

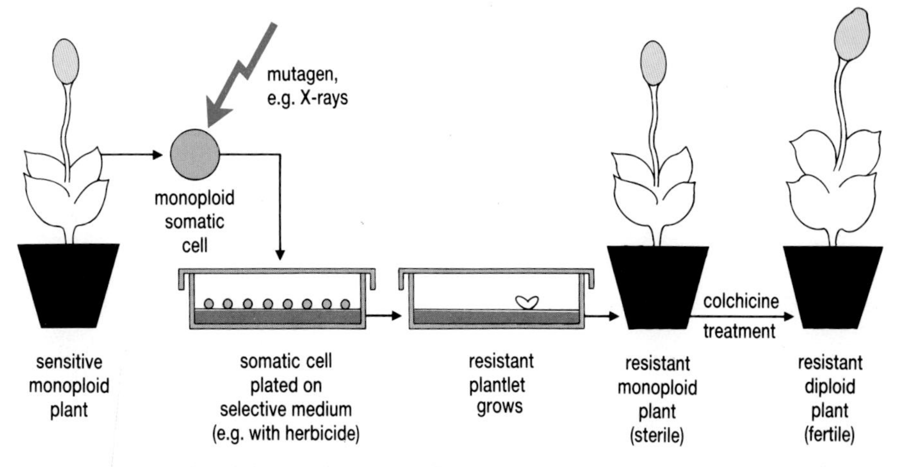

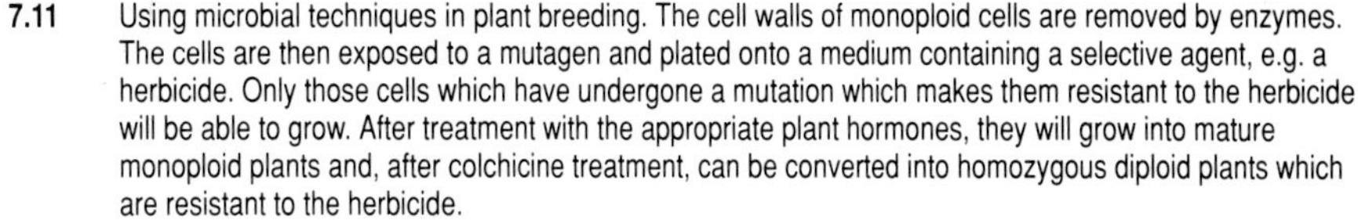

7.11 Using microbial techniques in plant breeding. The cell walls of monoploid cells are removed by enzymes. The cells are then exposed to a mutagen and plated onto a medium containing a selective agent, e.g. a herbicide. Only those cells which have undergone a mutation which makes them resistant to the herbicide will be able to grow. After treatment with the appropriate plant hormones, they will grow into mature monoploid plants and, after colchicine treatment, can be converted into homozygous diploid plants which are resistant to the herbicide.

Polyploids

You need to distinguish two types of polyploid, both of which are important in plant breeding. Imagine a diploid ($2x$) plant cell. For some reason the process of mitosis goes wrong so that cell division does not occur. As a result the cell ends up with four sets of chromosomes, that is, it is tetraploid ($4x$). This is an example of **autopolyploidy** (auto = same). Autopolyploids contain multiple sets of chromosomes all derived from the same species. By contrast, **allopolyploids** (allo = different) contain multiple sets of chromosomes derived from different species. The classic example of an allopolyploid is bread wheat, *Triticum*. Allopolyploidy has been very important in the evolution of plants and many modern crop plants are allopolyploids, for example many members of the genus *Brassica* .

Triploids ($3x$). These are usually autopolyploids derived from a cross between a tetraploid ($4x$) and a diploid ($2x$) as shown in Fig 7.12. Triploids are usually sterile but this can be useful for a plant grower. For example, commercially grown bananas are sexually sterile triploids which can be asexually propagated. The advantage of this is that the bananas you eat contain no seeds.

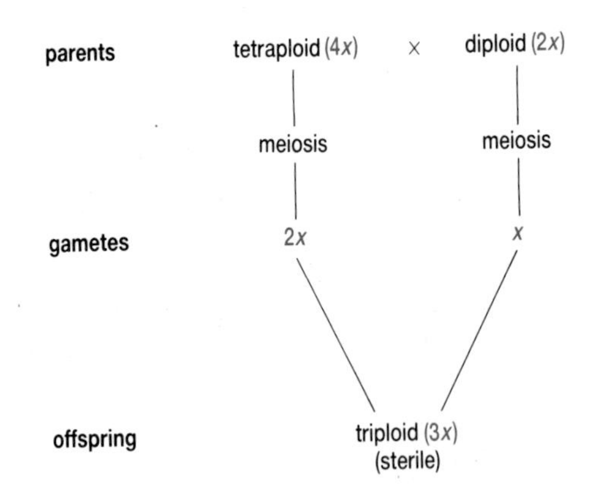

Fig 7.12 Producing a triploid.

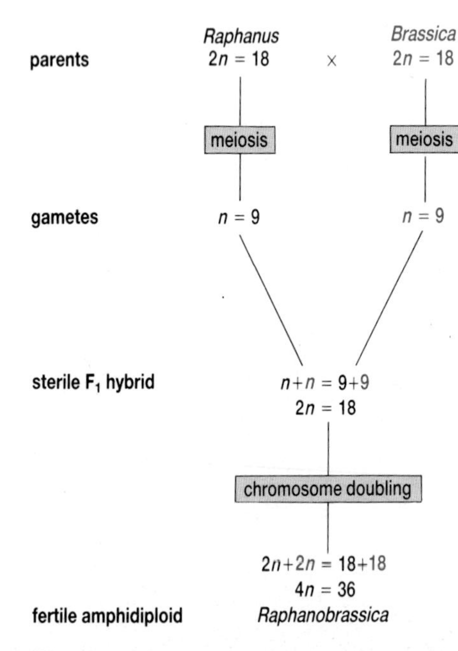

Fig 7.13 A summary of the experiments of Karpechenko. The fertile amphidiploid was produced as a result of an accidental doubling of the chromosome number rather than by treatment with colchicine, though the mechanism is the same in both cases.

Autotetraploids. These can occur naturally, as described above, or artificially using colchicine, a drug which disrupts the formation of the mitotic spindle in a dividing cell. As a result, the chromatids cannot separate during anaphase of mitosis, so producing a cell which contains twice the usual number of chromosomes. Some crop plants are natural autotetraploids. For example, most wild potatoes are diploid with 24 chromosomes but the principle cultivated species, *Solanum tuberosum* and *Solanum andigena*, are both tetraploid. The advantage of this lies in the fact that polyploids tend to have larger cells and hence the plants tend to be larger.

Artificial induction of autopolyploidy has now been tried with many crops. As a result practically all crop species have been examined as an autotetraploid to see whether they are superior to the diploid forms. Often autotetraploids are found to be more vigorous than diploids or they produce larger fruit, tubers and so on. For example, wild blackberries are usually diploid ($2x = 14$) while some cultivated blackberries are tetraploid, hexaploid or even octoploid and have larger fruit then their wild counterparts. Similarly, cultivated strawberries are octoploid

Allopolyploids – crossing the species barrier. Allopolyploids are potentially of great use to the plant breeder since they can be used to combine the advantages possessed by plants from different species and so extend the range of conditions under which a plant will grow. Natural allopolyploids are already widely used in agriculture, so a logical step would be to produce them artificially. An example will serve to illustrate the principles and pitfalls of producing artificial allopolyploids.

In 1928 a plant breeder, G Karpechenko, set out to produce a fertile hybrid between the cabbage (*Brassica*) and the radish (*Raphanus*) that would have the leaves of the cabbage but the roots of the radish which are not susceptible to club root, a fungal infection of cabbages. The two species both have 18 chromosomes and they are related closely enough to be crossed. Karpechenko's experiments are summarised in Fig 7.13.

Unfortunately, whilst both parents have 18 chromosomes they are sufficiently dissimilar to prevent pairing during meiosis (that is, the chromosomes are non-homologous). Consequently, the hybrid is unable to produce gametes and so it is sterile. However, Karpechenko was lucky. Some of his hybrid plants produced seed. These were planted and the plants which grew were fertile but they had 36 chromosomes. Clearly, an accident of cell division had occurred in the sterile hybrid which had resulted in a doubling of the chromosome number in cells which give rise to gametes (so-called germinal cells). So cells which had been $n_1 + n_2$ were now $2n_1 + 2n_2$. Meiosis can now occur because there is a homologous partner for each chromosome. The gametes produced are $n_1 + n_2$ which can fuse to give $2n_1 + 2n_2$ allopolyploid offspring which are also fertile. This kind of allopolyploid is called an **amphidiploid**. Unfortunately, Karpechenko's luck now ran out. All the amphidiploids he produced had the roots of a cabbage and the leaves of a radish!

The importance of Karpechenko's research was that it showed the way to make allopolyploids fertile. Providing the two parent plants are sufficiently closely related to cross in the first place then the hybrids can be made fertile by doubling the number of chromosomes in the germinal cells. Meiosis, gamete formation and reproduction can now occur. This chromosome doubling can be achieved using colchicine.

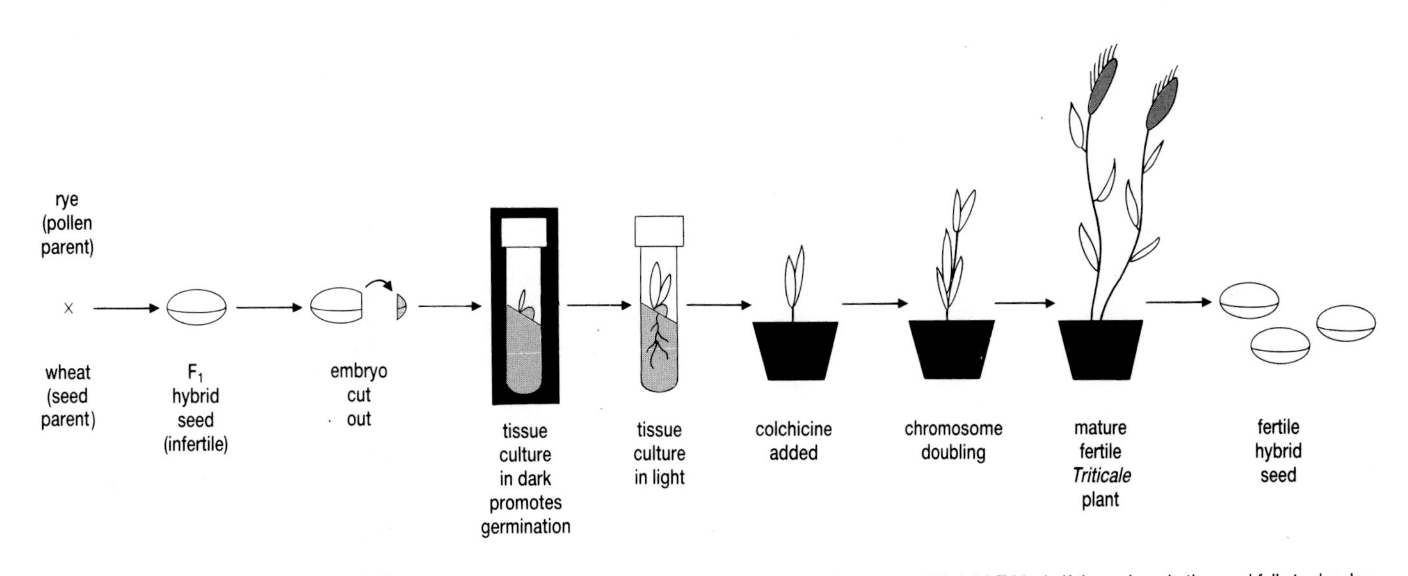

Fig. 7.14 Techniques for the production of the amphidiploid *Triticale*. If the embryo in the seed fails to develop, a common occurrence with wide crosses, then embryo culture can be used to produce a hybrid plant.

QUESTIONS

7.4 In maize, a single gene determines presence (*Wx*) or absence (*wx*) of amylose in the cell's starch. Cells that have *Wx* stain blue with iodine, and those that have only *wx* stain red. You are required to design a laboratory system for studying the frequency of rare mutations from *Wx* to *wx*.
(**Hint** : you might start by thinking which plant cell types are easy to grow in large numbers in the laboratory.)

7.5 **(a)** State three ways in which polyploids are useful to modern growers.

The plant genus *Brassica* (Cruciferae) contains a number of species from which some of our commonest vegetable and fodder plants, for example cabbage, swede and rape, have evolved. Hybridisation between the turnip *Brassica rapa*, $2n$ = 20) and the black mustard (*Brassica nigra*, $2n$ = 16) has produced the brown mustard (*Brassica juncea*, $2n$ = 36).

(b) What chromosomal change occurred during the formation of the brown mustard hybrid, and what term is used to describe such a hybrid?

(c) Why are plants of brown mustard able to produce fertile seed?

An example of an artificially produced amphidiploid which may be of great use in the future is *Triticale*, a polyploid hybrid which combines the hardiness of rye with the high yield of wheat. However, the seeds produced by such wide crosses often fail to germinate because the seed embryo dies as the endosperm fails to feed it properly. Another tissue culture technique, **embryo culture**, can overcome this problem (Fig 7.14). In addition to being used with *Triticale*, embryo culture has also proved successful in raising new forage crops based on crossing different species of clover.

7.4 PROTOPLAST FUSION

The techniques of allopolyploidy production described above depend upon using the plants' natural reproductive mechanisms to produce the hybrid. Often, however, the plants the breeder wants to hybridise are so dissimilar that they will not cross-fertilise. This is where **protoplast fusion** comes in. A protoplast (Fig 7.16) is a plant cell which has had its cell walls removed by an enzyme. If protoplasts are subjected to certain chemicals, for example calcium or polyethylene glycol (PEG), the cell membranes will fuse producing cells with two or more nuclei. The exact method used depends upon the plant species involved and is largely a matter of following a recipe. An example is shown in Fig 7.15.

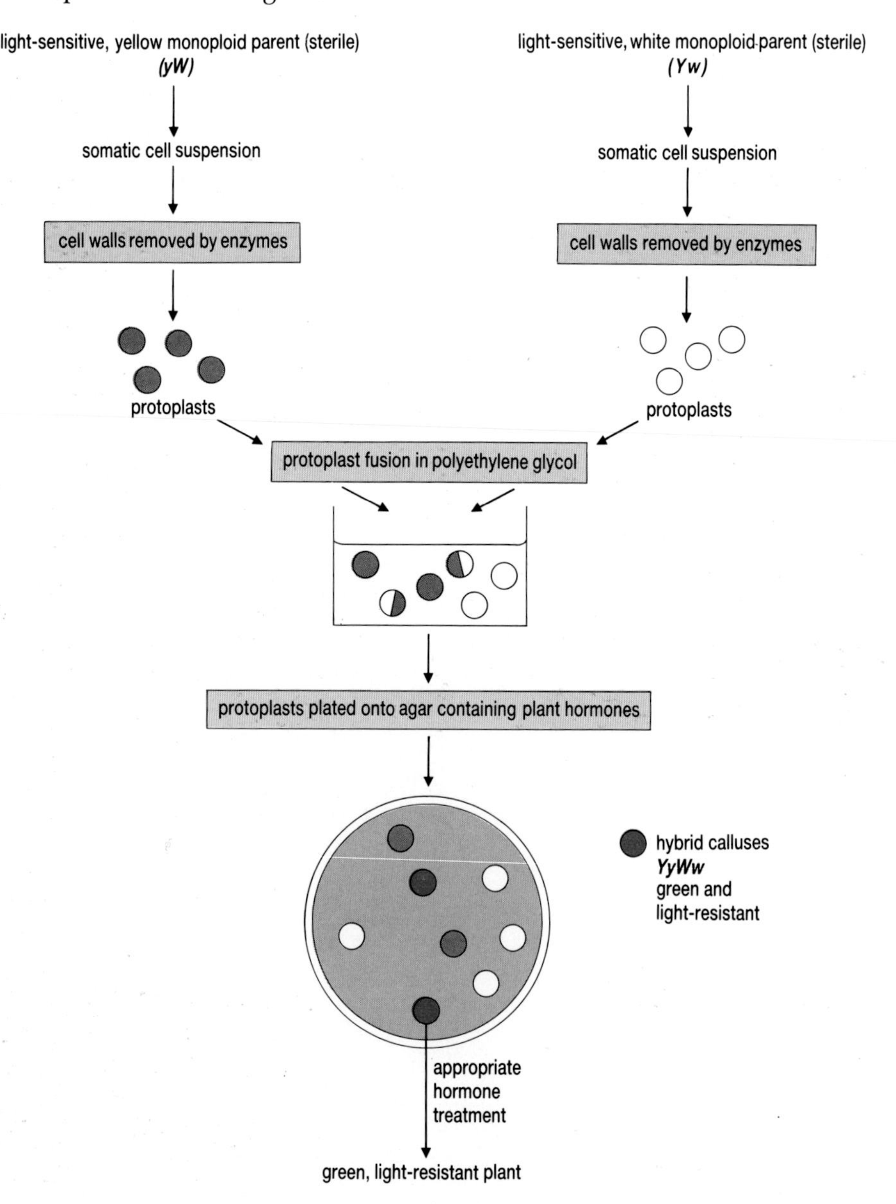

Fig 7.15 The use of protoplast fusion in the hybridisation of two strains of tobacco, *Nicotiana tabacum*.

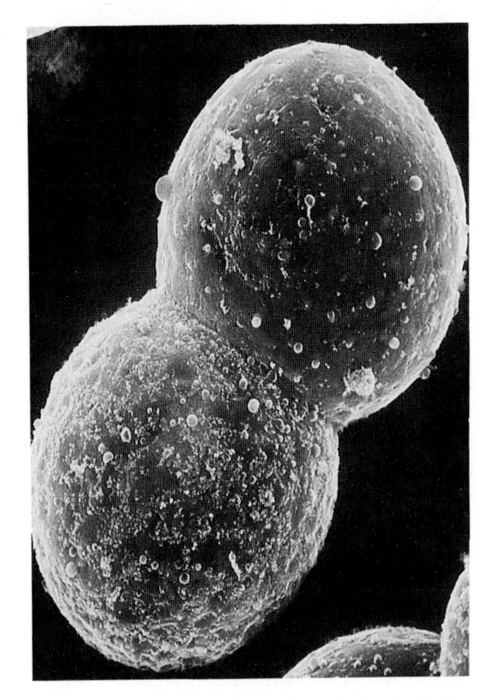

Fig 7.16 Scanning electron micrograph showing fusion of a pair of protoplasts produced from two tobacco leaf cells.

Interesting things can now happen when the nuclei divide. The chromosomes may become mixed up to produce an instant polyploid. If the two plants involved are from different but closely related species an allopolyploid may be formed. For example, petunia has been fused with tobacco. Since these plants are hybrids of somatic cells rather than sex cells they are called **somatic cell hybrids**.

Whilst the technique is not appropriate for exchanging genes between unrelated species of plant (only genetic engineering has the potential to do that), it is proving to be invaluable for transferring genes between closely related plants. For example, the wild potato *Solanum brevidens* contains genes that confer resistance to potato leaf roll virus, one of the most troublesome pests of commercial potatoes in Great Britain. A normal cross between *S. brevidens* and *S. tuberosum* is not possible because there are reproductive barriers between them, so plant breeders are trying to use protoplast fusion to create an appropriate hybrid.

An alternative form of somatic hybrid is being used to develop resistance to root wilt, a fungal disease, in coconuts. Here, just the nucleus of one variety of coconut is transferred into the cytoplasm of a cell from another variety of coconut or other closely related palms which are resistant to this disease. The reason for adopting this alternative procedure is that the genes for root wilt resistance are carried in the DNA found in mitochondria and not in the nuclear DNA. So it is hoped that implanting nuclei from high yielding varieties of coconut into the cytoplasm of resistant varieties will produce high yielding, resistant varieties. By adopting this procedure, the Indian scientists at Hindustan Lever may well be able to produce the new varieties within a matter of years rather than the decades it would take using conventional methods.

7.5 TECHNIQUES WITH ANIMALS

For the reasons given at the beginning of Chapter 5, little mention has been made of animal breeding techniques. Much of what has been said about outbreeding plants is also applicable to animals, but animal breeding does have its own unique set of problems. These include the long generation times of most agriculturally important animals, the small number of offspring that they produce and the ethical problems that genetic experiments with animals can present. Furthermore, the techniques of cell culture that are making such an impact on plant breeding are not widely applicable to animals at the moment. Nonetheless, animal breeding has made great strides in developing improved varieties and strains of domesticated animals, particularly poultry. In this final section we will briefly examine some of the techniques that animal breeders can use to improve and speed up their experiments. Most of these involve interfering with the natural reproductive cycle of the animal.

Artificial insemination (AI)

One of the major advances in animal breeding was the introduction of artificial insemination. This allows the sperm from one superior male to be used to fertilise a large number of different females. This is possible because each ejaculation or service produces enormous numbers of sperm which can be collected in an artificial vagina. The sperm is then tested for motility and stored in a special medium called an extender, for example milk and egg yolk mixed with citrate buffer. In addition to providing a suitable medium for storage, the extender also dilutes the sperm, making it go even further. The diluted semen is now divided into small vials, each of which contains enough for one insemination. The type of animal will determine the insemination procedure used. In large animals like cows, the sperm is placed directly into the uterus using a tube called a catheter. In smaller animals the sperm is placed into the vagina.

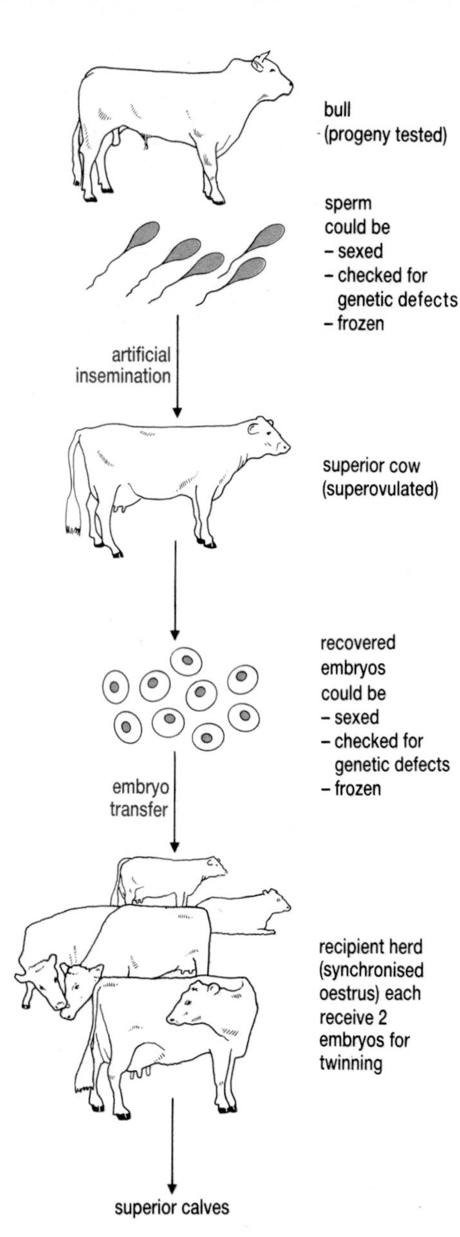

Fig 7.17 Modern technology and animal breeding.

Not all animals are suitable for use in AI programmes. For example, it may not make economic sense to use AI or it may be difficult to store the sperm. Short-term storage is possible using a simple refrigerator like the one in most kitchens. Long-term storage is usually achieved by freezing the sperm at –196 °C in liquid nitrogen. Unfortunately, storage at such low temperatures can damage the sperm, so more has to be used in the insemination process. Some sperm will not survive such long-term storage.

Embryo transplantation

An important limitation in animal breeding is the long gestation period and small numbers of offspring produced by agricultural animals. One way of improving this situation is to use techniques you are probably familiar with from the test-tube baby programme. One important technique is the use of surrogate mothers, which involves transferring the embryo from the real mother to a foster mother. There are a number of reasons for doing this.

- Increasing the reproductive rate of a superior individual. A particularly good animal may be treated with a hormone to increase egg production. A large number of eggs are fertilised and the extra ones can be transferred to less superior mothers who will then act as surrogates.
- Transport of embryos. It is sometimes more convenient and less stressful to a valuable animal to transfer an embryo in another animal. For example, sheep, cow and goat embryos have all been transferred in rabbits. This is analogous to taking seeds from one part of the world to another.
- Cloning. Early embryonic cells are totipotent like meristematic plant cells. For example, if you split an embryo into two separate cells after the first cell division, then each cell has the capacity to grow into a new individual. These individuals will be identical, that is, they are clones. This is the way identical twins are produced. In theory, the procedure can be repeated several times so that large numbers of identical individuals can be produced from just one egg. The embryos can then be grown in tissue culture until they reach the correct stage for implantation into the mothers. This is obviously an important way of producing large numbers of genetically superior animals.
- Testing. Transplanted embryos can be tested for sex before being implanted in their new mothers. New diagnostic tests based on recombinant DNA technology (see Theme 5) can be used to test for diseases at the early embryo stage. If the embryo is the wrong sex or it is abnormal then it can be discarded and the breeder can try again quickly.

The relationship between all these techniques is shown in Fig 7.17.

Selective breeding and the environment

It would be wrong to leave you with the impression that selective breeding alone has led to substantial increases in the yield and quality of agricultural products. Probably the greatest influence on plant yields in the last 50 years has not been the improved varieties that have become available, but the widespread availability and use of nitrogen fertilisers. By providing varieties which can use this extra nitrogen more efficiently, plant breeders have further improved yields. You should never forget that it is the subtle interplay of the environment and genotypes which produces phenotypes.

SUMMARY ASSIGNMENT

1. Make sure you have definitions and/or explanations of the following terms: gene bank, induced mutation, polyploidy, autopolyploid, allopolyploid, male sterility, self-incompatibility, tissue culture, anther culture, embryo culture, protoplast fusion, artificial insemination, embryo transplantation.
2. Discuss the importance of polyploidy with special reference to plant breeding.

Chapter 8

THE GENETICS OF PEST CONTROL

Pests are species which interfere with the activities of humans. Every year millions of tonnes of valuable food are lost through the action of a range of pests and pathogens which attack crops. Such pests can range from the very large, for example elephants in Africa destroying crops, to the minute viruses. In addition to the losses suffered in fields, large quantities of food are damaged whilst in store making them unfit for people to eat. Given that much of this damage takes place in the poorer countries of the world, there is an urgent need to develop methods of pest control. Two methods are used widely. The first involves killing the pest, for example through the use of toxic chemicals called pesticides or biological control using the pest's natural enemies. The second involves trying to breed resistance into plants by transferring genes from resistant species or varieties to susceptible ones. Both of these techniques require an understanding of genetics if they are going to be applied properly.

LEARNING OBJECTIVES

After completing the work in this chapter you will be able to:

1. explain the origin of resistance to pesticides;
2. propose methods for reducing the build up of resistance in pest species;
3. define and describe the use of the gene-for-gene hypothesis;
4. explain the difference between specific and general resistance;
5. appreciate the problems faced by breeders in their attempts to produce resistant plants.

8.1 BASIC STRATEGIES OF PEST CONTROL

A pest species can be considered to be under control when it is not causing excessive economic damage, and uncontrolled when it is. The point at which a pest becomes uncontrolled will depend upon the particular pest involved. For example, an insect that destroys 5 to 6 per cent of an apple crop may be causing an insignificant amount of biological damage but may be destroying the grower's profit margin. On the other hand, an insect pest of forest trees, for example a moth like spruce budworm, may actually defoliate (remove all the leaves or needles) over large areas of a forest without adversely affecting the timber industry.

Two basic pest control strategies can be identified: the pesticide strategy and biological control strategy.

Pesticides

Pest control in most agricultural systems is achieved by the use of poisonous chemicals called pesticides. The range of pesticides available and their use is described in detail in another book in this series, *Applied Ecology*. A basic glossary of terms is given in Table 8.1.

Table 8.1 A glossary of terms

Resistance	any inherited characteristic of an organism which reduces the effect of an adverse environmental factor, e.g. a pest, a pesticide or a physical factor such as salinity
Susceptibility	the opposite of resistance
Pesticide	a chemical substance used for destroying pests
Herbicide	a substance which is poisonous to plants and which can be used to destroy or inhibit the growth of weeds. The three major types are: • phenoxyacetic acids (e.g. 2,4,5-T) • triazines (e.g. simazine, atrizine) • ureas
Insecticide	a substance which is poisonous to insects. The major types include: • chlorinated hydrocarbons (e.g. DDT, dieldrin) • organophosphates (e.g. malathion) • carbamates (e.g. propoxur) • synthetic pyrethroids
Fungicide	substance which is poisonous to fungi
Systemic	pesticides which enter a plant through the roots, stems or leaves and are then translocated into the rest of the plant tissues

A major problem associated with the use of pesticides is that the pest species often become resistant to pesticides which previously killed them. For example, the insects which attack cotton have developed resistance to so many insecticides that it is practically impossible to grow cotton in some parts of Mexico, Central America and southern Texas.

The pest species does not usually become totally immune to the pesticide. Rather it takes more pesticide to kill the pest than before (see case study 2). The additional cost of the extra pesticide treatments may reduce the farmer's profit to such an extent that the treatment becomes uneconomic. In addition, high levels of pesticide application can damage the crop they were designed to protect and may present enormous ecological problems.

The exact biochemical mechanisms involved in the development of pesticide resistance depend upon the pest species involved, but they include storage of toxic material, chemical or physical barriers to the entry of the toxin, the production of enzymes which break down the pesticide, avoidance of metabolic pathways affected by the pesticide or even alteration in the behaviour of the pest species so that it avoids the insecticide.

Since genes are likely to play a part in all these mechanisms, it is not surprising to find that the genetic basis of pesticide resistance also varies between pest species and even between populations of the same pest species. In most cases, the genetic basis of resistance is unknown. In those examples which have been investigated, such as warfarin resistance in rats (case study 1), it often appears that resistance is controlled by just one or two gene loci, that is, resistance is under major gene control. However, this is likely to be an oversimplification. Remember, genes are constantly interacting with each other through the interaction of the products which they encode. So the precise level of resistance of a particular organism to a pesticide will depend not only on the possession of major resistance genes but also on the genetic background in which these genes are operating. In other words, the major genes for resistance may be expressed differently in different individuals.

One estimate suggests that banning the use of all pesticides would reduce agricultural output in the United States of America by 30 per cent.

Given that crops have to be protected from pests, then new pesticides will have to be developed. If these new pesticides are to be successful, it is important to understand how a pest population becomes resistant to pesticides. This is where the geneticist comes into the war against pests. The development of pesticide resistance is discussed further in section 8.2.

QUESTION

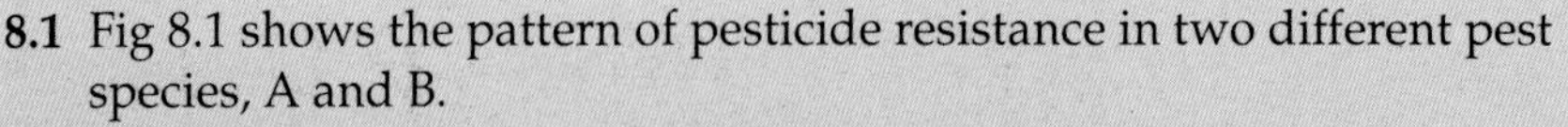

8.1 Fig 8.1 shows the pattern of pesticide resistance in two different pest species, A and B.
(a) For each species, explain whether resistance is under major gene control or polygenic control.
(b) Explain why the resistance levels of the susceptible and resistant individuals of species A are not all the same.

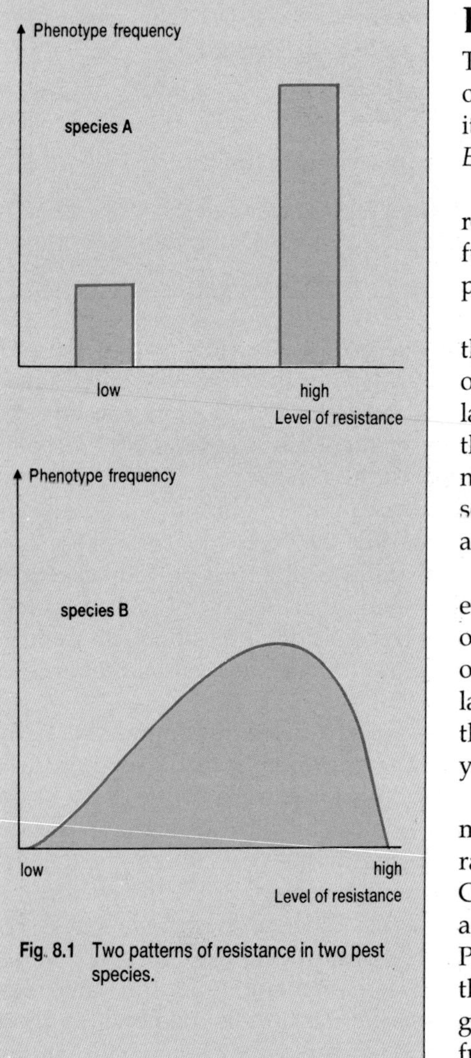

Fig. 8.1 Two patterns of resistance in two pest species.

Biological control

This term covers a wide range of different strategies, ranging from the use of a pest species' natural enemies to the use of a pest's hormones to disrupt its breeding cycle. Again, these are described in more detail in *Applied Ecology*.

The geneticists' contribution to the biological control of pests is in their role as selective breeders. Plant varieties can be made resistant to insect, fungal and viral pests by incorporating resistance genes into the cultivated plants. The process is described in section 8.3.

An alternative genetic control strategy is to alter the genetic make-up of the pest species. The simplest genetic manipulation that can be carried out on a pest species is to make it sterile. This can be achieved by releasing large numbers of males of the pest species which have been made sterile through the use of chemicals or radiation. These sterilised males mate with normal females but their sperm is unable to fertilise any eggs. Consequently, the number of offspring in the next generation is greatly reduced and the pest is brought under control.

The sterile male technique has had a number of successes. The classic example is the control of the screw worm. This is the larval (maggot) stage of the fly *Cochliomyia hominivorax* which lays its eggs in wounds on cattle, other domestic animals and sometimes people. The eggs hatch and the larvae then start to eat the host animal. Economic losses from this pest in the southern United States of America were estimated at $120 million per year.

In a pilot experiment, screw worm flies were reared in captivity and the males were then sterilised at the pupal stage by irradiation with gamma rays. These sterile males were then released in large numbers on the small Caribbean island of Curaçao. The females of this species only mate once, so any female that mates with a sterilised male will not lay any fertilised eggs. Providing the sterile males outnumber the wild males and then mate with the wild females, the pest population should decrease in size. In a few generations, about one year, the screw worm fly was completely eradicated from the island.

The experiment was then repeated in the United States of America. At a cost of $60 million, the screw worm was eradicated from the southern states. However, since the fly still lived in Mexico, the liberation of sterile males had to be continued in order to prevent reinvasion of the screw worm fly.

8.2 THE DEVELOPMENT OF PESTICIDE RESISTANCE

If the rose bushes in your garden were covered in aphids, you would usually reach for the insecticide and spray them. But the insecticides we use today are very different from those available 30 or even 20 years ago. The evolution of a new type of insecticide spray and of new methods of pest management stem directly from the evolution of resistance to insecticides in the pests themselves.

A pest population becomes resistant to a particular pesticide because it already contains alleles for resistance before the pesticide is applied. The pesticide does *not* cause the mutations which will produce the resistance alleles. Rather, these alleles are already present in the gene pool of the pest population, albeit at a low frequency.

By applying a pesticide to a population of pests, you are applying an extraordinarily powerful selection pressure. This can lead to dramatic changes in the allele frequency of the pest population's gene pool such that the frequency of the resistance alleles increases, which can result in the population becoming resistant to the pesticide.

Notice we are not talking about individuals developing pesticide resistance, but populations. The mechanisms involved in the development of pesticide resistance are examined in the two case studies below.

Case study 1 – Warfarin resistance in rats

Rats (Fig 8.2) are a major health hazard and a serious pest in stored food crops. In an attempt to control the rat population, the poison warfarin was introduced in the 1950s. Warfarin is an excellent rat poison because of its low toxicity to other animals. Warfarin kills rats because it interferes with the normal mechanism of blood clotting.

Fig 8.2 Rats do an enormous amount of damage as well as spread disease.

In 1958 some rats caught just to the east of Glasgow in Scotland were found to be resistant to warfarin. Even when fed quite high doses of warfarin, the blood of these rats still clotted normally. However, a really high dose of warfarin killed them. By 1972, 12 areas in Britain (Fig 8.3) were known to have warfarin-resistant rats.

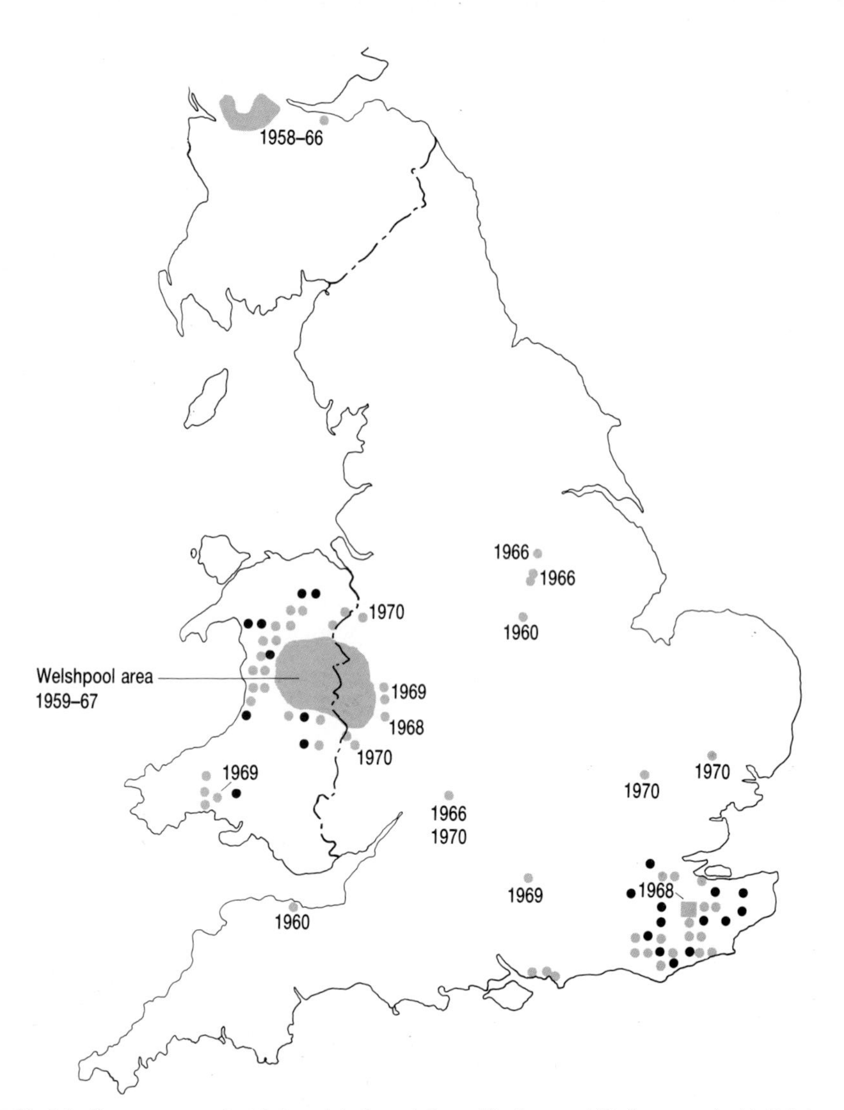

Fig 8.3 The occurrence of warfarin-resistant populations of the brown rat (*Rattus norvegicus*) in Britain. Undated samples are from rats tested in 1972. ● = resistant sample; ● non-resistant sample. (After Greaves & Rennison, 1973.)

The difference between warfarin-resistant and warfarin-susceptible rats can be determined in the laboratory by testing how long it takes for the rats' blood to clot. A blood sample is taken from a rat, usually from the large blood vessel in the tail, the sample is mixed with warfarin and subjected to a standard clotting test carried out using standard laboratory equipment. The blood of the resistant rats clots in just a few seconds whilst the blood of susceptible rats may take more than a minute to clot.

Warfarin resistance in rats is usually determined by a dominant allele at one locus. Animals carrying the resistance allele require larger amounts of vitamin K in their diets than the susceptible rats. Vitamin K is an essential part of the blood clotting mechanism in mammals. Warfarin somehow seems to interact with the vitamin K and stops it from playing its normal role in blood clotting. This leaves the susceptible rats open to the possibility of a fatal haemorrhage. Resistant rats apparently do not use vitamin K in the same way and thus their blood clots normally even when they have eaten large amounts of warfarin.

The specific mechanisms of warfarin resistance, however, vary between populations. For example, the resistant Scottish rats are killed by lower doses of warfarin than those found around Montgomery in Wales. The Welsh rats also appear to need more vitamin K than the Scottish rats in order to maintain their resistance. It seems probable that different alleles at the same locus determine resistance in the two areas.

Interestingly, the number of resistant rats rose to about 50 per cent in each population and then stayed at that level. This suggests the genetics of warfarin resistance is a little more complicated than was first suspected. The three possible genotypes and phenotypes are shown below:

Rw^SRw^S normal rats susceptible to warfarin
Rw^RRw^S heterozygotes, resistant to warfarin with a slightly increased requirement for vitamin K
Rw^RRw^R warfarin-resistant but having a 20 fold increased requirement for vitamin K

So the heterozygotes enjoy a selective advantage over both homozygotes. They are resistant to warfarin, but they do not require the massive amounts of vitamin K that places the dominant homozygotes at a selective disadvantage. This state of affairs is called **balanced polymorphism**. The susceptibility to warfarin of the recessive homozygotes (Rw^SRw^S) is balanced by the reduction in fitness of the dominant homozygotes (Rw^RRw^R) due to their increased demand for vitamin K. The implications of this are that the normal recessive allele can never be eliminated from the population as there will always be a pool of Rw^S alleles hiding from selection in the superior heterozygotes.

Warfarin is used to treat people with diseases where blood clots may form which block blood vessels, e.g. deep vein thrombosis. If these blood clots break loose they can become lodged in blood vessels in the lungs causing pulmonary embolism, which is often fatal. A dominantly inherited pattern of resistance to warfarin has been detected in people, though fortunately the allele responsible seems to be rare.

Many populations of mice are also becoming resistant to warfarin. However, the genetic mechanism is different to that in rats. Again, a major gene is involved, the dominant allele of which confers resistance. However, the effect of this major gene is modified by other genes so that the genetic basis of warfarin resistance in mice is best described as being polygenic. Warfarin resistance in rats is also affected by modified genes but these have a relatively minor effect.

QUESTIONS

8.2 Give a detailed description of the experiments and crosses you would carry out to prove that warfarin resistance is due to a dominant allele at one locus.

8.3 Explain the origin of warfarin resistance in the Scottish and Welsh rat populations.

8.4 With the aid of a gene pool model like the one used in section 5.3, explain how the resistance allele spread through the population. Remember, rats are diploid and the heterozygote enjoys a selective advantage over both homozygotes.

8.5 Why will the warfarin-susceptible allele never be totally eliminated from the rat gene pool?

8.6 Geneticists investigating warfarin resistance in rats around Welshpool (Fig 8.3) found that the level of warfarin resistance decreased from 60 per cent to 40 per cent after the use of warfarin was halted.
(a) Account for this observation.
(b) Suggest why resistance alleles normally occur at low frequencies in populations not exposed to pesticides.

8.7 Find out how rat populations are controlled now. Your local Environmental Health Officer will be able to help you find this information. They are usually more than willing to help. Would you expect rats to develop resistance to new poisons?

8.8 **Extension question**
Using the Hardy-Weinberg equation, calculate the frequency of each allele and each genotype in a population where 50 per cent of the rats are resistant to warfarin.

Case study 2 – Cotton in Peru

The Cañete valley is Peru's most successful cotton growing area. The landowners are very wealthy, which has enabled them to buy the latest in crop management technology. In particular, the farmers rapidly adopted the use of organochlorine insecticides, for example DDT, developed during the Second World War to combat malaria. Initially the chemicals worked like magic, killing off the insect pests to which cotton is particularly prone. However, the farmers soon found they had to use more and more insecticide as the pests began to develop resistance. It was not that the insects were becoming immune to the insecticides, rather it was taking more insecticide to kill them.

By the mid 1950s the Cañete farmers were spending about 30 per cent of their production costs on organochlorine insecticides, with some farmers spraying as many as 40 times a year. Entomologists from Peru's National Agrarian University suggested a series of measures to remedy this ecological disaster. Fields are now checked regularly to estimate pest levels and the levels of the pests' predators (these are killed by the wide-spectrum organochlorine insecticides). Farmers cut their use of insecticides to only one or two sprays per year and they only spray when the pests in a field have reached a level which economically justifies the use of an expensive insecticide. The insecticides used are selective (narrow-spectrum), slow-acting and short-lived. They cause only limited damage to the pests' natural enemies.

QUESTIONS

8.9 What do you understand by the terms 'narrow-' and 'broad-spectrum' insecticide?

8.10 You have been approached by a group of farmers who are suffering from similar problems to those experienced by the people in the Cañete valley. Explain why the intensive use of broad-spectrum insecticides has caused their problems and how the remedies suggested by the Peruvian entomologists will help them. Remember, you are talking to farmers not scientists, so you will need to keep your explanation simple, though academically honest, and try to use diagrams wherever possible.

Case study 3 – Herbicide resistance in the Groundsel, *Senecio vulgaris*

Of the biological factors (insect pests, fungal infection and so on), weeds are probably the single most important cause of crop losses worldwide: *average* losses to weeds in arable crops are thought to be of the order of 10 per cent. In 1981 the cost of herbicides exceeded £100 million in the United Kingdom and the expenditure on herbicides continues to increase substantially each year. For example, in the United Kingdom weeds of cereal crops are often sprayed three times a year.

Three major groups of chemicals are used as herbicides, phenoxyacetic acids, triazines and ureas. Herbicide resistance first appeared in weeds found in Hawaiian sugar cane plantations in the early 1950s. By 1983 about 50 species worldwide were resistant to 11 herbicides, mostly in the triazine group. One of the best examples is provided by the groundsel, *Senecio vulgaris*.

This plant is a major pest. For example, it infests over a quarter of a million hectares of arable land in Washington, USA. The plant was controlled successfully by triazine herbicides. These herbicides act by binding to the thylakoid membranes of the chloroplasts, inhibiting photosystem II and thus preventing photosynthesis. Resistant plants are thought to possess mechanisms which prevent the triazine herbicides from binding to the thylakoid membranes. By 1982 seven different groundsel populations in the United Kingdom were known to be resistant to triazine, the genetic basis of resistance being different in all seven populations.

QUESTIONS

8.11 Herbicide resistance is much less of a problem than insecticide resistance. At least 360 insect pests are now known to be resistant to at least one insecticide. By comparison, only 50 species of plant are resistant to herbicides.

(a) Account for this observation. You will need to think about the factors which affect the rate at which a resistance allele will increase in frequency in a pest population's gene pool.

(b) Of the plants which have become resistant to herbicides, the majority are annuals, produce a large number of seeds and are often self-fertile. How do these characteristics contribute to the rapid increase in the frequency of resistance alleles in the weed's gene pool?

8.12 Fungal infections of crops represent another major loss to farmers. Fungi are often controlled using systemic fungicides which often affect a single step in a single metabolic pathway in cells. Over 50 species of fungi are now known to be resistant to systemic fungicides. By contrast, resistance to general fungicides, based on mercury or copper, has rarely been observed. Account for these observations.

A final note

It would be wrong to leave you with the impression that all pesticides are bad. The use of DDT to control mosquitoes, which carry the parasites causing both malaria and yellow fever, has undoubtedly saved millions of lives. The problem is that these strong chemicals produce a very strong selection pressure which will rapidly change the gene pool of a pest species. A system where spraying is reduced to one or two times a year will reduce that selection pressure, so resistance is not so strongly selected for. Furthermore, varying the insecticide used will ensure that one particular part of the population does not become resistant to all insecticides. The use of selective insecticides means that you only hit the pest which is the current problem. This means potential pest species cannot then develop resistance. The sensible use of pesticides in a system of integrated pest management (IPM) is what is needed, not the prohibition of all pesticides. You can read more about IPM in another book in this series called *Applied Ecology*. A general model for the development of pesticide resistance is shown in Fig 8.4.

8.3 NATURAL RESISTANCE

In addition to insect pests, crop plants are attacked by a whole variety of other parasites. These include fungal diseases like smuts, rusts and mildews, and viruses. Some examples are shown in Fig 8.5.

It was common knowledge, even 100 years ago, that some varieties of a crop plant were more resistant than others to these plant parasites. Indeed, in 1905 one scientist had shown that resistance to yellow rust, caused by a fungus called *Puccinia striiformis*, was inherited. A cross between Rivet wheat (resistant) and a susceptible wheat had produced a pattern of inheritance which followed Mendel's laws. Since this pioneering work, many other investigations into the genetics of host resistance have been carried out. Such research is vital if selective breeders are going to transfer genes for resistance between plants. Questions like how many genes are involved, is resistance dominant or recessive, do alleles interact, for

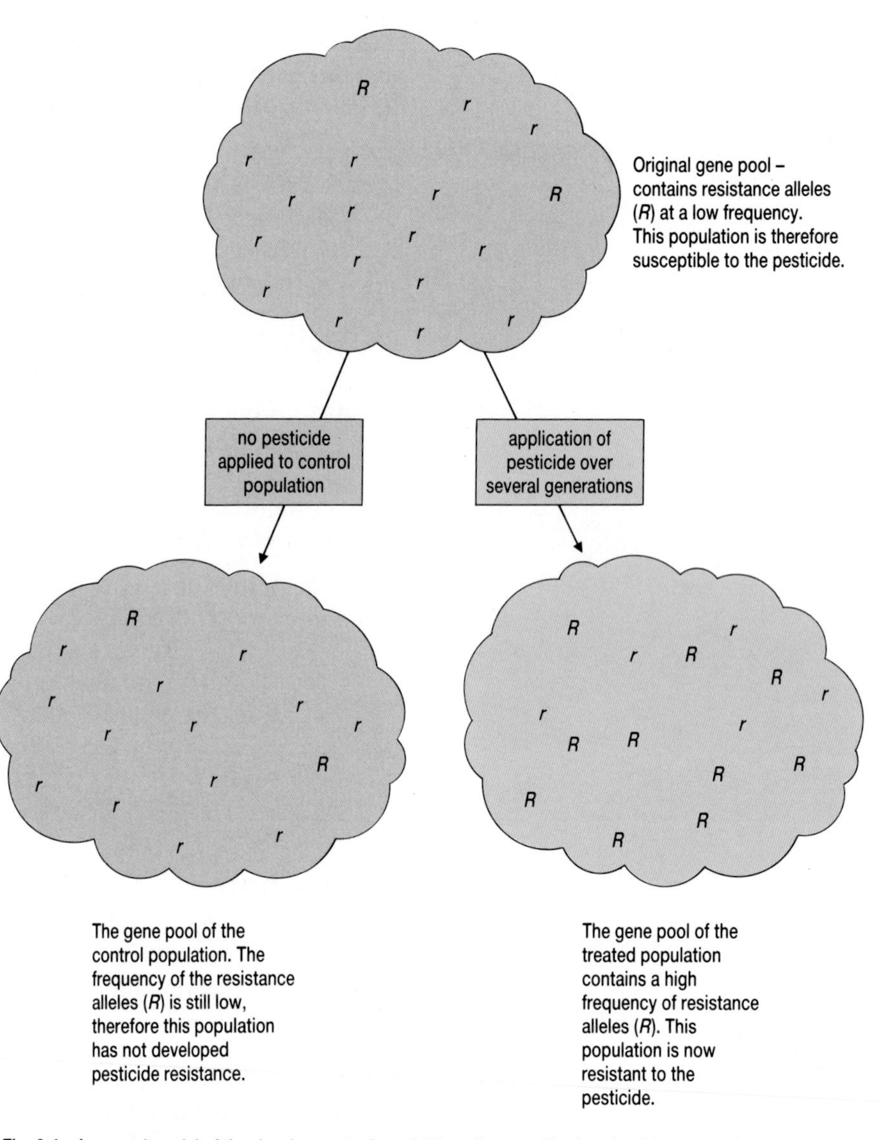

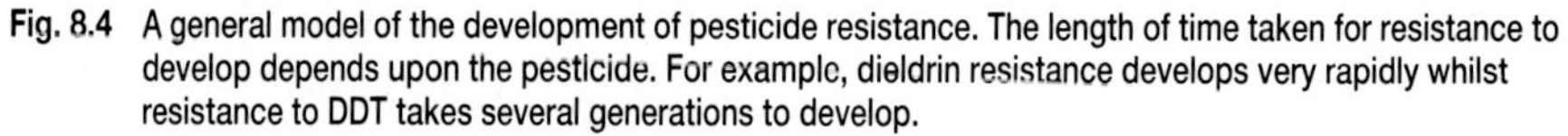

Fig. 8.4 A general model of the development of pesticide resistance. The length of time taken for resistance to develop depends upon the pesticide. For example, dieldrin resistance develops very rapidly whilst resistance to DDT takes several generations to develop.

example through complementary gene action or epistasis, all need answering if the selective breeder is going to put together the correct sequence of crosses to transfer the resistant phenotype. This research has revealed that the genetics of resistance are complex, but some consistent patterns have emerged. It is these we are now going to examine.

Some useful terminology

Before we start to explore the genetics of plant resistance you need to know a few terms used by plant pathologists. Organisms which are parasitic on plants are divided into **physiologic races** on the basis of their ability to infect different varieties of the same plant. The exact range of cultivars which the pest can attack will depend on the virulence genes they contain. If a physiologic race contains a large number of genes for virulence it is said to be highly **virulent**. On a susceptible host some physiologic races may grow and/or reproduce faster than others. Such races are said to be **aggressive**.

(b)

(a)

(c)

Fig. 8.5 **(a)** Adult Colorado beetles and their eggs. Strict quarantine rules have ensured that this devastating pest of potatoes has not become established in Great Britain.
(b) A severe infestation of powdery mildew on wheat. Whilst the fungus may not kill the plant, it severely reduces both the yield and the quality of the grain.
(c) Virus infection of potato plant showing the characteristic leaf curling. Since plant viruses are often spread by aphids, seed potatoes are usually grown in cold areas, such as Scotland, where aphids are rare.

An experiment – resistance to rust in flax

Flax is a plant used to supply fibres for the rope and linen industry. Like most other plants it is attacked by a range of parasites, one of which, *Melamspora lini*, causes flax rust. Look at Table 8.2. This shows the results of a cross between two varieties of flax, Ottawa and Bombay. Ottawa is resistant to race 24 of *M. lini* but susceptible to race 22 whilst the situation is reversed for Bombay. If Bombay and Ottawa are crossed, the resultant F_1 progeny are resistant to both varieties. If the F_1 are now selfed, the F_2 are

Table 8.2 Resistance to races 22 and 24 of *Melamspora lini* in a cross between the flax cultivars Ottawa and Bombay (S = susceptible, R = resistant) (After Flor, 1946.)

Race	Ottawa	Bombay	Flax F_1	Flax F_2			
22	S	R	R	R	S	R	S
24	R	S	R	R	R	S	S
		Observed F_2 ratios:		110	32	45	9
		Expected F_2 ratios (9:3:3:1):		109	36	36	12

resistant and susceptible in a ratio consistent with the 9:3:3:1 ratio of a dihybrid cross.

In a second experiment the two rust races, 22 and 24, were crossed. The results of this experiment are shown in Table 8.3. The rust F_1 generation was unable to attack either of the flax cultivars, in other words both flax cultivars are resistant (R). If the rust F_1 is now selfed, a different pattern emerges in the F_2. Now some of the rusts can attack the two flax cultivars. How can we interpret this data? We need to make three assumptions.

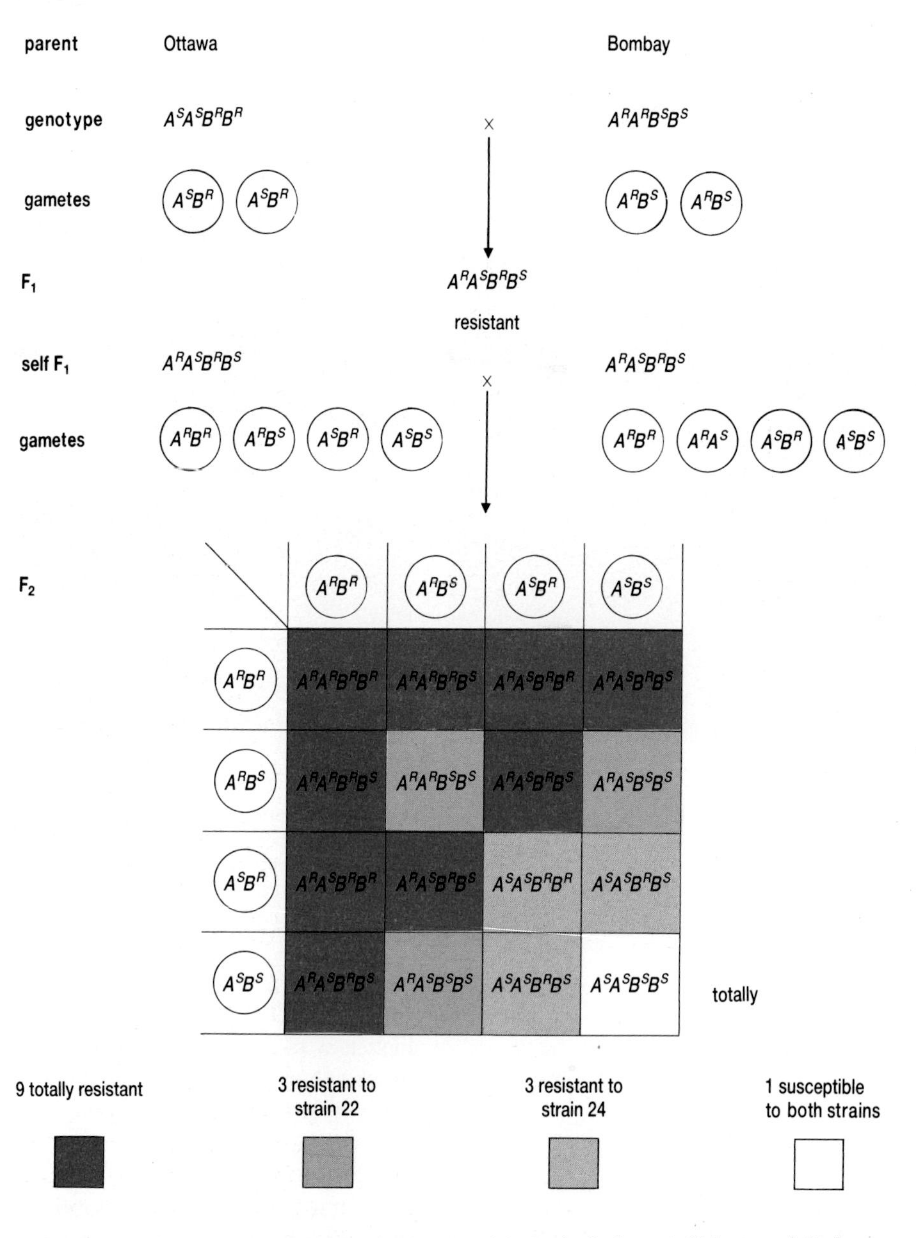

Fig 8.6 A model to show the inheritance of resistance to infection by the flax rust, *Melamspora lini*, in flax.

Table 8.3 Virulence to two cultivars of flax, Ottawa and Bombay, in a cross between races 22 and 24 of the flax rust *Melamspora lini* (S = susceptible, R = resistant) (Flor, 1946).

Cultivar	Race 22	Race 24	Rust F_1	Rust F_2			
Ottawa	S	R	R	R	S	R	S
Bombay	R	S	R	R	R	S	S
		Observed F_2 ratios:		78	27	23	5
		Expected F_2 ratios (9:3:3:1):		75	25	25	8

Assumption 1. Ottawa and Bombay each have two loci which determine resistance to attack by the two physiologic races of *M. lini*. Let us call the locus which controls resistance to race 22 locus A and the locus which controls resistance to race 24 locus B.

Assumption 2. At each locus there are just two possible alleles which we will call *R* for resistance and *S* for susceptibility.

Assumption 3. *R* is dominant to *S*.

We also know that the two varieties are true breeding. This means that they are homozygous at the two loci. So using this knowledge and our assumptions, we can now model the cross shown in Table 8.2.

The model is shown in Fig 8.6 and you should recognise this as a straightforward dihybrid cross. Instead of using flower colour or pea shape as the phenotypic characters we are interested in, we are using susceptibility or resistance to two races of pathogen. The results of our model are consistent with the results shown in Table 8.2. So we can conclude that two loci control resistance in the flax plants and the resistance allele at each locus (*R*) is dominant to the susceptible allele (*S*).

QUESTION

8.13 Using the same method as above, that is, setting out your assumptions first, construct a model to explain the inheritance of virulence in *M. lini*. The two races, 22 and 24, are pure-breeding. You may get the answer wrong at first but don't give up. Try a different model or modify your assumptions. If you get really stuck then talk with a friend. After all, science is a cooperative venture.

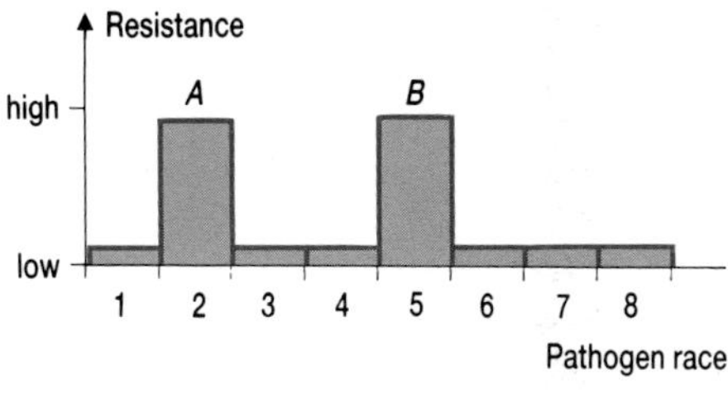

Fig 8.7 Specific or vertical resistance. This plant has two genes encoding specific resistance, *A* and *B*, but has low general resistance.

The gene-for-gene concept

Using the results from experiments like the one you have just analysed, a major theory of host–pathogen interaction evolved. This is called the gene-for-gene concept and was defined by Flor (the scientist who did the research on flax rust) as follows: **for each gene conditioning rust reaction in the host, there is a specific gene conditioning pathogenicity in the parasite.** What this means is that resistance and susceptibility are the result of specific combinations of corresponding alleles in both organisms. Rather like a lock and a key, the pathogen has got to be carrying the right virulence alleles to infect the host plant. This suggests that there is an intimate genetic relationship between a host and its pathogens which has been built through the course evolution.

The validity of this concept has now been tested using a variety of parasites and host plants other than flax. The parasites include rusts, smuts, powdery mildews, nematodes, insects, viruses and bacteria. So far, the experiments support the theory. This theory now forms the basis of much of the interpretation of the genetics of different host–parasite systems in plants.

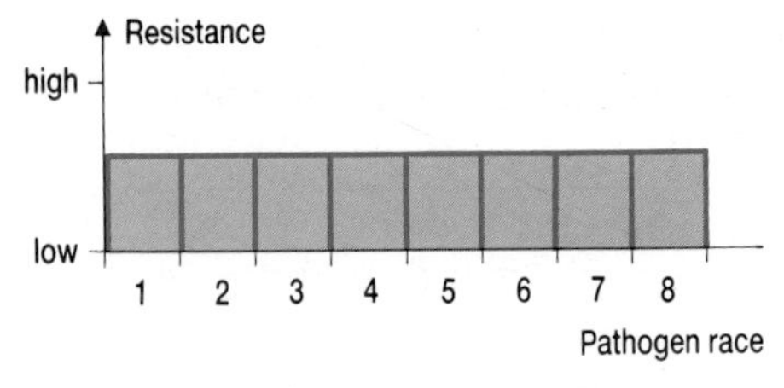

Fig 8.8 General or horizontal resistance. This plant has no specific resistance but a large amount of general resistance which is controlled by many genes, i.e. polygenically.

Specific and general resistance

Look at Fig. 8.7. This shows a cultivar of a crop plant which has a high level of resistance to just two physiologic races of a pathogen, but is susceptible to other physiologic races. This plant is said to show **vertical** or **specific resistance**. Such a pattern of resistance is usually controlled by just one or two major genes. The two varieties of flax described above are showing specific resistance.

Now look at Fig 8.8. This shows another crop cultivar exhibiting **horizontal** or **general resistance**. Here the crop is reasonably resistant to a whole range of pathogen races but is not specifically resistant to any particular race. General resistance is usually under polygenic control, that is, resistance in the host is produced by a large number of loci with the effects of the alleles at each locus being additive. In other words, general resistance is a quantitative character whilst specific resistance is a qualitative character.

QUESTIONS

8.14 Fig 8.9 shows two other patterns of resistance. Interpret these patterns in terms of specific and general resistance.

8.15 Using the gene-for-gene concept, explain why even a single allelic change, for example due to a mutation, would result in the breakdown of specific resistance so that a plant which was resistant to a particular physiologic race becomes susceptible. Why will this not happen with a plant showing general resistance?

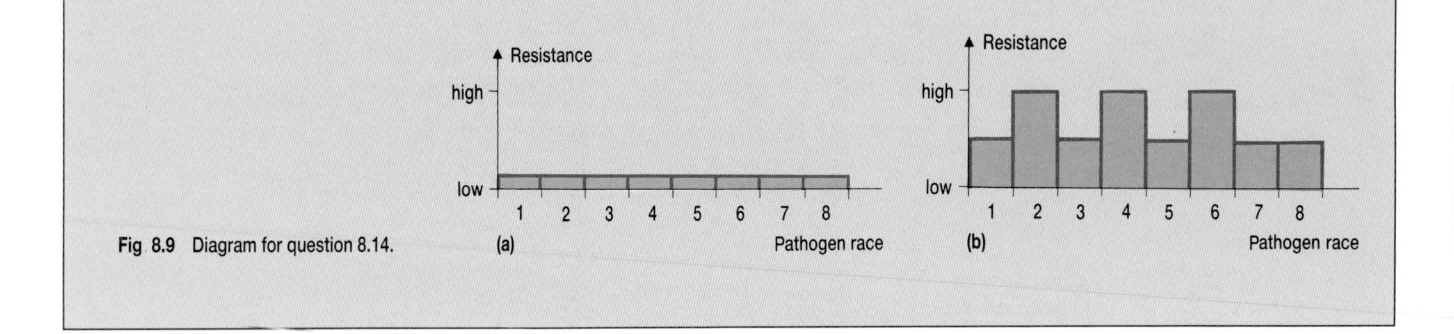

Fig 8.9 Diagram for question 8.14.

8.4 USING RESISTANCE

Having established the pattern of inheritance for resistance to a particular pathogen, how can breeders use this information? There are basically two strategies – the variation strategy and the breeding strategy.

The variation strategy

In Chapter 5 we looked at the concept of background genes. These were the genes which adapted a particular group of plants to a particular locality. Now a land race, which is a locally adapted population, will contain a whole host of these background genes adapting them to a local environment. One aspect of that environment will be the local pests and parasites. So if a farmer was to maintain a good variety of different genotypes in a crop, the chances of a particular pathogen eliminating all the plants in one year are small. However, if the farmer planted a high yielding but highly inbred line, one pathogen could destroy the whole crop in one year. This was demonstrated only too clearly during the potato famine in Ireland during the 1840s. The potato farmers in Ireland used only one variety, the so-called lumper potato – which was almost totally eradicated by the potato blight fungus, *Phytophthora infestaris*, shown in Fig 8.10.

Throughout the potato famine, when tens of thousands of people died of starvation or emigrated to the United States of America, Ireland was a net exporter of wheat. The average person in Ireland could not, however, afford to buy wheat flour.

Fig 8.10 Potato tubers infested by potato blight fungus which makes them unfit for human consumption.

This lack of variation in modern crop plants is still a problem today. For example, the discovery of male sterility genes in the variety of maize called Texas June meant that by the late 1960s most of the maize varieties in the United States of America contained genes from this one single variety. Now, whilst maize is nowhere as genetically uniform as an inbreeding crop can be, this concentration of genes from just one variety still represents a major reduction in the variability of the maize gene pool. This became evident in the summer of 1970 when just one strain (called strain T) of southern leaf blight fungus, *Helminthosporium maydis*, attacked the maize crop. Starting in the southern states of the American corn belt the fungus spread northwards at the astonishing rate of 150 km per day. Maize breeders are currently very concerned to recover some of the lost variability of their crop.

QUESTIONS

8.16 Where might maize breeders look to find some of their lost variability, and how might they return it to the gene pool of the maize crop?

8.17 One project at the Plant Breeding Institute at Cambridge involved producing 'varieties' of potatoes suitable for the Third World. Rather than sending tubers to the farmers, the PBI sent seed which does not breed true. Why is this a sensible strategy to use in a Third World country which may not have enough foreign currency to spend on pesticides?

Breeding for resistance

Breeding for resistance seems like a good idea. But remember that selective breeding is expensive, so it only makes economic sense to breed plants resistant to pathogens which cause real economic damage. For example, suppose a variety of wild potato was found which was resistant to a particular fungus. You might think that it would be a good idea to transfer the resistance genes to a cultivated variety. But think about all the years of backcrossing that you would have to carry out, all the trials and so on. Then you realise that the level of damage being caused by the fungus to commercial potatoes is so small that it does not make economic sense to embark on the breeding programme.

The second design feature that the plant breeder also has to bear in mind is whether there is sufficient genetic variability for the trait in question. If resistance is only controlled by one or a few genes then transfer might not be too difficult. But what about polygenic control? Here the breeder will need to estimate the amount of variation in the population and assess the heritability of the trait before the difficult task of transferring the resistance genes can start. It is worth bearing these constraints in mind as you read through the three case studies on resistance breeding in potatoes given below. The case studies illustrate the pitfalls and the successes of breeding for resistance.

Case study 4 – Potato blight

A POTATO'S LOT IS NOT A HAPPY ONE
Worldwide, the catalogue of potato pests includes 128 insects, 68 nematodes, 38 fungi, 23 viruses and 6 bacteria. The list is still growing.

Potato cultivars produced from *Solanum tuberosum*, the potato you eat, are susceptible to practically all physiologic races of the blight fungus *P. infestans*. However, a wild Central American potato, *S. demissum*, possesses a number of major genes, called *R* genes, conferring specific resistance to particular races of *P. infestans*. A great deal of effort was put into transferring these *R* genes into cultivars of *S. tuberosum* which were commercially acceptable. However, as fast as the breeders produced cultivars containing more and more *R* genes, so the pathogen produced more and more races with correspondingly larger numbers of virulence genes capable of overcoming the *R* genes. The breeders and the fungus were locked in a genetic arms race and the potato blight was winning.

Physiologic races are numbered according to the virulence they contain, so a race capable of overcoming *R* genes 1 and 4 will be numbered race 1,4. The breeders now struck back. They devised a strategy in which they introduced cultivars which contained several resistance genes. In 1961, Pentland Dell, a variety of potato which contains three *R* genes R_1, R_2 and R_3 was introduced. This remained blight free until 1967 but it was then heavily attacked all over southern Britain. In 1968 it was heavily attacked throughout the whole of Britain. In these two years 23 races of the fungus were isolated from Pentland Dell. More worrying for the breeder, the fungus had cracked the multiple resistance gene strategy. Some of the 23 races contained up to nine virulence genes.

By 1969, at least 11 *R* genes, numbered R_1 to R_{11}, had been identified, as had the corresponding virulence genes in the potato blight. Also, in addition to the races which contained single virulence genes, many races containing multiple virulence genes had been isolated. One from Britain had the whole set – 1, 2, 3, 4, 5, 6, 7, 8, 9, 10, 11. Game, set and match to the fungus, but the war goes on.

Case study 5 – Potato cyst nematode

The major growing area for potatoes in Britain is in the Fenlands of Cambridgeshire and Lincolnshire. In the 1950s, practically the entire Fenland potato industry was wiped out by an attack of the potato cyst nematode, *Globodera*. This animal attacks and destroys the potato tubers and roots. Three wild species of potato show resistance: *S. vernii*, *S. cruzianam* and *S. andigena*, the ancestor of *S. tuberosum*. A cross between *S. tuberosum* and *S. andigena* in 1955 eventually led to the production of a resistant variety, Maris Piper, released to farmers in 1966. This variety is now the mainstay of the Fenland crop and the commonest variety of potato eaten in England.

Case study 6 – Scaring off the aphids

This case study is included to show you that there are many ways of defeating a pest problem and sometimes you need to be a little subtle and not attack the pest itself. Viruses are a major problem for potato growers and they are particularly difficult to control. However, viruses are usually carried from plant to plant by aphids. So an alternative strategy to attacking the virus directly would be to make the plants resistant to aphids. A wild potato, *S. berthaultii*, has hairy, sticky leaves which aphids apparently don't like. In addition to its hairiness, *S. berthaultii* has a chemical defence system too. Some of the leaf hairs produce a substance called (E)-beta-farnesene (pronounced far-knee-seen). This substance is chemically very similar to an aphid alarm pheromone. This alarm pheromone is a substance produced by aphids which alerts other aphids to danger. So by producing (E)-beta-farnesene, *S. berthaultii* is literally scaring the aphids away.

Geneticists at the Rothampstead Research Station and Plant Breeding Institute are trying to breed varieties of potato containing the 'aphid scaring' genes from *S. berthaultii*. The programme highlights one of the problems involved in crossing a domesticated and a wild species. *S. berthaultii* is a plant adapted to a semi-arid climate by having stolons that spread out for a long distance. Stolons are the underground stems of the potato which swell to produce the tubers we eat. *S. berthaultii* produces small tubers at the end of stolons which are 2 to 3 metres in length. Such a habit will result in very low yields from a potato field because each plant would take up so much room. The breeders are now faced with a long programme of backcrossing to dilute out the *S. berthaultii* genes that they do not want.

QUESTION

8.18 An inbred strain of maize, strain A, is resistant to a particular race of the pathogenic fungus *Puccinia sorghi*, whereas a different inbred maize, strain S, is entirely susceptible to attack. The results of crosses between these two lines are shown in Table 1.

Table 1

Cross	Progeny	Result
A × S	F_1 (A × S)	all resistant
F_1 (A × S) × S	backcross (A)	50% resistant 50% susceptible

(a) (i) You are told that strains A and S are both inbred. What does this suggest about the genotypes of the two strains?

(ii) Explain how the results in the table suggest that the difference in resistance between strain S and strain A is due to a single gene with two alleles, a dominant one for resistance and a recessive one for susceptibility.

(iii) What results would be expected from a backcross between F_1 (A × S) and strain A and why?

Another inbred strain of maize, strain B, is also resistant to *Puccinia sorghi* and when crossed with strain S gives similar results to those in Table 1, that is, as in Table 2.

Table 2

Cross	Progeny	Result
B × S	F_1 (B × S)	all resistant
F_1 (B × S) × S	backcross (B)	50% resistant 50% susceptible

(continued)

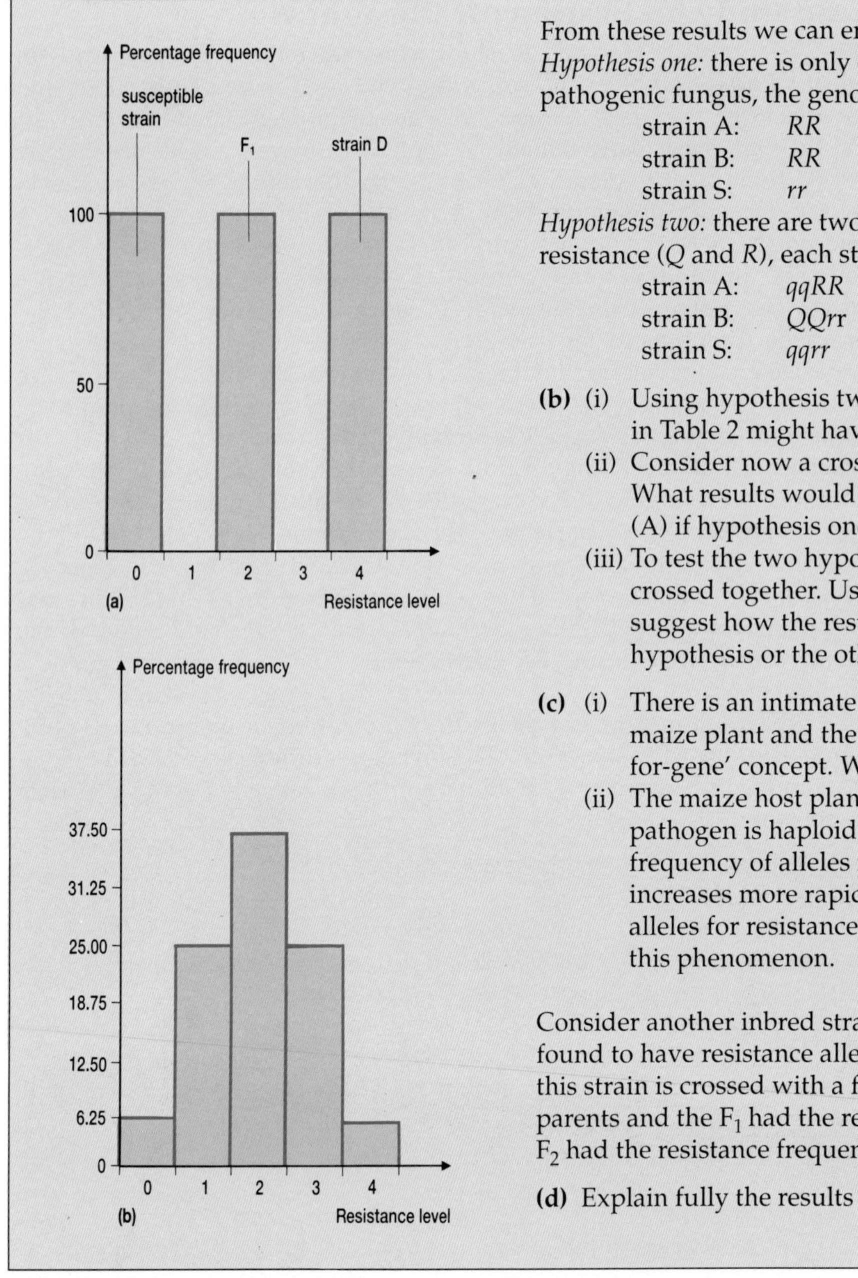

From these results we can erect two hypotheses:

Hypothesis one: there is only one gene involved in resistance to the pathogenic fungus, the genotypes of the different strains being:

strain A:	*RR*	(resistant)
strain B:	*RR*	(resistant)
strain S:	*rr*	(susceptible)

Hypothesis two: there are two genes at different loci involved in resistance (*Q* and *R*), each strain having a different genotype:

strain A:	*qqRR*	(resistant)
strain B:	*QQrr*	(resistant)
strain S:	*qqrr*	(susceptible)

(b) (i) Using hypothesis two, show how results from the crosses in Table 2 might have been obtained.

(ii) Consider now a cross between strain A and strain B. What results would be expected in the F_1 (A × B) plants (A) if hypothesis one is true, (B) if hypothesis two is true?

(iii) To test the two hypotheses the F_1 (A × B) plants were crossed together. Using the genotypes given above, suggest how the results in the F_2 might support one hypothesis or the other. Show your reasoning.

(c) (i) There is an intimate genetic relationship between the maize plant and the pathogenic fungus, called the 'gene-for-gene' concept. What is meant by this term?

(ii) The maize host plants are diploid, whereas the fungal pathogen is haploid. When the two organisms interact, the frequency of alleles for virulence in the fungus often increases more rapidly than the frequency of dominant alleles for resistance in the maize. Suggest *two* reasons for this phenomenon.

Consider another inbred strain of maize, strain D. This has been found to have resistance alleles at two gene loci (*G* and *H*). When this strain is crossed with a fully susceptible strain (*gghh*) the parents and the F_1 had the resistance levels shown in Fig (a). The F_2 had the resistance frequencies shown in Fig (b).

(d) Explain fully the results from the F_1 and F_2.

(JMB, 1988)

SUMMARY ASSIGNMENT

1. Your answers to the questions in this chapter will serve as a summary of the work in this chapter.
2. Make sure you have definitions and, where appropriate, examples of the following terms: gene-for-gene concept; horizontal resistance; general resistance.
3. Make a copy of Fig 8.4 to serve as a summary of the development of pesticide resistance.

Theme 3
VIRUSES AND BACTERIA

Electron micrograph of bacteriophages heavily infecting a bacterium, *Escherichia coli.*

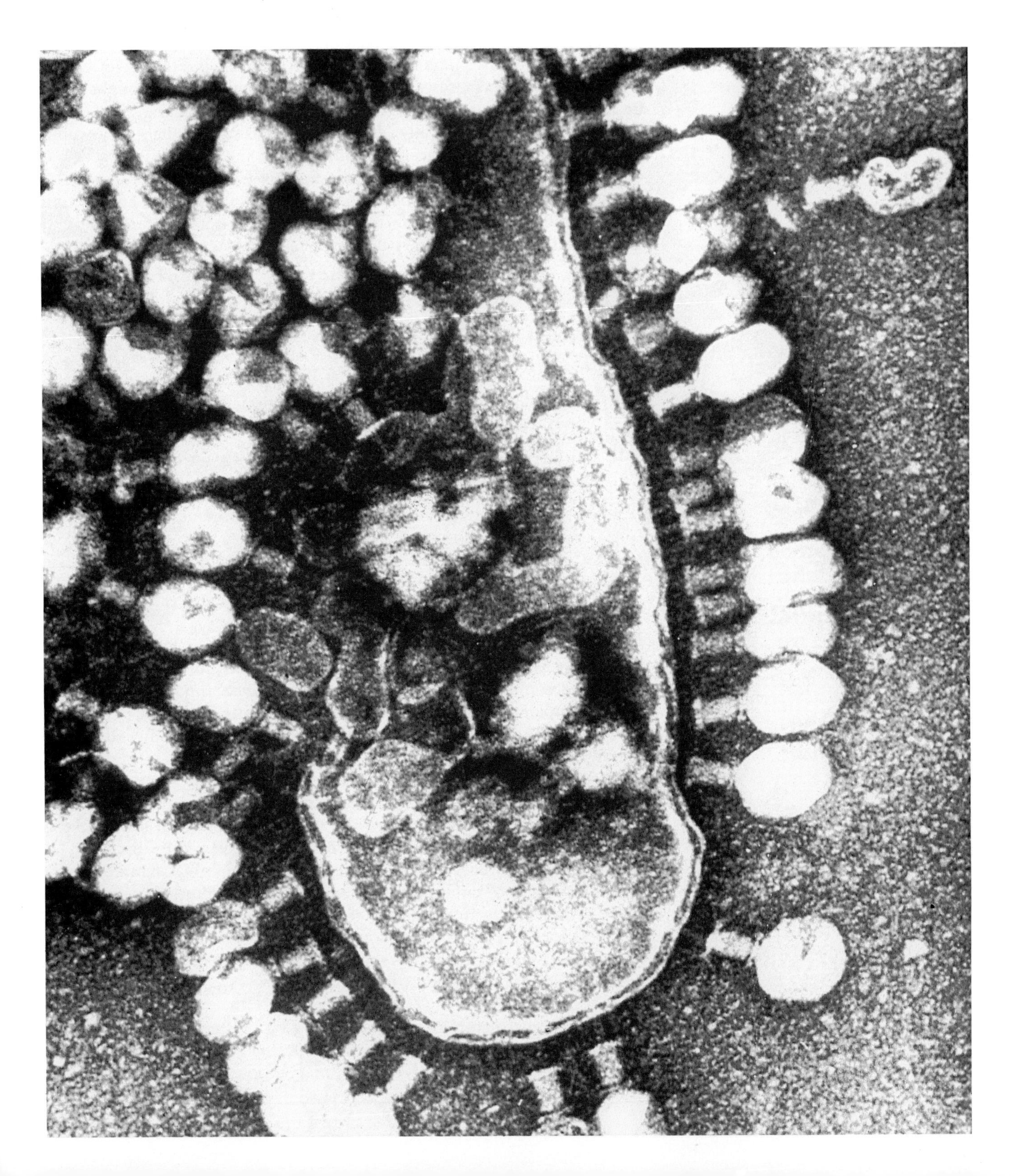

Chapter 9

VIRUSES

Sinister, invisible, untreatable agents of disease causing epidemics and death. This is the frightening picture of viruses and bacteria commonly painted by the press and television. Certainly some viruses and bacteria do cause serious diseases in people, plants and animals. But the use of viruses and bacteria as genetic research tools has led to major breakthroughs in our understanding of genetic processes. Now, by manipulating viruses and bacteria, we can genetically engineer organisms so that they could, potentially at least, produce large amounts of the very drugs needed to combat viral and bacterial diseases. In this theme you are going to investigate some aspects of the biology of viruses (Chapter 9) and bacteria (Chapter 10). This basic information is essential if you are going to understand how viruses and bacteria can be used by genetic engineers.

LEARNING OBJECTIVES

After completing the work in this chapter you will be able to:

1. describe the nature of viruses;
2. understand how the presence of bacteriophages (viruses which infect bacteria) is detected;
3. outline the life cycles of virulent bacteriophages, temperate bacteriophages and retroviruses;
4. explain how studies with viruses have increased our understanding of cancer.

9.1 WHAT ARE VIRUSES?

The discovery of viruses

The role of viruses as infectious agents was recognised long before their true nature was understood. The term virus (Latin for poison) was first used to describe the infectious nature of filtered fluids which caused diseases in plants, for example tobacco mosaic virus (TMV). In 1898, it was shown that the agent which caused foot-and-mouth disease in cattle could be transmitted by a cell-free filtrate.

These findings paved the way for the recognition of many other viral agents of infectious diseases. Soon the tumour-producing ability of viruses was indicated by the discovery of the viral transmission of fowl leukosis and of chicken sarcoma (both forms of cancer). The discovery of **bacteriophages** (viruses which infect bacteria) was of great significance for the development of virology as a science because it afforded an important model system for the investigation of basic virology.

In 1933 bacteriophages were shown to be composed of protein and DNA, while tobacco mosaic virus was shown to be made up of RNA and protein. Gradually it became evident that different viruses contained *either* DNA *or* RNA, but *not both*.

The physical structure of viruses remained a mystery until the development of the electron microscope in the 1930s. Even now, when we can 'see' at least some viruses with an electron microscope, we still depend on the

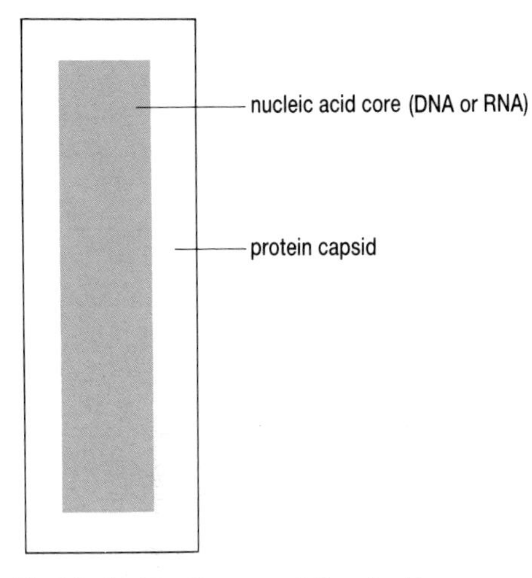

Fig 9.1 A schematic representation of a virion.

symptoms of experimentally induced virus infections to investigate the biology of viruses.

The basic characteristics of viruses

The pioneering investigations referred to above showed that:

- viruses are extremely small, infectious agents which can only reproduce inside a specific host cell;
- the basic viral particle, a **virion** (see Fig 9.1), consists of **nucleic acid** (either RNA or DNA) surrounded by a protein coat called the **capsid**. The capsid protects the nucleic acid, transmits the virion between host cells and sometimes starts viral replication. In some viruses, for example tobacco mosaic virus (TMV), the capsid is made up of distinct protein subunits called **capsomeres**;
- viruses exhibit a range of different structures from simple rods, through icosahedrons to complex. Some of these are shown in Fig 9.2;
- viruses can have a variety of life cycles but they are all totally dependent on living cells for their reproduction.

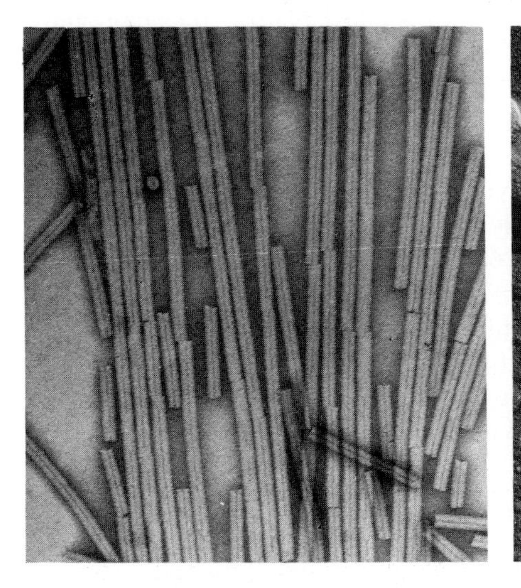

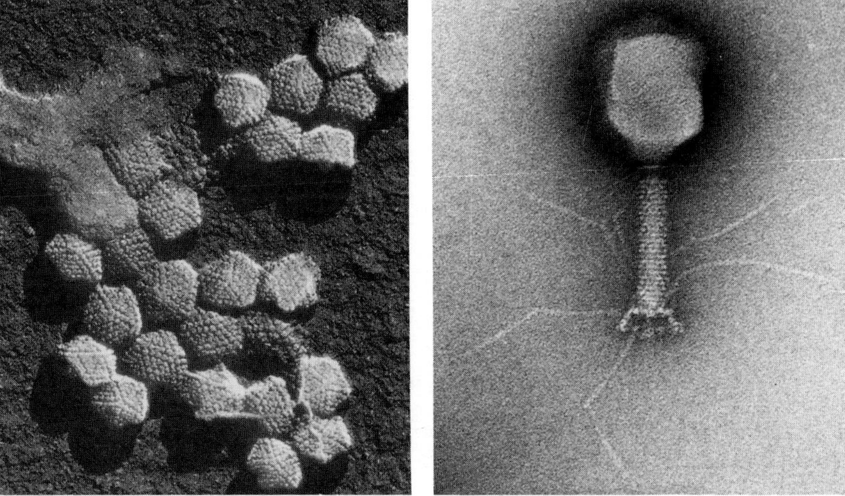

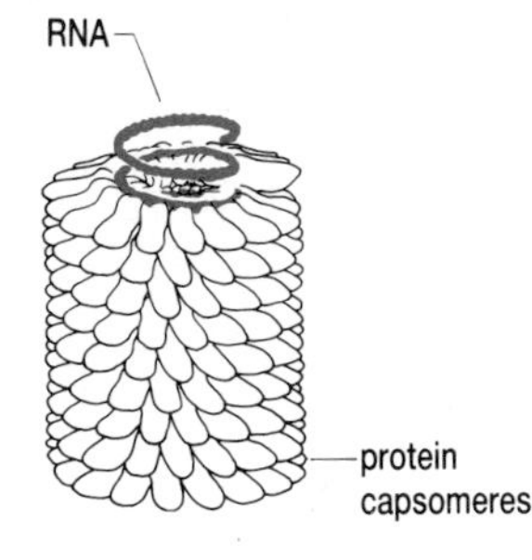

(a) helical viruses, e.g. tobacco mosaic virus

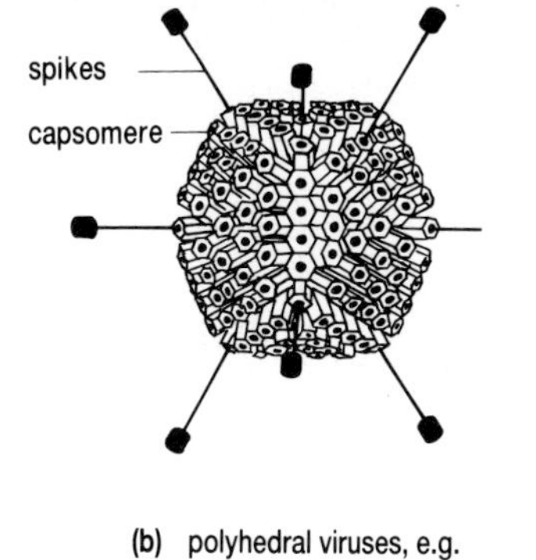

(b) polyhedral viruses, e.g. adenovirus

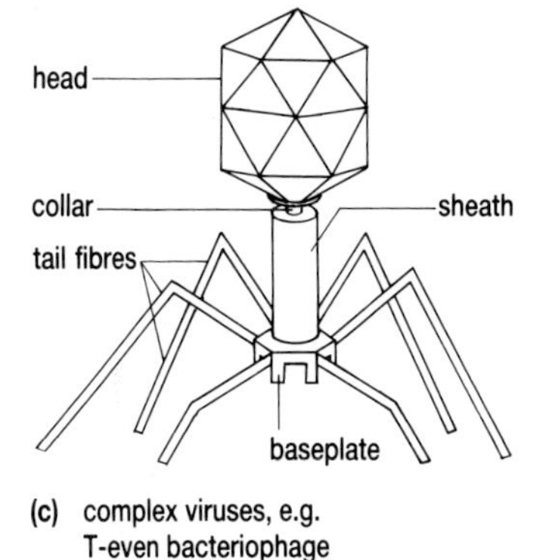

(c) complex viruses, e.g. T-even bacteriophage

Fig 9.2 (a) Electron micrograph of tobacco mosaic virus (×200 000) illustrating helical symmetry.
(b) Electron micrograph of adenovirus (×280 000) illustrating polyhedral symmetry.
(c) Electron micrograph of T4 phage (×420 000) illustrating complex symmetry.

Naming viruses

Viruses are not named like other organisms. Instead, each virus is given a unique series of letters and numbers which describe the hosts they infect and their properties, for example simian virus is called SV, human immunodeficiency virus number 1 is called HIV-1. Whilst such a system of classification is haphazard, it is widely used. A more formal classification of viruses which reflects their biological relationship may evolve in due course from the work of the International Committee on Taxonomy of Viruses.

Viruses and genetics

In addition to their medical importance, the study of viruses has made a significant contribution to our understanding of genetics. In particular, the group of viruses which infect bacteria, the bacteriophages (usually abbreviated to phages), are widely used. Phages are particularly suitable for use in genetic studies for the following reasons:

- When a phage particle infects a bacterial cell, the cell lyses (bursts) after about 20 minutes, releasing several hundred phages into the environment. Thus phages have a very rapid rate of multiplication.
- For this reason phages can be used to observe and to score rare genetic events, such as mutation and recombination. For example, if a cell is infected with two genetically different strains of a particular phage then the two parental types, together with both types of recombinant, can be recovered after only 20 minutes. By plating these progeny phages on a suitable host strain of bacteria, the parental and recombinant types may be recognised after 24 hours or less. A similar experiment with *Drosophila* would take at least 3–4 weeks, and over a year with peas.
- The average phage genome is very small. For example, the genome of ϕX174 (ϕ is pronounced fie) has only got nine genes whilst that of λ phage has less than 60. By comparison, a bacterium like *Escherichia coli*, a harmless (normal) component of the intestinal flora of both humans and agricultural animals, probably has several thousand genes. This simplicity means that it is possible to identify practically all the genes within a viral genome. As a result, we can investigate and understand the organisation and regulation of an entire genome.

QUESTIONS

9.1 What types of molecules do viruses contain and how are these molecules organised in a virion?

9.2 Use the magnifications provided in Fig 9.2 to calculate the size of each virus particle in nm. You will need a drawing of a phage to go in your notes.

9.3 Bacteriophages are often used as a 'model system' in viral genetics. What do you understand by the term **model system** in this context, and why are bacteriophages so widely used in genetics?

9.4 Viruses are a simple form of life but do you think they are primitive or highly evolved organisms?

9.2 VIRAL LIFE CYCLES

Viruses infect practically all cell types, bacterial, plant and animal, and then use the host cell's metabolic machinery to make new viruses. Since viruses are totally dependent on living cells for their reproduction, they can be considered obligate parasites. The basic life cycle of a virus can be divided into a number of stages.

- The virion, or just its nucleic acid, enters a host cell.
- The early viral proteins, which are involved in the replication of the viral genetic material, are synthesised using the information present in the viral genome.
- The viral nucleic acids are then synthesised.

- The late viral proteins, including the viral coat proteins, are produced and the viral nucleic acid is packaged inside the capsid.
- The virus is then released from the host cell.

Virulent viruses kill their host cells in the process of reproduction, for example T4 phage (T = type) which infects *E. coli*. Others do not destroy their hosts, but instead permit them to continue to grow and divide while still producing and secreting new viruses. Another type, for example λ phage, simian virus 40 (SV40) and HIV-1, may actually become recombined into the DNA of the host cell and behave as if they were part of the chromosome of the host cell. They are then replicated passively as part of the DNA of the host cell. The three life cycles examined below have been chosen to reflect this diversity of reproductive styles and to introduce you to viruses which are important in genetic research, genetic engineering and medicine respectively.

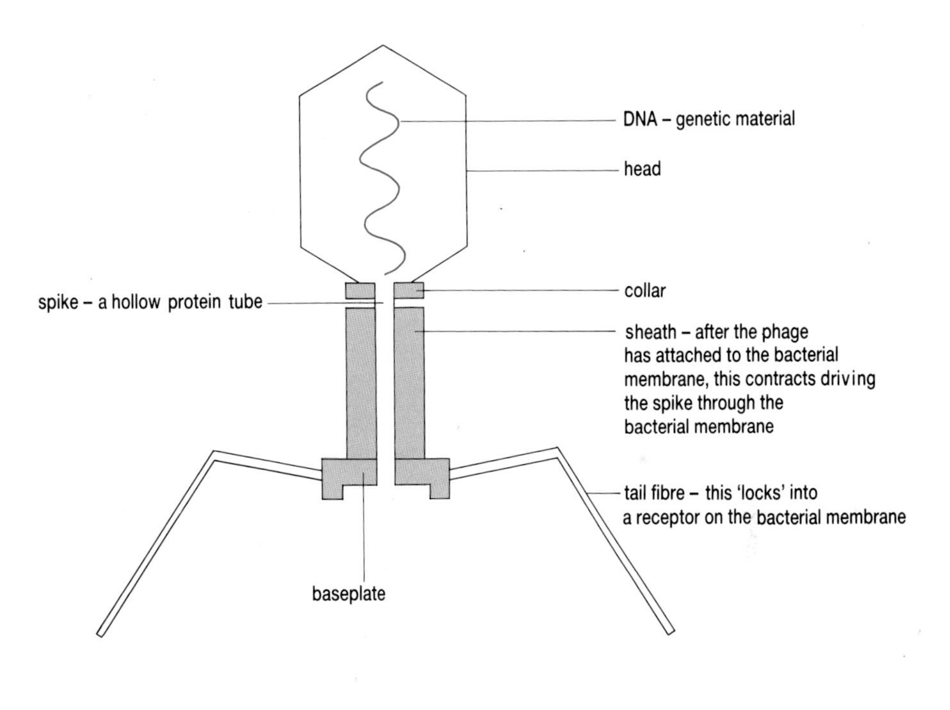

Fig 9.3 A cross-section through T4 phage.

The lytic life cycle of virulent bacteriophages

The life cycle of T4 phage is used as a model for the development of all virulent phages. The sequence of events involved in producing new phages has now been unravelled using a variety of techniques, for example electron microscopy. The stages in the life cycle of T4 bacteriophage (Fig 9.3) are shown in Fig 9.4. Notice that only the nucleic acid, not the whole phage, enters the host cell.

For the geneticist, the exciting aspect of such a life cycle is that it consists of a number of steps. Each step is controlled by one or more specific gene products, for example enzymes. Mutants of T4 phage which produce defective gene products can be isolated and their effect on development can be analysed. Such analysis helps reveal how viral genomes are organised and function. This knowledge is essential if we wish to understand how viruses cause disease and how we can use them as tools for genetic engineering. The techniques used to isolate and grow bacteriophages are described in Box 9.1.

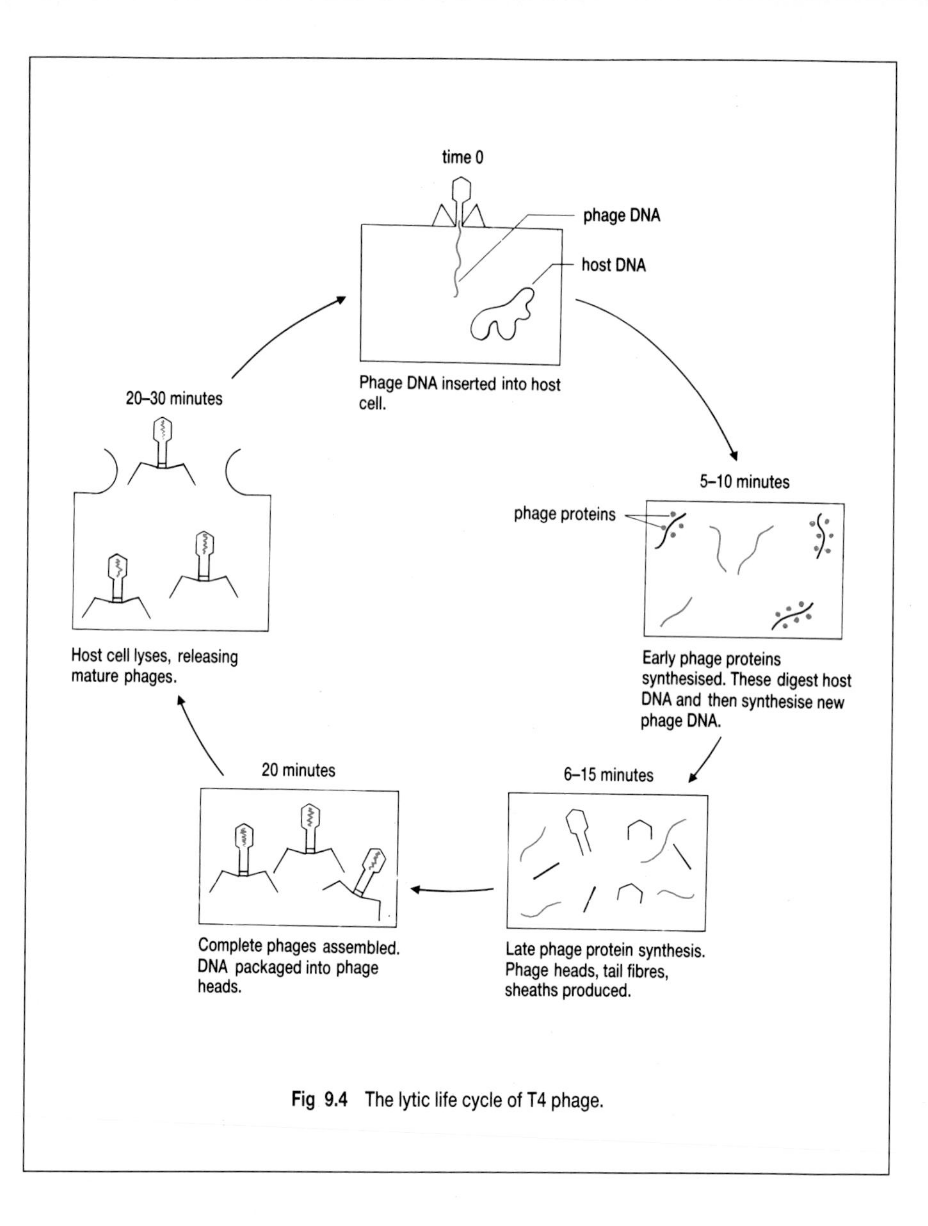

Fig 9.4 The lytic life cycle of T4 phage.

BOX 9.1

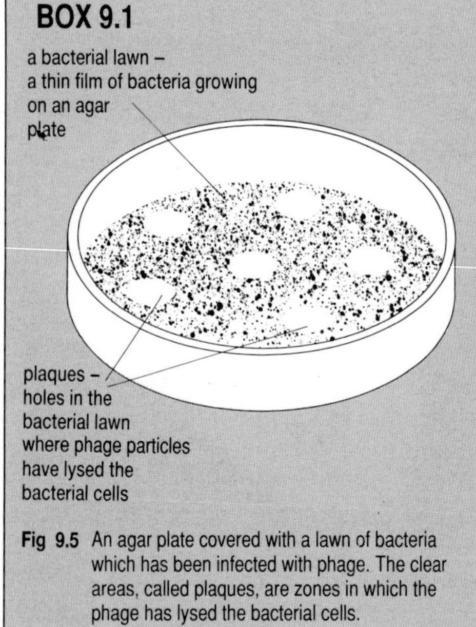

Fig 9.5 An agar plate covered with a lawn of bacteria which has been infected with phage. The clear areas, called plaques, are zones in which the phage has lysed the bacterial cells.

Isolating and cultivating bacteriophages

Phages can be isolated and grown in young, actively growing cultures of bacteria in broth or on agar plates. In liquid cultures, lysing of the bacteria may cause a cloudy culture to become clear, whereas in agar-plate cultures, clear zones or plaques become visible to the naked eye.

The best and most usual source of new strains of phage is the host habitat. For example, coliphage or other phages which infect bacteria found in the intestinal tract can best be isolated from sewage or manure. This is done by centrifugation or filtration of the source material and the addition of chloroform to kill the bacterial cells. A small amount, say 1 ml, of this preparation is mixed with the host organism, for example *Escherichia coli,* and spread on an agar plate. Growth of the phage is indicated by the appearance of plaques in the otherwise opaque growth of the host bacterium, as shown in Fig 9.5. Each plaque is the result of one phage particle infecting a single bacterial cell. The phages released from this infected cell then infect the surrounding bacterial cells. Repetition of this process results in a zone of lysed bacterial cells surrounding the cell which was originally infected, so producing a plaque.

Life cycle of temperate (lysogenic) phages

In addition to the **lytic life cycle** shown by the virulent phages, some phages can adopt an alternative life style. These are the **temperate phages,** for example λ phage, which do not always cause the lysis of their host. Members of this group of phages can enter a **lysogenic phase** in which they become an integral part of the bacterial genome.

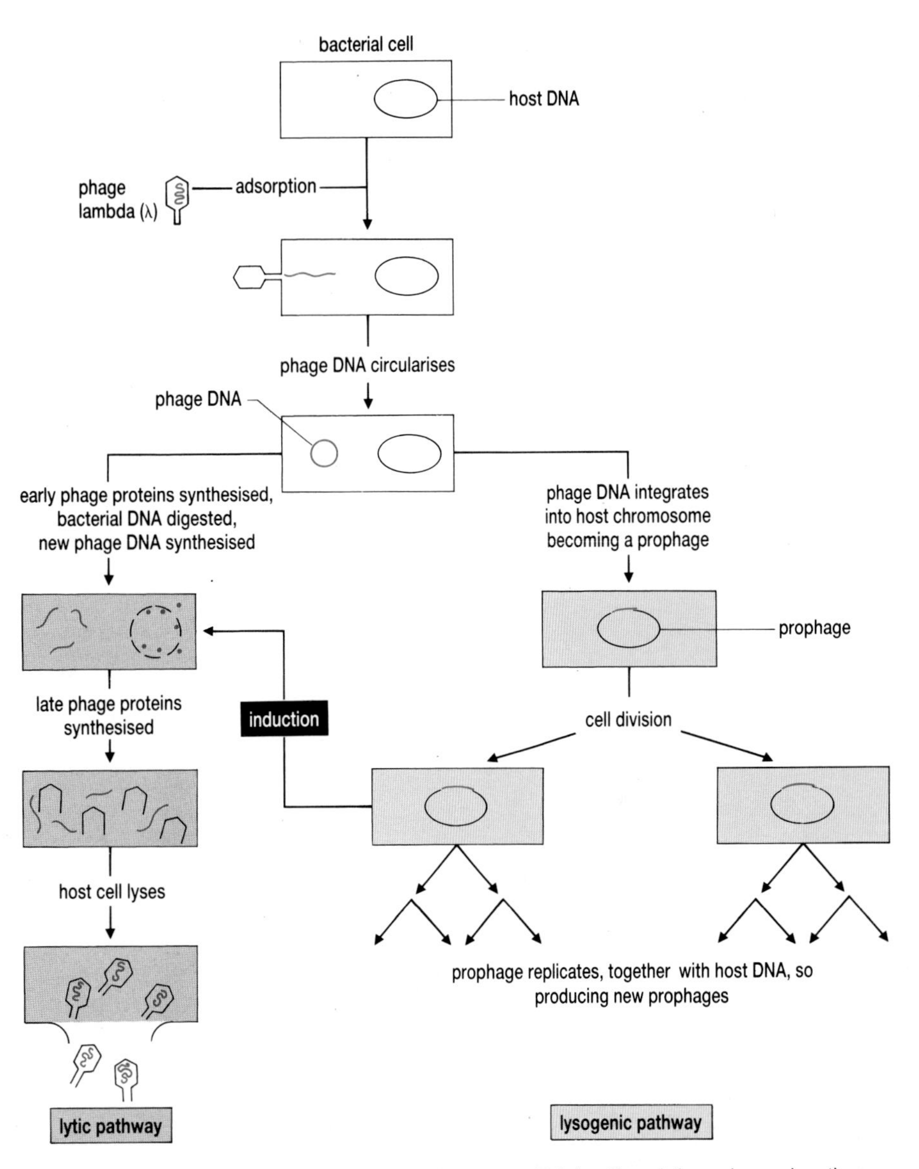

Fig 9.6 The life cycle of λ phage. As shown, the phage can multiply by either a lytic or a lysogenic pathway.

The process of **lysogenisation**, shown in Fig 9.6, involves suppression of phage virulence usually followed by integration of the phage DNA into the bacterial genome. The integrated phage DNA, called a **prophage**, is then replicated every time the bacterial chromosome is duplicated prior to cell division. Each daughter cell then receives not only a copy of the parental chromosome but a copy of the prophage DNA as well.

Occasionally the prophage will undergo a process called **induction** and re-enter the lytic cycle of development. In a population of lysogenic bacteria, about one in a million cells per generation will spontaneously undergo induction. The rate of induction can be increased by a variety of agents which damage the DNA of the host bacterium, for example ultraviolet light. After induction, the phage starts to replicate rapidly and eventually the host cell lyses, releasing progeny phage.

Life cycle of retroviruses

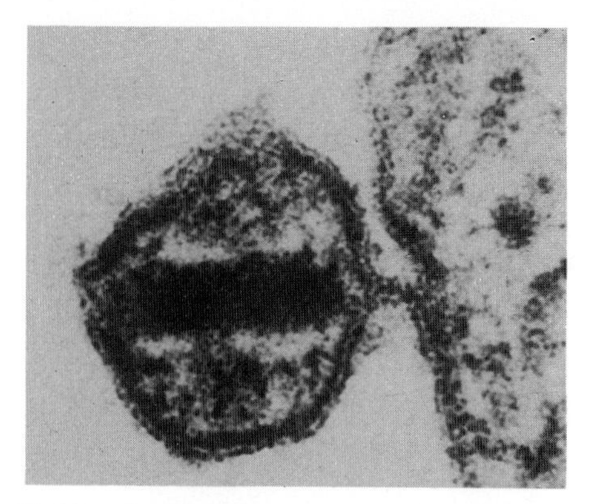

Fig 9.7 Electron micrograph of HIV-1, the virus which causes AIDS. The protein core contains the genetic material, in this case RNA. The HIV-1 virion is approximately 100 nm in diameter.

This group of viruses, which includes HIV-1, the cause of AIDS (see Fig 9.7), differs from the phages in three important ways. Firstly, they are viruses which infect animal cells not bacteria. Secondly, the whole virion enters the cell not just the genetic material. Thirdly, the genetic material is RNA not DNA. The life cycle of HIV-1 is shown in Fig 9.8.

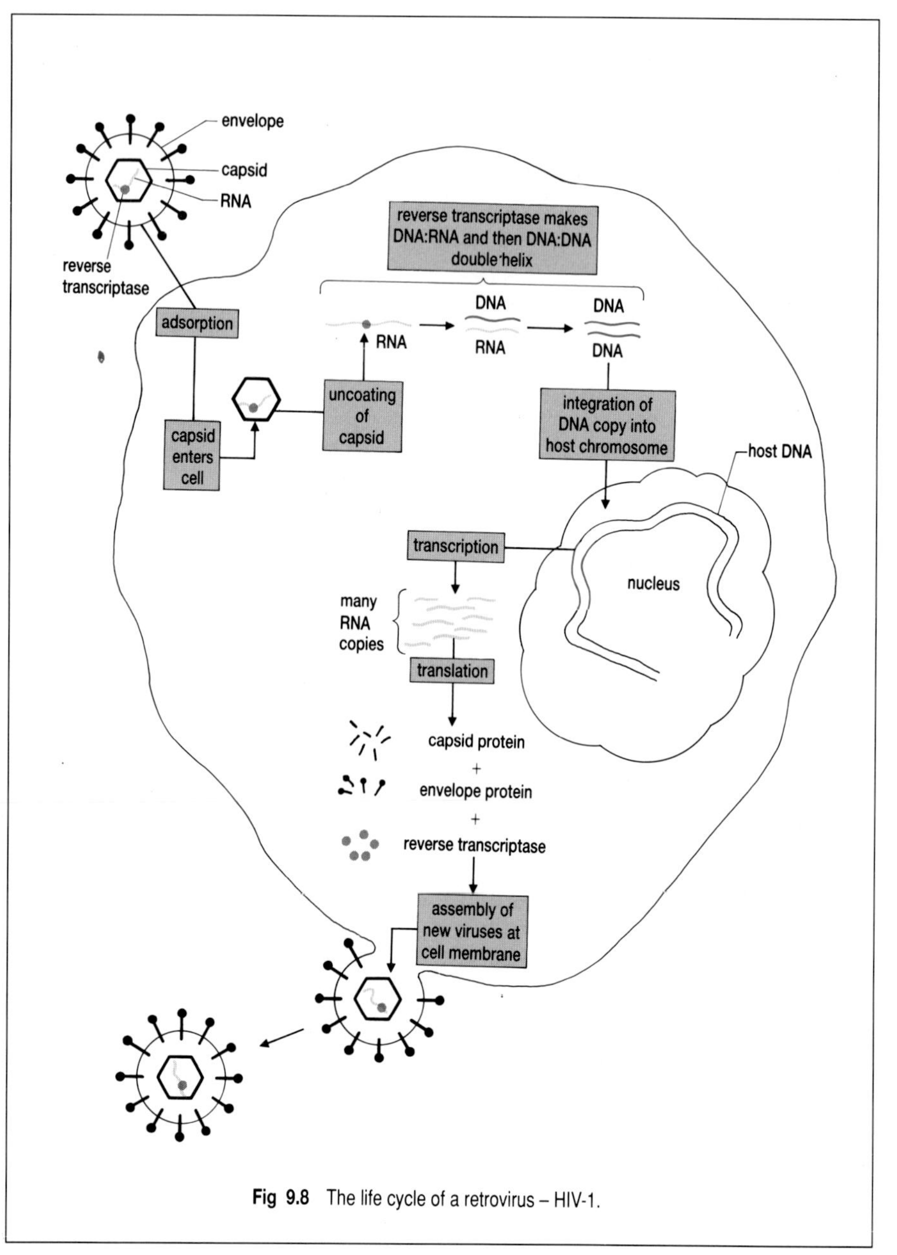

Fig 9.8 The life cycle of a retrovirus – HIV-1.

HIV has an outer lipid coat which surrounds the protein core containing the genetic material (see Fig 9.7). This lipid coat fuses with the cell membrane of the target cell, releasing the viral RNA into the cell along with an enzyme, **RNA reverse transcriptase**. This enzyme, which is unique to the retroviruses, makes the infected cell turn the viral RNA into viral DNA. This DNA copy, called the provirus, can insert itself into one of the chromosomes in the nucleus of the cell. The provirus can now remain dormant, the latent stage of infection, for months, perhaps even years. Eventually, for some unknown reason, the provirus is 'switched on' and starts to direct the synthesis of new viruses. The new viruses are then budded off from the cell membrane to infect new cells.

9.3 EXTENSION: VIRUSES, ONCOGENES AND CANCER

What is cancer?

Cancer is not a single disease. More than 100 distinct types of cancer are recognised, each having a unique set of symptoms and requiring a specific course of therapy. Most of them can be grouped into four categories:

Leukaemias. Abnormal numbers of white blood cells (leucocytes) are produced by the bone marrow.

Lymphomas. Abnormal numbers of lymphocytes (a type of leucocyte) are produced by the spleen and lymph nodes.

Sarcomas. Solid tumours growing from tissues derived from embryonic mesoderm, for example connective tissues, cartilage, bone, muscle and fat.

Carcinomas. Solid tumours which grow from epithelial tissues; the most common form of cancer. Epithelial tissues are the internal and external body surface coverings and their derivatives, for example skin, glands, nerves, breasts and the linings of the respiratory, gastrointestinal, urinary and genital systems.

Cancer has three major characteristics:

Hyperplasia. The uncontrolled growth of cells.

Anaplasia. The structural abnormality of cells.

Metastasis. The ability of a malignant cell to detach itself from a tumour and establish a new tumour at another site within the organism.

These changes suggest that in cancer cells the factors which regulate normal cell differentiation have been altered.

Is cancer a genetic disease?

Some forms of cancer have a clear-cut genetic determination following a pattern of single gene inheritance. For example, retinoblastoma, which starts during early childhood as a tumorous growth in one or both eyes that rapidly spreads to the brain resulting in death if left untreated, is inherited as a dominant trait. Although there are many kinds of cancer which follow a simple mode of inheritance, the incidence of each is low and they only affect a small proportion of the people suffering from cancer. Most common forms of cancer do not have a clear-cut genetic determination, though there is some evidence that susceptibility to some cancers may be inherited.

At the cellular level, however, cancer is clearly a genetic disease. Cancer cells transmit their **neoplastic** (cancerous) characteristics to daughter cells; that is why cancerous cells continue to proliferate. The transformation of normal into cancerous cells must, then, be due to genetic alterations. So a change in the DNA of a cell causes it to become cancerous. But what sort of change?

(**Note**: Transformation is the term used for the conversion of a normal cell into a cancerous one. This is quite distinct from the usual genetic meaning which refers to the incorporation of free DNA into a cell, as discussed in section 10.2.)

Cancer and viruses

For a long time biologists thought that cancer might be caused by viruses. Certainly, early experiments associated viruses with both leukaemias and sarcomas in chickens (see section 9.1). But for many years these discoveries were not thought to be relevant to humans. However, the discovery that some retroviruses, so-called 'acute' transforming viruses, were able to produce cancer in infected animals and transform cultured animal cells from normal into neoplastic ones has changed this point of view. About 20 acute transforming viruses have been isolated from rats (for example, Harvey sarcoma virus), mice (for example, Maloney sarcoma virus), monkeys, and chickens (Rous sarcoma virus). In 1982, the first retrovirus which caused cancer in people, human T-cell leukaemia virus (HTLV), was isolated by Robert Gallo in the United States of America.

The following article explains how the study of acute transforming viruses has helped to understand what causes cancer. Read through the article and then answer the questions below.

Cancer and the Viral Trigger

When a retrovirus infects a cell it can sometimes "capture" a cellular gene and incorporate it into the viral DNA. Exactly how this happens is unclear: the mechanism may vary, but the results can be devastating. It can turn a harmless retrovirus into an "acute transforming retrovirus", which turns infected cells into cancer cells. Acute transforming retroviruses are rare and very unusual. They appear to be created within an individual animal when a suitable cellular gene is incorporated into a retroviral genome: the viruses then cause cancer and die with the animal. So, unlike normal viruses, these "transformers" are not transferred from one animal to another. Moreover, most of them are incapable of multiplying without the help of normal retroviruses, because incorporation of the cellular gene into their own genome damages the virus.

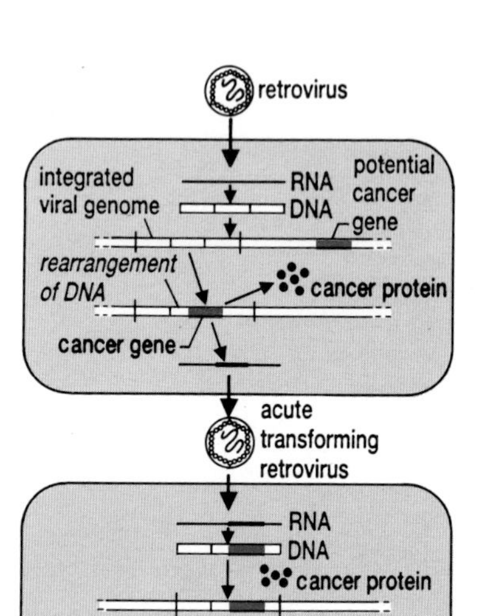

Although these viruses are rare, their study has revealed a central secret of all cancers. In 1983, two research teams, one led by Russell Doolittle at the University of California, San Diego, and the other by Michael Waterfield of the Imperial Cancer Research Fund in London, announced that the cellular gene that had apparently been "captured" by one acute transforming retrovirus coded for a "growth factor" — a protein involved in switching on cell growth. Since then, scientists have found that many other acute transforming retroviruses have captured genes for other proteins crucially involved in cell growth. So the idea developed that acute transforming retroviruses had captured cellular genes which were once normal genes essential to their growth. These cellular genes, when incorporated into a virus and either changed slightly or made unusually active, became "cancer genes" (oncogenes), capable of transforming cells into cancer cells. This became more plausible with the discovery that other retroviruses — the "slow transforming retroviruses" — could cause cancer when the viral genome simply integrated next to or even within the cellular genes in a way that altered them or made them more active.

The general theory of cancer that has emerged from the studies says that all cells contain potential cancer genes (proto-oncogenes), which are essential for cell growth and multiplication, but which can cause cancer if they are altered slightly, or go out of control and make too much of the protein they code for or make it at the wrong time. Viruses can turn the potential cancer genes into true cancer genes, but so might many other things, such as chemical carcinogens, radiation and other causes of change and mutation of DNA.

New Scientist May 1988
In an article called VIRUSES WORK TO IMPROVE THEIR IMAGE by Andrew Scott.

QUESTIONS

9.5 What is an oncogene?

9.6 How do retroviruses become acute transforming viruses?

9.7 How has the study of acute transforming and slow transforming viruses revolutionised our understanding of cancer?

SUMMARY ASSIGNMENT

1. Produce an essay plan to compare and contrast the life cycles of virulent phages, temperate phages and retroviruses.
2. Using your essay plan, try to produce the completed essay in about 40 minutes.

Chapter 10

BACTERIAL GENETICS

Bacteria affect every aspect of our daily lives. They cause disease, provide us with foods and medicines and dispose of our wastes. They are of vital importance to the industrial microbiologist and they are essential to the genetic engineer. Bacteria are crucial research tools in biochemical genetics, being used to formulate biological principles applicable to a wide variety of organisms including people. Indeed, more is known about the detailed biology of *Escherichia coli* (Fig 10.1) than any other organism.

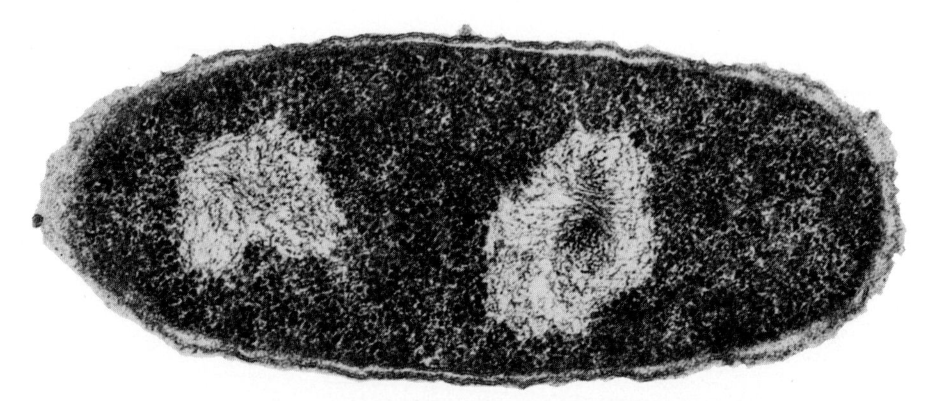

Fig 10.1 Transmission electron micrograph of a section of the bacterium *Escherischia coli* (×120 000). The DNA appears as fibrils within the two white patches. The darkly staining area consists mainly of ribosomes. The cell wall is clearly visible running around the outside of the cell.

Whilst bacteria are widely used in genetics, they do have a number of unusual features. You will investigate some of these in this chapter.

LEARNING OBJECTIVES

After completing the work in this chapter you will be able to:

1. describe the various types of genetic element found inside bacterial cells;
2. describe how bacteria transfer DNA molecules between themselves;
3. account for the origin and rapid spread of antibiotic resistance among bacteria.

10.1 THE GENETIC ELEMENTS IN THE BACTERIAL CELL

Four inherited genetic elements can be found inside a bacterial cell (Fig 10.2). An important point to notice from Fig 10.2 is that the DNA in bacterial cells is in the form of closed circles. Linear sequences, prophages and transposable elements, are integrated into one of the circles. This circularity evidently has some function. If, for example, the bacterial chromosome is opened out (linearised) by cutting it with an enzyme then it cannot replicate properly and will eventually be broken down by enzymes. A recently discovered exception to this rule are the large, linear plasmids found in some members of the *Streptomyces*, a bacterial genus which produces many antibiotics.

Bacteria are haploid, but occasionally a fragment of chromosomal DNA is transferred from one bacterial cell (the **donor**) to another (the **recipient**). The mechanisms involved in this transfer of DNA between cells are discussed later. Such a DNA transfer can produce a partially diploid cell (Fig 10.3).

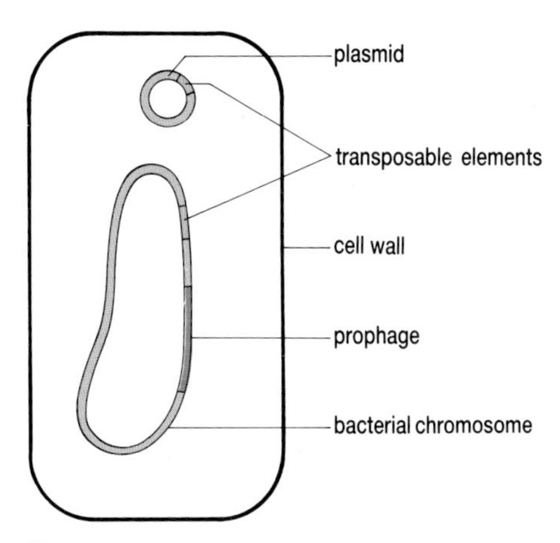

Fig 10.2 The genetic elements in a bacterial cell.

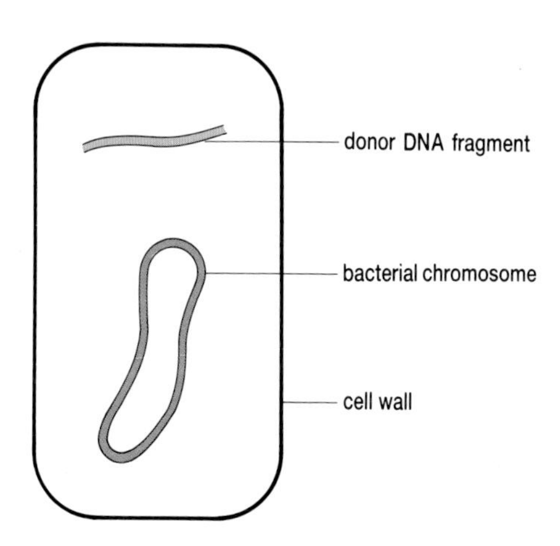

Fig 10.3 A partially diploid bacterial cell.

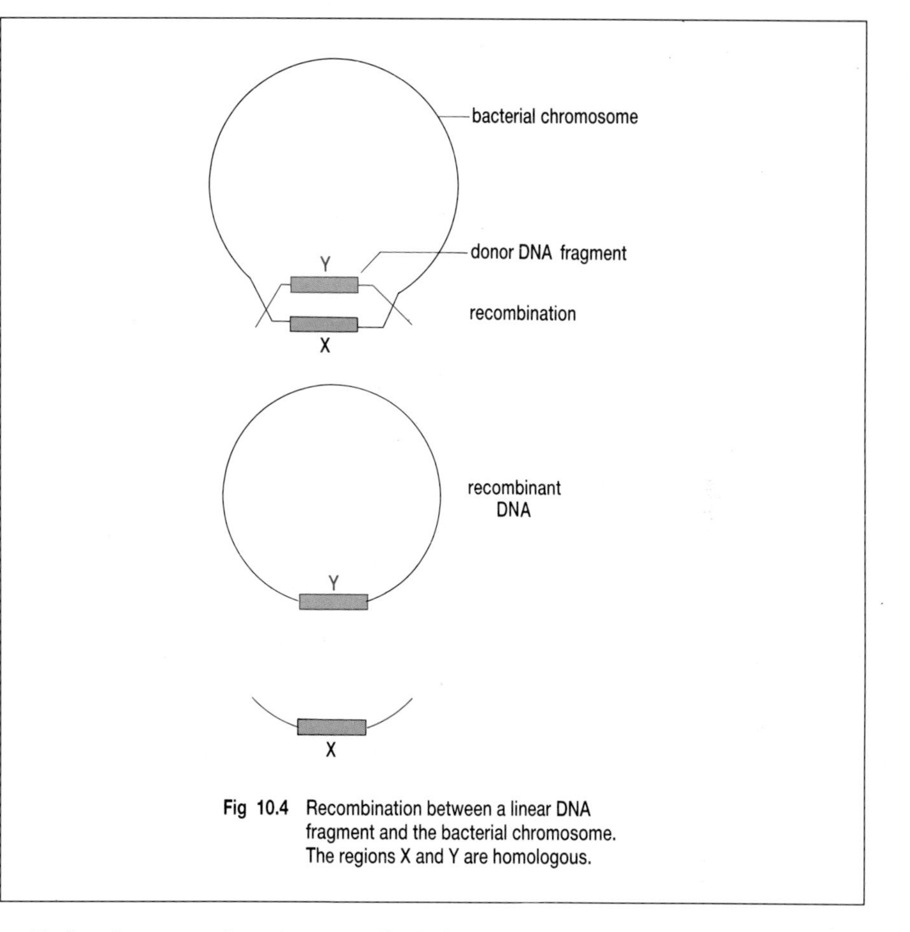

Fig 10.4 Recombination between a linear DNA fragment and the bacterial chromosome. The regions X and Y are homologous.

If the donor and recipient cells belong to the same or closely related species then the sequence of DNA bases in the transferred fragment will be very similar (**homologous**) to a corresponding part of the recipient bacteria's chromosome. Two DNA molecules which are homologous can pair up and recombination can occur between them (Fig 10.4). Notice that an even number of crossovers are needed between a circle and a linear DNA fragment to produce recombinant cells which contain a closed circle of DNA.

INVESTIGATION 10.1

You have been given two test tubes. One contains millions of bacteria that are sensitive to penicillin, the other contains pure DNA extracted from a different strain of the same bacterium which is resistant to penicillin. You mix the contents of the test tubes together and incubate the mixture for a period of time. This allows the bacteria to take up the DNA fragments from the incubation mixture. Devise a procedure which would enable you to count the number of recombinant individuals in the incubation mixture which contain the antibiotic-resistance gene at the end of the incubation period. To help you develop your experiment you will need to find out: (i) how bacteria are grown; (ii) how bacteria growing in a culture are counted; (iii) how **selective media** can be used to isolate bacteria.

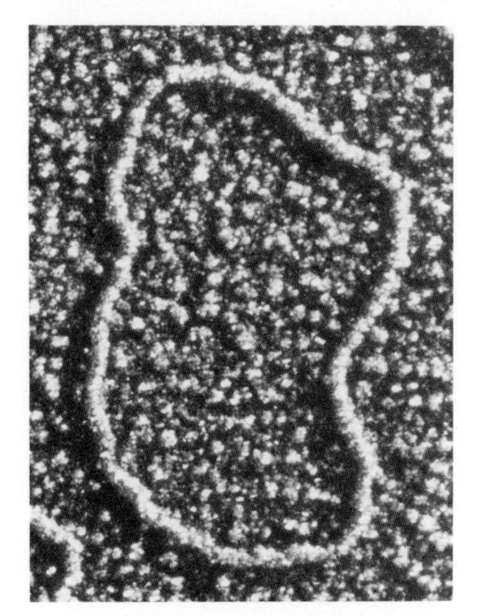

Fig 10.5 Transmission electron micrograph of pBR 322 a bacterial plasmid commonly used in genetic engineering (× 129 200). Notice that the plasmid is a closed loop of DNA.

The bacterial chromosome

All the essential genes required for the normal functioning of a bacterium are carried within one double-stranded, circular DNA chromosome. This is in contrast to your genes which are carried on a number of chromosomes, each of which is made up of double-stranded, linear DNA molecules.

Additional genetic elements

In addition to the main chromosome, a bacterium may contain the three additional types of genetic element shown in Fig 10.2. It is important that you realise that these are not distinct categories, but show a considerable degree of overlap. For example, some types of **prophage** exist as **plasmids**, whilst **transposable elements** are found in practically all bacterial chromosomes and plasmids, and in some phage genomes as well. Bearing this in mind, the three different types are now considered in turn.

Plasmids. These are circular, double-stranded, extrachromosomal DNA molecules (Fig 10.5) which vary in size from below 1 kb to over 300 kb (see Box 10.1 if you are unfamiliar with this way of measuring the size of DNA molecules). As a general rule, the small plasmids occur in multiple copies per cell, that is they have a high **copy number**, whilst the large plasmids have a low copy number. In addition to being found in a wide variety of bacterial cells, plasmids are a normal component of yeast genomes and have also been observed in other eukaryotes.

BOX 10.1

Measuring the size of DNA Molecules

The length of DNA molecules and DNA sequences is measured in terms of base pairs. A base pair is simply two nitrogenous bases which have paired up by hydrogen bonds. One base pair is indicated by the red box on Fig 10.6. One kilobase (kb) is a length of DNA which contains 1000 base pairs.

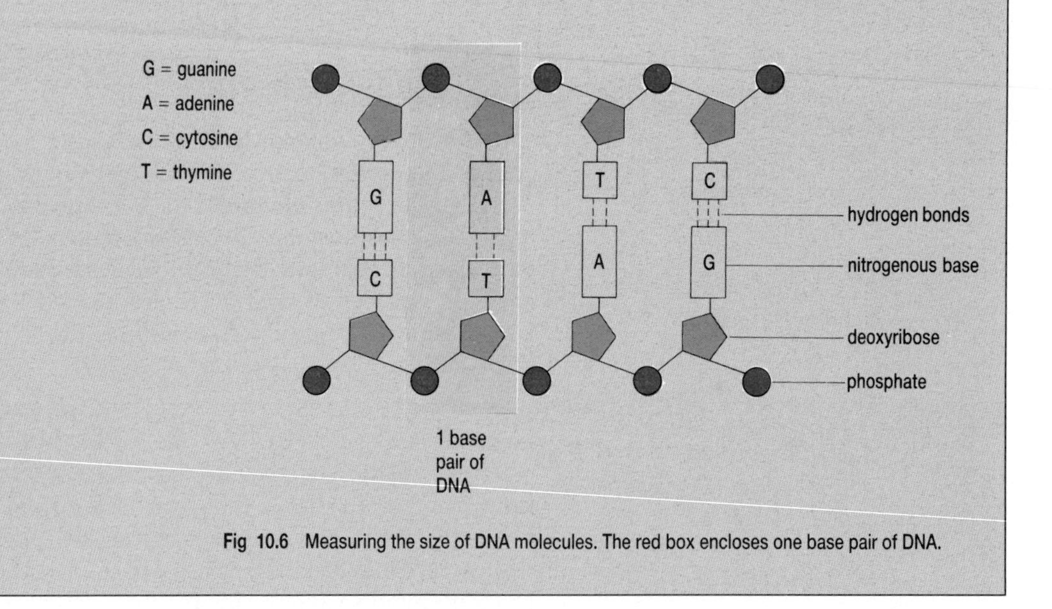

Fig 10.6 Measuring the size of DNA molecules. The red box encloses one base pair of DNA.

In some respects plasmids are like phages in that they cannot replicate outside a bacterial host. Plasmids include the DNA sequences necessary to initiate their own reproduction and they can replicate in the cytoplasm of their bacterial host independently of the bacterial chromosome. In particular, each plasmid has a sequence of bases which serves as the **origin of replication**. Without this sequence the plasmid cannot replicate. Some

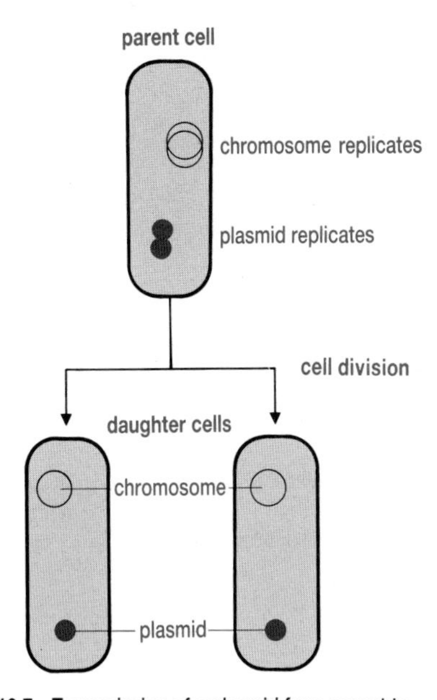

Fig 10.7 Transmission of a plasmid from parent to daughter cells, a process called **vertical transmission.**

plasmids can only replicate in specific bacteria whilst others can replicate in a wide range of hosts.

Plasmids can enter new cells in two ways. Look at Fig 10.7. This shows a bacterial cell containing a plasmid undergoing cell division. In addition to passing a copy of the bacterial chromosome to the daughter cells, the parental cell has also transferred a copy of the plasmid. This is called **vertical transmission**. Alternatively, in **horizontal transmission** the plasmid transfers a copy of itself from the host bacterial cell to a different and sometimes unrelated cell. Indeed, the **recipient cell** does not even have to be the same species as the **donor cell**. The process by which the transfer usually occurs is called **conjugation** and those plasmids carrying the genes which enable this transfer process to take place are called **conjugative plasmids**. Conjugation is discussed in more detail below.

In addition to carrying genes which control their replication and promote their transfer, some plasmids may also contain genes which code for special functions which may confer an advantage on the host bacterium. Some examples of these functions are given below.

Bacteriocinogenic factors. These are plasmids which carry genes that determine the formation of **bacteriocins** – protein toxins which kill related strains of bacteria. The bacteriocins of *E. coli* are called **colicins**; those of *Yersinia pestis* (the causative agent of bubonic plague) are called **pestins** and so on. Bacteriocins have proved useful for distinguishing between certain strains of the same species of bacteria in medical bacteriological diagnosis.

Catabolic plasmids. These carry genes which specify the enzymes necessary to use a wide range of unusual carbon sources. For example, some strains of a soil-living bacterium called *Pseudomonas* can use, among other substances, camphor and octane as energy sources.

Virulence factors. Some strains of the bubonic plague bacterium contain a plasmid which improves the ability of the bacterium to infect new hosts. The plasmid does not make the bacterium **pathogenic**, it simply improves the efficiency of infection. Many of you may have suffered from traveller's diarrhoea when you have gone away on holiday. This can be caused by strains of *E. coli* which contain virulence factors that carry genes coding for the production of **enterotoxins**. You may be relieved to know that it is these protein poisons which cause the diarrhoea, not the bacterium.

R or resistance plasmids. These are conjugative plasmids which can also confer resistance to a number of **antibiotics**. Some forms of resistance are due to a gene whose product is an enzyme that destroys a specific antibiotic. The genetics and implications of antibiotic resistance for people are explored in more detail in section 10.3.

Whilst some of the functions coded for by plasmid genes may prove useful to people, many do not. Indeed some, like antibiotic resistance, are a positive menace to health. However plasmids, particularly those which carry antibiotic-resistance genes, lie at the heart of genetic engineering. Paradoxically, then, we can use the plasmids which confer antibiotic resistance to make new antibiotics, as well as increasing the availability of compounds like insulin and AIDS-free factor VIII, so essential to haemophiliacs.

Some plasmids, described as being **cryptic**, have not yet been associated with any genes encoding a special function.

Prophages. As you have seen with virulent phages, such as T4 phage, entry of the phage nucleic acids into the bacterium sets off a chain of events which inevitably leads to cell lysis and the release of new phages. The genome of a virulent phage cannot really be considered as an extra genetic element of the infected bacterial cell since it is present for such a short period of time. However, the other group of bacterial viruses, the temperate phages, which have the potential to lysogenise the host cell (see section 9.2) can become additional genetic elements, called prophages. Most prophages, such as λ, become part of the bacterial chromosome (Fig 10.8); others, for example P1 coliphage found in *E. coli*, exist as plasmids.

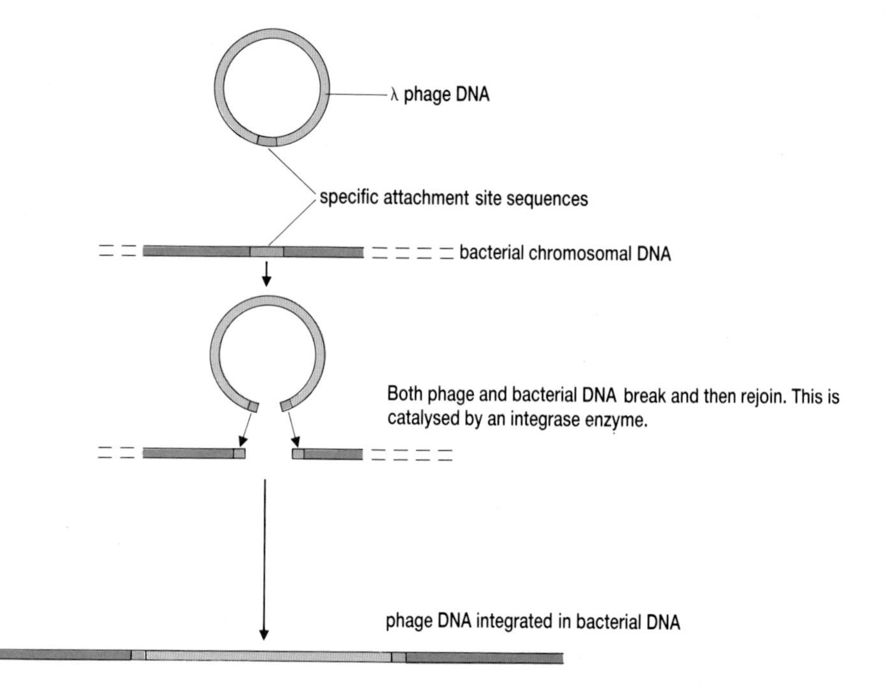

Fig 10.8 The integration of λ phage DNA into a bacterial chromosome.

Like plasmids, prophages may carry genes which code for some special function which is then expressed by the bacterium. For example, a bacterium which contains a prophage may become immune to further attack by related phages. More seriously for us, prophages may change the biology of some bacteria so that they become pathogenic and cause a disease. For example, a prophage of the bacterium *Corynebacterium diphtheriae*, which causes diphtheria in humans, carries a gene which encodes the diphtheria toxin. In the absence of the prophage the bacterium is harmless.

On the positive side, temperate phages are extremely useful agents for transferring DNA from one bacterium to another. This process, called transduction (see section 10.2), is an important research tool in bacterial genetics.

Transposable genetic elements. You probably think of DNA as being a relatively stable molecule and, therefore, the genome of an organism as very static. In the late 1960s, this view was radically altered by the discovery of transposable genetic elements in bacteria. These are pieces of linear DNA which have the ability to move (**transpose**) around the bacterial genome either between plasmids, or between a plasmid and the chromosome, or between either of these and a resident prophage. Unlike plasmids, transposable elements cannot exist as independent entities in the bacterial cytoplasm. They must be integrated into a self-replicating molecule of DNA such as the bacterial chromosome or a plasmid.

The first class of transposable elements to be discovered in bacteria were called **insertion sequences**. These are short pieces of DNA which only carry the genes that allow them to jump around the bacterial genome. Insertion sequences are relatively small but the second group of transposable elements, **transposons**, are much larger and they carry

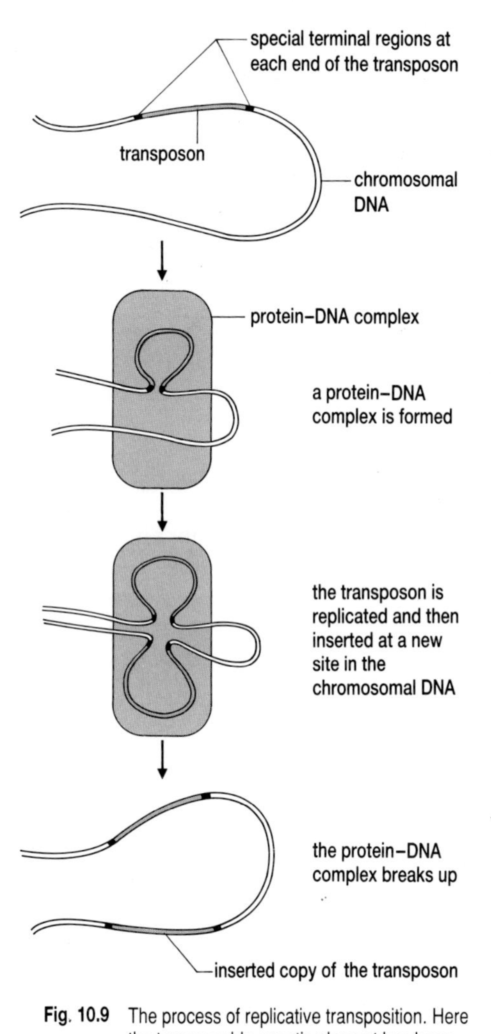

Fig. 10.9 The process of replicative transposition. Here the transposable genetic element has been replicated, leaving one copy at the original site and a new copy at the new site. The black bands represent the two identical DNA sequences commonly found at the ends of transposable genetic elements.

additional genes which are totally unrelated to transposition. These genes can, however, encode functions which may be of considerable benefit to the bacterium, for example heavy metal and antibiotic resistance. Indeed, it now seems probable that many of the special functions encoded by plasmids and prophages are actually carried by transposons.

The process of transposition can take two forms, replicative and non-replicative (Fig 10.9). Though transposition involves the movement of linear pieces of DNA within a bacterial cell, it differs from the normal process of recombination between homologous regions of DNA in two important ways.

1. Transposition does not require extensive DNA homology between the transposable element and the target DNA. In other words, a transposable element can insert anywhere in the cell's genome, though certain parts of the genome, termed hot spots, seem to be more prone to invasion than others.

2. Recombination between homologous regions of DNA is under the control of the bacterial *rec* (short for recombination) genes. Transposition is controlled by genes in the transposable elements themselves (Fig 10.10).

The first transposable elements were described in corn plants by Barbara McClintock in 1951. It is now apparent that transposable elements occur both in prokaryotes and eukaryotes. Even you have got some jumping genes!

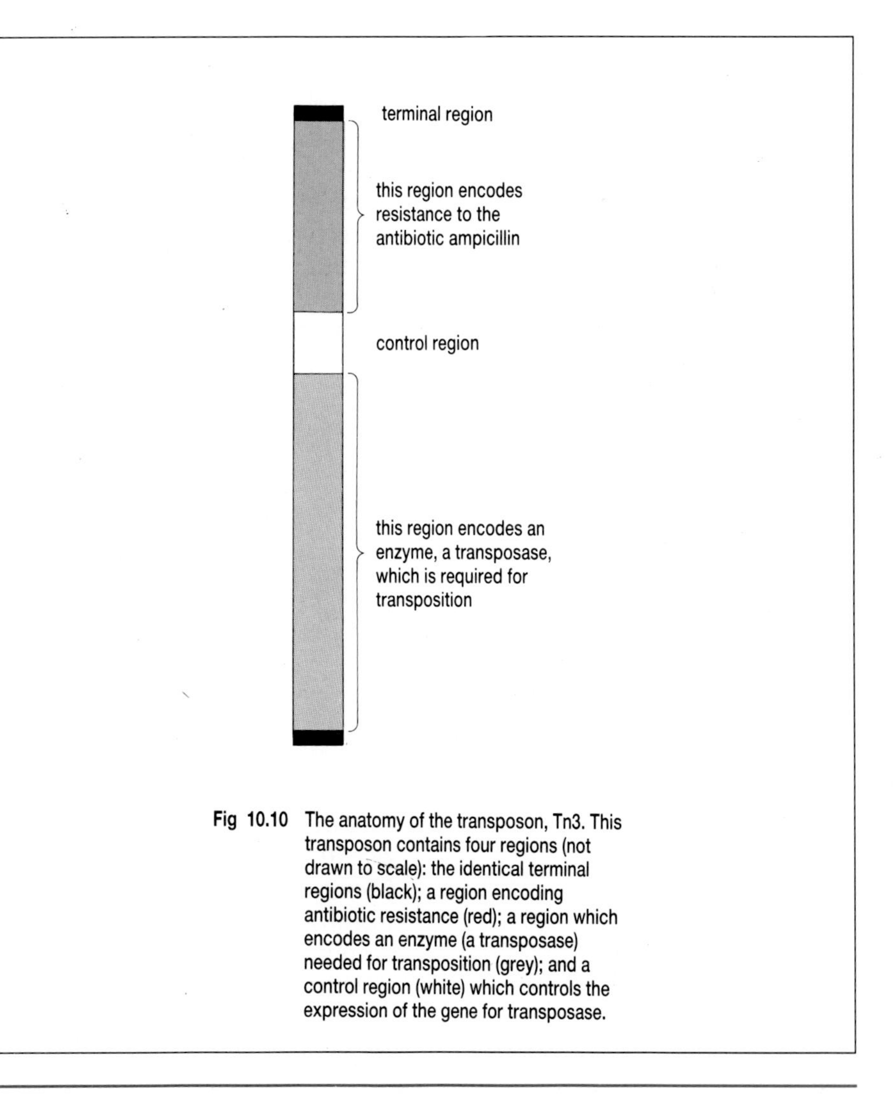

Fig 10.10 The anatomy of the transposon, Tn3. This transposon contains four regions (not drawn to scale): the identical terminal regions (black); a region encoding antibiotic resistance (red); a region which encodes an enzyme (a transposase) needed for transposition (grey); and a control region (white) which controls the expression of the gene for transposase.

QUESTIONS

10.1 Compare and contrast the process of recombination in bacterial and eukaryotic cells.

10.2 What advantages would the possession of an R factor confer on the host bacterial cell?

10.3 The process of transposition can result in a mutation if the insertion occurs within a gene. Explain this observation.

10.2 DNA TRANSFER BETWEEN BACTERIAL CELLS

DNA can be transferred between bacterial cells in three ways: transformation, transduction and conjugation.

Transformation

This involves the uptake of DNA, both chromosomal fragments and entire plasmids, from the environment. Following entry into the cell, the transforming DNA may recombine with host DNA or may replicate independently as a plasmid or prophage. Cells which are able to take up DNA are described as being **competent**. Some bacteria, for example *Haemophilus influenzae* and *Bacillus subtilis*, are naturally competent. Others, for example *E. coli*, must be subjected to various treatments to induce the cell to take up the DNA. Transformation is an essential step in most genetic engineering experiments.

Transduction

This involves the transfer of bacterial DNA from one cell to another by means of a phage particle. Look at Fig 10.11. This shows the process of generalised transduction. Here, a piece of the bacterial DNA, produced by the digestion of the host cell's genome by the early phage proteins, has been packaged into a phage particle. Notice that the phage particle contains no phage DNA. The bacterial DNA can be transferred to a new cell when it is infected by the phage particle. Occasionally, an entire plasmid can become packaged into a phage and then transferred to a recipient cell. Indeed, this seems to be a natural mechanism of plasmid transfer in members of the genus *Staphylococcus*.

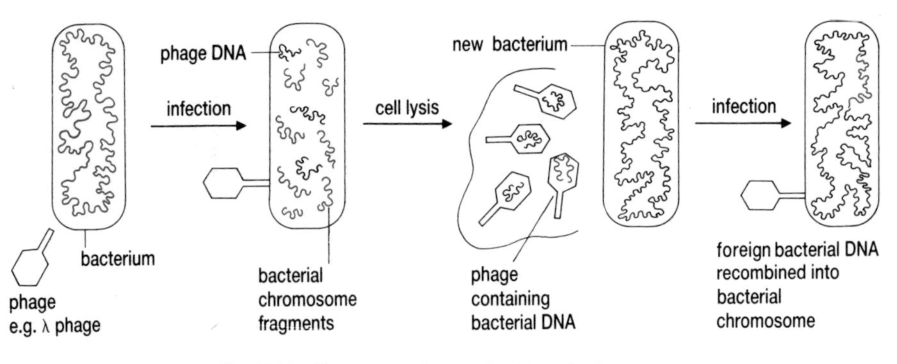

Fig. 10.11 The process of generalised transduction.

A different kind of transduction, specialised transduction, is shown in Fig 10.12. Here the transducing phage particle contains both phage and bacterial DNA. The bacterial genes transferred are those which flank the integration sites of prophages. Usually the excision of a prophage from a bacterial chromosome, prior to its reproduction to form new phages, is very precise. Occasionally something goes wrong and a piece of bacterial DNA is included with the phage DNA, so producing a specialised transducing phage.

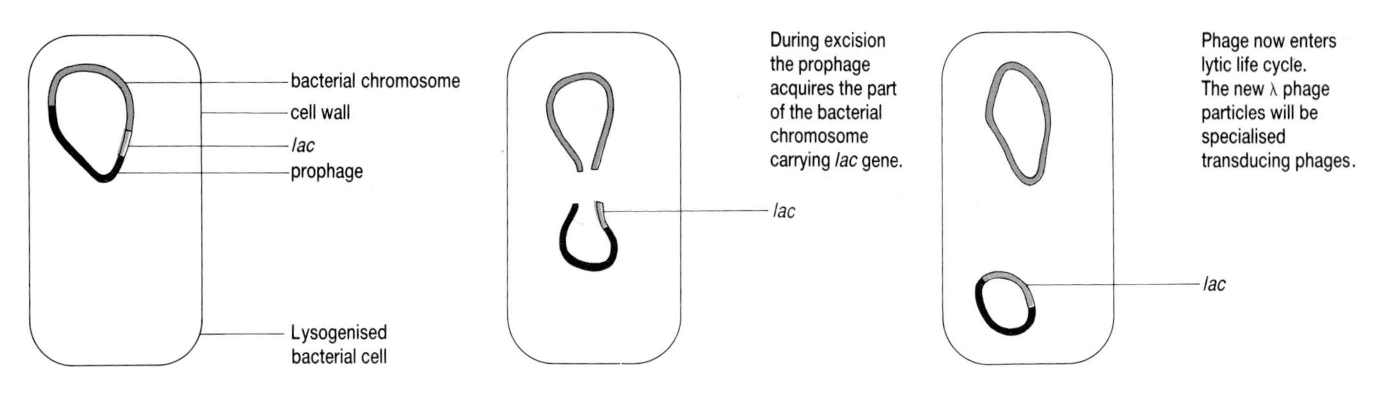

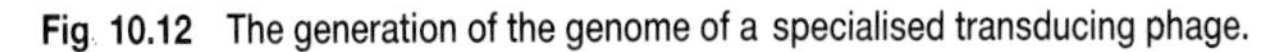
Fig. 10.12 The generation of the genome of a specialised transducing phage.

INVESTIGATION 10.2

The *gal* gene in *E. coli* encodes the enzyme galactase which breaks down the sugar galactose. In the presence of high concentrations of galactose the *gal* gene becomes activated. *E. coli* which contain the normal *gal* gene (gal^+ strains) can live in a medium where galactose is the only energy source. By contrast, *E. coli* containing a mutant *gal* gene (gal^- strains) which encodes non-functional galactase cannot survive on a medium which contains galactose as the only energy source. The *gal* gene flanks the insertion site of prophage λ. You are provided with a test tube of nutrient broth containing a gal^+ strain of *E. coli* which is lysogenised by λ. In another tube you have a non-lysogenised gal^- strain of *E. coli*.

(a) How could you increase the rate of induction of the prophage (see Chapter 9)?
(b) How could you separate the phages, produced as a result of induction, from the bacterial cells in the test tube?
(c) How would you check to see if any of the phages in your pure culture were specialised transducing phages? Give as complete a set of experimental instructions as you can.
(d) Draw a diagram to show the sequence of events involved in the specialised transduction of a gal^- bacterium to a gal^+ bacterium. Think carefully about what will happen to the transducing DNA once it has been injected into the recipient cell.

Conjugation

This involves the transfer of a plasmid between cells and differs from the other two methods of DNA transfer in that it involves direct contact between the donor and recipient cells through a cytoplasmic bridge (Fig 10.13). The process is illustrated in Fig 10.14 for a plasmid called F (for fertility) which was one of the first conjugative plasmids to be studied in detail. Notice that the plasmid is not lost from the donor, F^+, cell since only a single strand of plasmid DNA is transferred to the recipient, F^-, cell. Subsequent replication restores the DNA duplex in each cell.

It is now apparent that many plasmids have the ability to initiate conjugation and the phenomenon occurs in a number of different bacterial species, for example the streptococci, the *Streptomyces* and all the members of the family Enterobacteriaceae to which *E. coli* and *Salmonella typhimurium* belong. Other plasmids in these species, whilst not being able to initiate conjugation, can be passed to another bacterium when conjugation is initiated by a conjugative plasmid in the same cell. This process is called **mobilisation**.

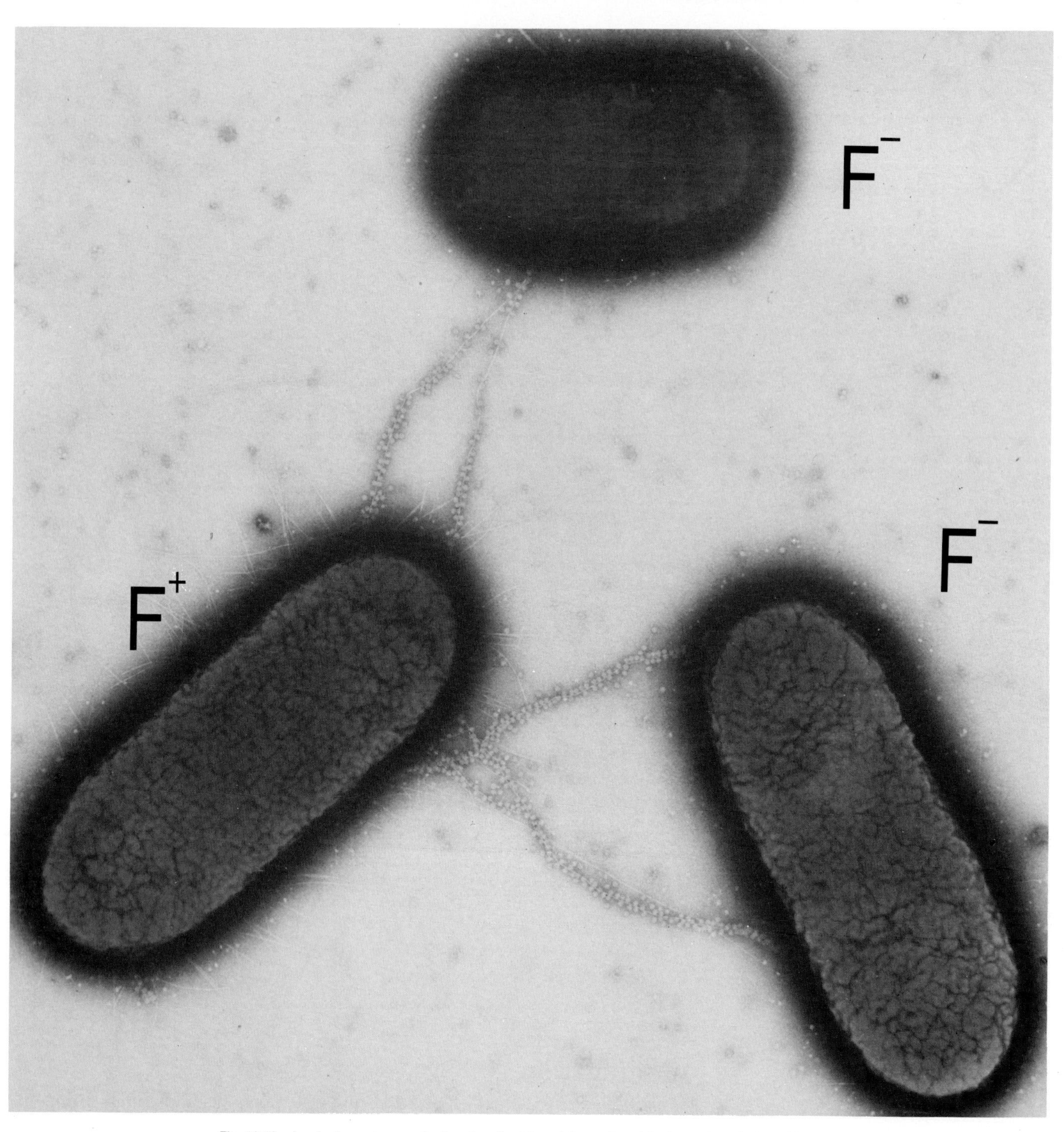

Fig 10.13 An electron micrograph of conjugating F^+ and F^- strains of the bacterium *Escherichia coli*.

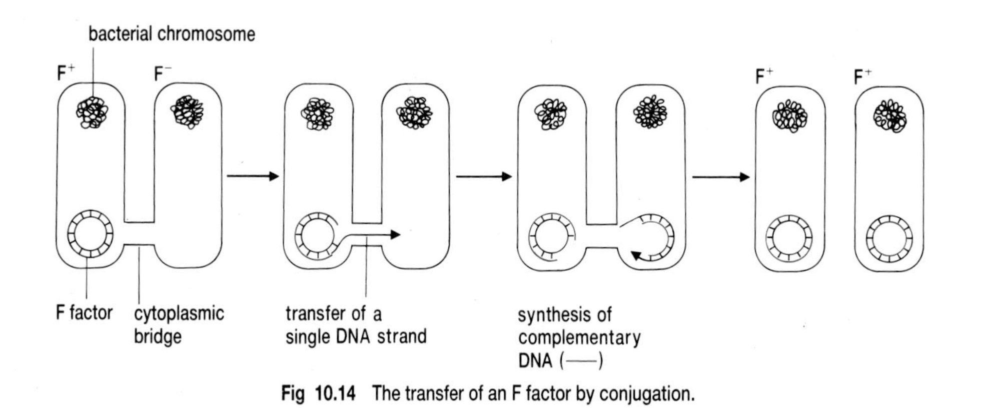

Fig 10.14 The transfer of an F factor by conjugation.

Some bacterial genera, for example *Staphylococcus* and *Bacillus*, do not appear to have conjugative plasmids, and plasmid transfer in these groups is dependent upon transduction and transformation.

The Hfr system. The F plasmid, found in some strains of *E. coli*, has the unusual ability to transfer fragments of bacterial chromosome (occasionally the whole chromosome) from donor to recipient cell. The process, shown in Fig 10.15, is a consequence of the ability of this plasmid to integrate stably with the chromosome. Strains of *E. Coli* in which F is integrated into the bacterial chromosome are called Hfr (high frequency of recombination) strains. When an Hfr strain establishes contact with an F^- cell, chromosomal DNA can be transferred along with DNA from the integrated F factor.

The F factor can be excised from the Hfr chromosome by an exact reversal of the integration process. Occasionally, during the excision process, the plasmid will pick up a piece of bacterial DNA, for example the lactase gene (lac^+) in Fig 10.16 so producing an F ' factor.

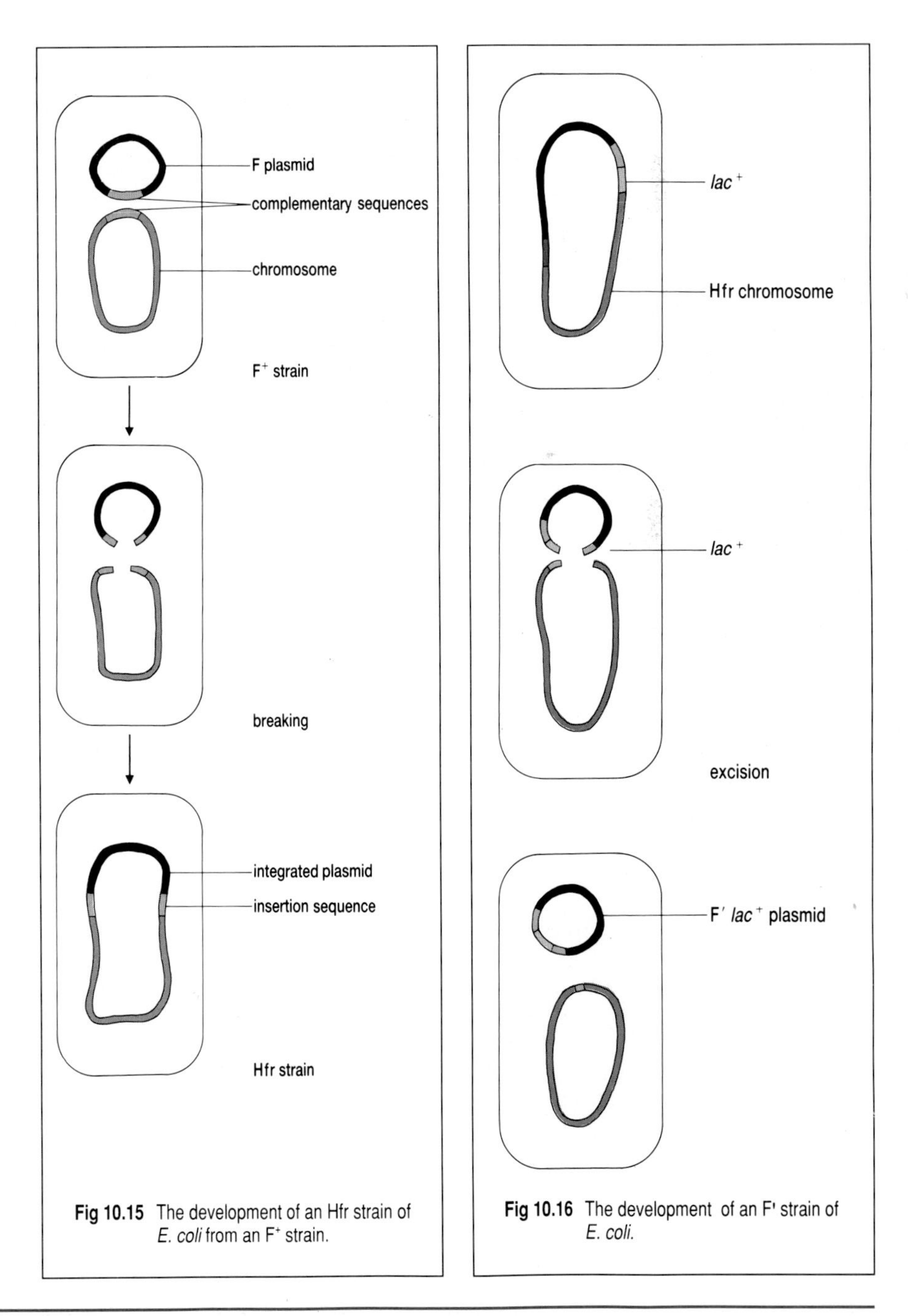

Fig 10.15 The development of an Hfr strain of *E. coli* from an F^+ strain.

Fig 10.16 The development of an F' strain of *E. coli*.

QUESTIONS

10.4 Using a series of annotated drawings,

(a) explain the difference between transformation, transduction and conjugation.

(b) How does the Hfr system found in some strains of *E. coli* differ from conjugation in other bacteria.

10.5 In 1928 Frederick Griffith, an English bacteriologist, was investigating a bacterium which causes pneumonia, *Streptococcus pneumoniae*. He found that virulent strains of this bacterium produced shiny, smooth colonies when they were grown on nutrient agar. For this reason they are called S (for smooth) strains. Occasionally, Griffith isolated mutant strains which produced colonies with a rough surface. These are called R (for rough) strains. The R strains are non-pathogenic. So if a mouse is injected with an R strain, it does not contract pneumonia. If exactly the same mouse is injected with an S strain, it contracts pneumonia and invariably dies. Griffith performed the following experiment. He injected a mouse with a mixture of an R strain of *S. pneumoniae* and a heat-killed S strain. The mouse contracted pneumonia and died. Upon autopsy Griffith found that the mouse contained live S strain *S. pneumoniae*. He then injected a mouse with just the heat-killed S strain. The mouse remained alive and healthy. Give a detailed explanation of Griffith's results.

10.3 R PLASMIDS AND ANTIBIOTIC RESISTANCE IN BACTERIA

So far we have looked at the genetic structure and mechanisms of DNA transfer in bacteria. In this section you will explore the implications of these two aspects of bacterial biology by considering the problem of antibiotic resistance. There are a number of different mechanisms which result in antibiotic resistance in bacteria, but most antibiotic resistance genes are carried on R plasmids. Under normal circumstances bacteria carrying R plasmids will not be at a selective advantage. However, the application of antibiotics immediately creates a selective pressure which will favour bacteria carrying the appropriate R plasmid (Fig 10.17).

Fig 10.18 Electron micrograph of a pathogenic strain of the bacterium *Shigella* (× 66 000) which causes dysentry. The bacterium is spread by flies, direct contact and water contaminated by faeces containing the bacillus.

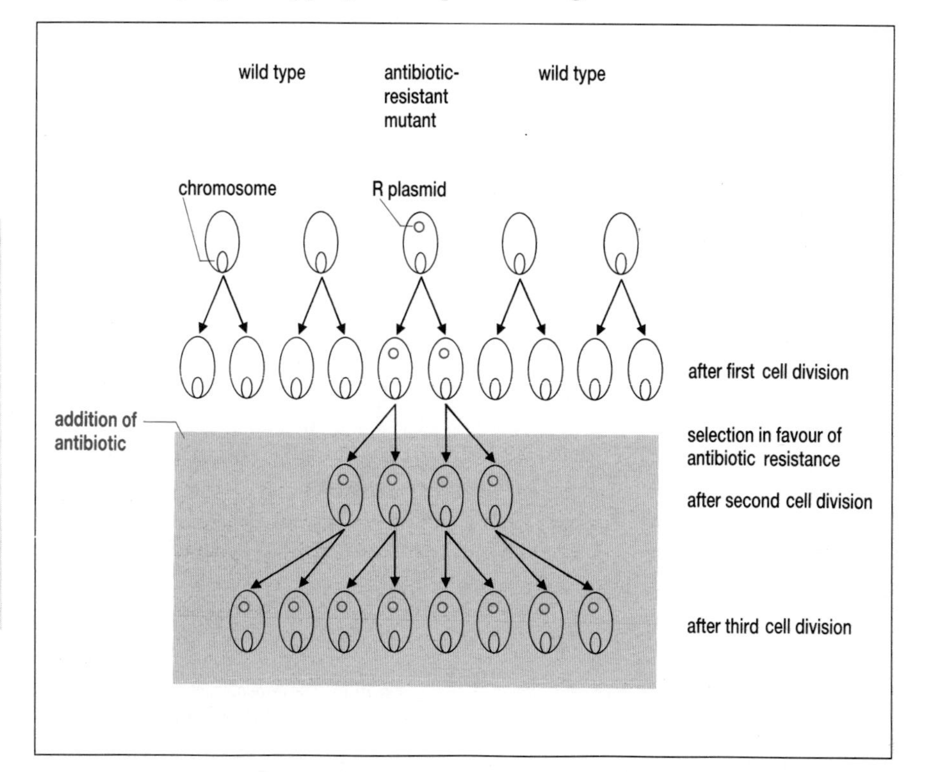

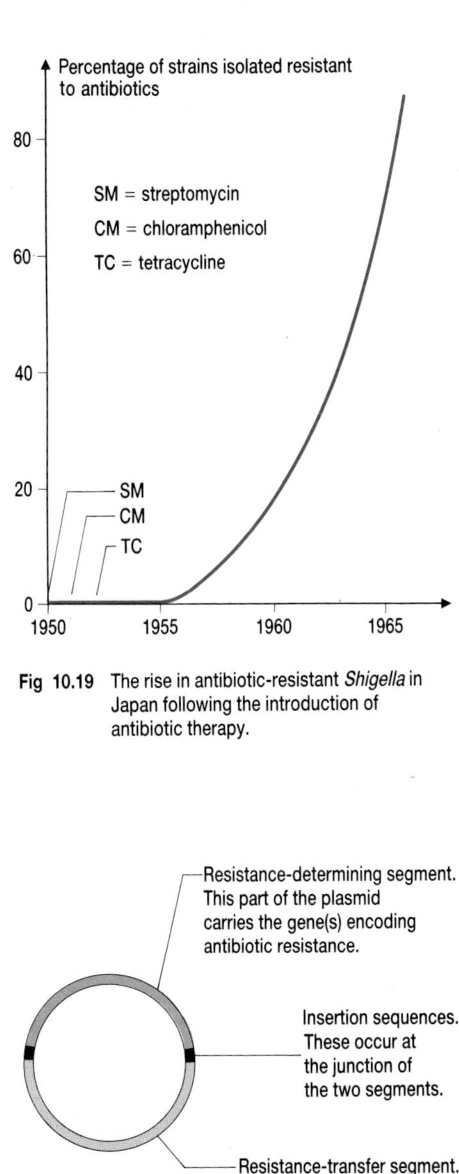

Fig 10.19 The rise in antibiotic-resistant *Shigella* in Japan following the introduction of antibiotic therapy.

Fig 10.20 The anatomy of an R plasmid. Two regions can be recognised: one encodes antibiotic resistance (red) whilst the other controls the transfer of the plasmid to other cells by conjugation (grey). The two regions are joined by insertion sequences like those found at the end of transposons (black).

The introduction of antibacterial drugs must rate as one of the greatest advances in the fight against disease. Unfortunately, the nature of antibiotics has been misunderstood with the result that many strains of bacteria are becoming resistant to them. In an antibiotic-rich society like ours, where these powerful drugs are too readily prescribed, bacterial resistance is becoming an increasing problem. Three case studies will be used to explain the origin of plasmid-encoded antibiotic resistance and how the problems posed by antibiotic resistance may be tackled.

Case study 1 – Multiple antibiotic resistance in *Shigella*

Shigella (Fig 10.18), belonging to the group of bacteria called the Enterobacteriaceae causes bacillary dysentery (severe diarrhoea) in humans.

Tsutomu Watanabe, an eminent Japanese bacteriologist, was the first to note the increasing number of *Shigella* infections which were becoming antibiotic resistant. The time scale is shown in Fig 10.19. In particular, Watanabe noted that in outbreaks of bacillary dysentery from 1956 onwards there was an increasing number of *Shigella* strains which showed multiple resistance to four drugs: streptomycin, chloramphenicol, tetracycline and sulphonamide, which had, up to then, been used successfully to treat *Shigella* infections. How had this multiple antibiotic resistance originated and then spread so rapidly?

The first step in solving this puzzle was to find out how the drug resistance was being transferred from bacterium to bacterium. In 1959 Watanabe demonstrated that conjugative plasmids could transfer antibiotic resistance from one bacterial cell to another. Since then a large number of plasmids, called R plasmids, which carry antibiotic-resistance genes, have been isolated. Not all these plasmids are conjugative. For example, the R plasmid which confers penicillin resistance on strains of *Staphylococcus aureus* is transferred between cells by transduction. Furthermore, not all cases of antibiotic resistance in bacteria are encoded by genes carried on plasmids. However, in the case of *Shigella*, the antibiotic-resistance genes were definitely located on conjugative R plasmids. Such plasmids have the general structure shown in Fig 10.20.

How does such an R plasmid originate? The generally accepted model involves a cryptic plasmid which evolves into an R plasmid by the acquisition of transposons that carry antibiotic-resistance genes (Fig 10.21). A recently documented example is the development of an R plasmid in

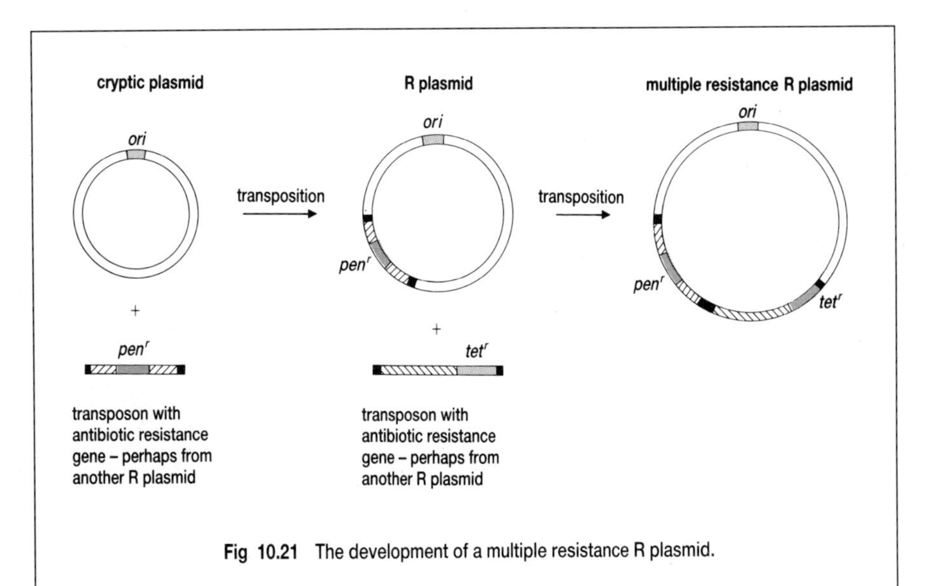

Fig 10.21 The development of a multiple resistance R plasmid.

Providentia stuartii from a cryptic plasmid which acquired two copies of the transposon called Tn1. This particular transposon carries a gene encoding ampicillin resistance. An R plasmid encoding multiple drug resistance could therefore be built up by the acquisition of several different transposons by a cryptic plasmid. If the original cryptic plasmid contained, or subsequently acquired, the genes necessary to initiate conjugation then the resulting R plasmid would be able to move between cells by conjugation.

How can we explain the rapid spread of the antibiotic resistance in *Shigella*? Further experiments showed that conjugative R plasmids could be transferred between all members of the large family of bacteria known as the Enterobacteriaceae and to other genera outside this family, for example *Pseudomonas* and *Proteus*. So an enteric bacterium which acquires a resistance plasmid can quickly pass it on to a member of the same species or to a different species, a process called infectious spread of drug resistance. In the case of *Shigella*, it appears that the resistance plasmids may have been transferred from *E. coli*, a normal component of the gut flora of humans, to *Shigella*.

It is now apparent that resistance (R) factors can spread between a large number of different types of bacteria. This has serious consequences. For example, urinary tract infections can be caused by a number of different bacteria, such as *E. coli*, *Proteus mirabilis*, *Klebsiella* and *Pseudomonas aeruginosa*. All these bacteria can take part in the great R factor transfer game. As a result, antibiotic resistance in urinary tract infections is now widespread.

Case study 2 – *Salmonella*

This enteric bacterium causes food poisoning in people. One of the worst outbreaks occurred in Stanley Royd Hospital in Wakefield in 1984 when 19 people died. In Great Britain, most *Salmonella* strains are now resistant to antibiotics. This alarming state of affairs has arisen partly through the misuse of antibiotics in agriculture.

Bacterial infections of livestock were a major problem before the introduction of antibiotics. Unfortunately, farmers did not simply use antibiotics **therapeutically** to treat diseased animals. They also used them **prophylactically**. So if one animal caught a disease, all the animals in the herd or flock would be treated with an antibiotic to prevent them catching the disease. Even more alarmingly, antibiotics like chloramphenicol, tetracycline and penicillin were given to farm animals as growth promoters.

The main reservoir of *Salmonella* is in farm animals like pigs, cattle and chickens. Since these animals had been given antibiotics as part of their diet, bacteria which occur naturally in their guts, like *E. coli*, which contain R plasmids, have been at a selective advantage compared to those strains which did not contain such plasmids. The effects of this can be seen among young calves which are being raised in intensive beef production systems. Under these conditions the calves are continually contracting diseases, like enteritis, which are treated with antibiotics. As a result up to 60 per cent of the *E. coli* in the guts of such animals are resistant to antibiotics. Whilst the *E. coli* are, in the main, harmless, they can transfer the R plasmids they contain to pathogenic bacteria like *Salmonella*. The implications of this are spelt out in the article on page 133. The authors were investigating an outbreak of diaorrhoea, caused by *Salmonella newport*, in Minnesota, USA.

People mainly catch *Salmonella* from contaminated agricultural products like meat which have not been cooked or stored correctly. For example 60 per cent of all frozen chickens are contaminated with *Salmonella*. These do not appear to pose a health hazard provided the chickens are stored and cooked properly.

Bacteria isolated from all the patients were resistant to the antibiotics ampicillin, carbenicillin and tetracycline and all carried the same plasmid. A survey of *S. newport* strains recently isolated from livestock turned up the same resistance pattern and plasmid in bacteria that had caused diarrhoeal disease in cows on a dairy farm in South Dakota several months before.

A beef feedlot adjoins the dairy farm. The cattle there had been fed tetracycline. A shipment of beef from the herd could be traced to supermarkets where it had been ground to make hamburger. Most of the affected patients had eaten hamburger bought at one or another of those supermarkets during the week before they became sick. Three others had got beef directly from the feedlot. All the remaining cases could be connected in some way to a hamburger-linked case. Both the epidemiological and the laboratory data thus point to specific beef cattle as the source of resistant *Salmonella*.

Interviews with the patients revealed that twelve of them had taken an antibiotic (one to which the *S. newport* strain was resistant) for a minor non-diarrhoeal illness just before developing salmonellosis. The implication is that what would have been an asymptomatic salmonella infection was rendered virulent by "selective pressure": the antibiotic killed off harmless intestinal bacteria that would normally have competed with the *Salmonella*, and in doing so it allowed the pathogens to proliferate.

In 1969, the Swann Committee recommended that penicillin and tetracycline, two of the most widely used antibiotics, should no longer be used for growth promotion in Great Britain. Subsequent legislation has tightened the use of antibiotics in agriculture considerably throughout the world. One of the results of this is that there is now a thriving black market in antibiotics for agricultural use.

Case study 3 – the *Klebsiella* scandal

In 1966 a neurosurgical unit in Great Britain suffered an increasing incidence of post-operative wound sepsis due to staphylococcal and enterobacterial infections. Patients were given large quantities of ampicillin and cloxacillin prophylactically to prevent this problem. The result was that whilst bacteria sensitive to these antibiotics were eradicated, highly virulent, antibiotic-resistant *Klebsiella* took their place. As a consequence of this change in the bacterial flora the incidence of *Klebsiella*-related pneumonia in the unit rocketed. Between October 1968 and May 1969, of 228 patients admitted to the intensive care unit, one in three had *Klebsiella*.

The bacterium was attacked with large amounts of antibiotic between July and September 1969. Whilst this tactic benefited individual patients, it made no difference to the numbers of *Klebsiella* in the environment. Faced with the possible closure of the unit, the use of all antibiotics was suspended between October 1969 and January 1970. *Klebsiella* disappeared rapidly, and surprisingly, no patient suffered as a result of the withdrawal of antibiotic therapy. Since that time the use of antibiotics in the unit has been strictly reserved for the treatment of a limited number of life-threatening infections.

QUESTIONS

10.6 What are the dangers of using antibiotics prophylactically?

10.7 In small groups, produce a set of rules which doctors should follow when prescribing antibiotics.

10.8 Why would you persuade a friend not to take antibiotics for a mild sore throat?

10.9 Hospital sewage contains a higher proportion of *E. coli* which are resistant to antibiotics than normal domestic sewage. Explain this observation.

SUMMARY ASSIGNMENT

1. What are the distinguishing features of the major types of genetic element in bacteria?
2. Keep a copy of question 10.4 as a summary of mechanisms of DNA transfer.
3. Construct an argument to convince a parliamentary committee not to restore the use of antibiotics as growth promoters in chickens. Remember, you are writing a report for politicians so you will need to explain the relevant aspects of bacterial genetics so that they can understand your argument.

Theme 4

RECOMBINANT DNA TECHNOLOGY

Biology stands at the threshold of a new era because of the techniques of recombinant DNA technology or genetic engineering. What is recombinant DNA? Why is it important? How can we use this new technology? Is it safe? These are some of the questions addressed in this theme.

PREREQUISITES

To complete the work in this theme successfully you will need to have a working knowledge of DNA, RNA and their role in protein synthesis, and to have completed Theme 3.

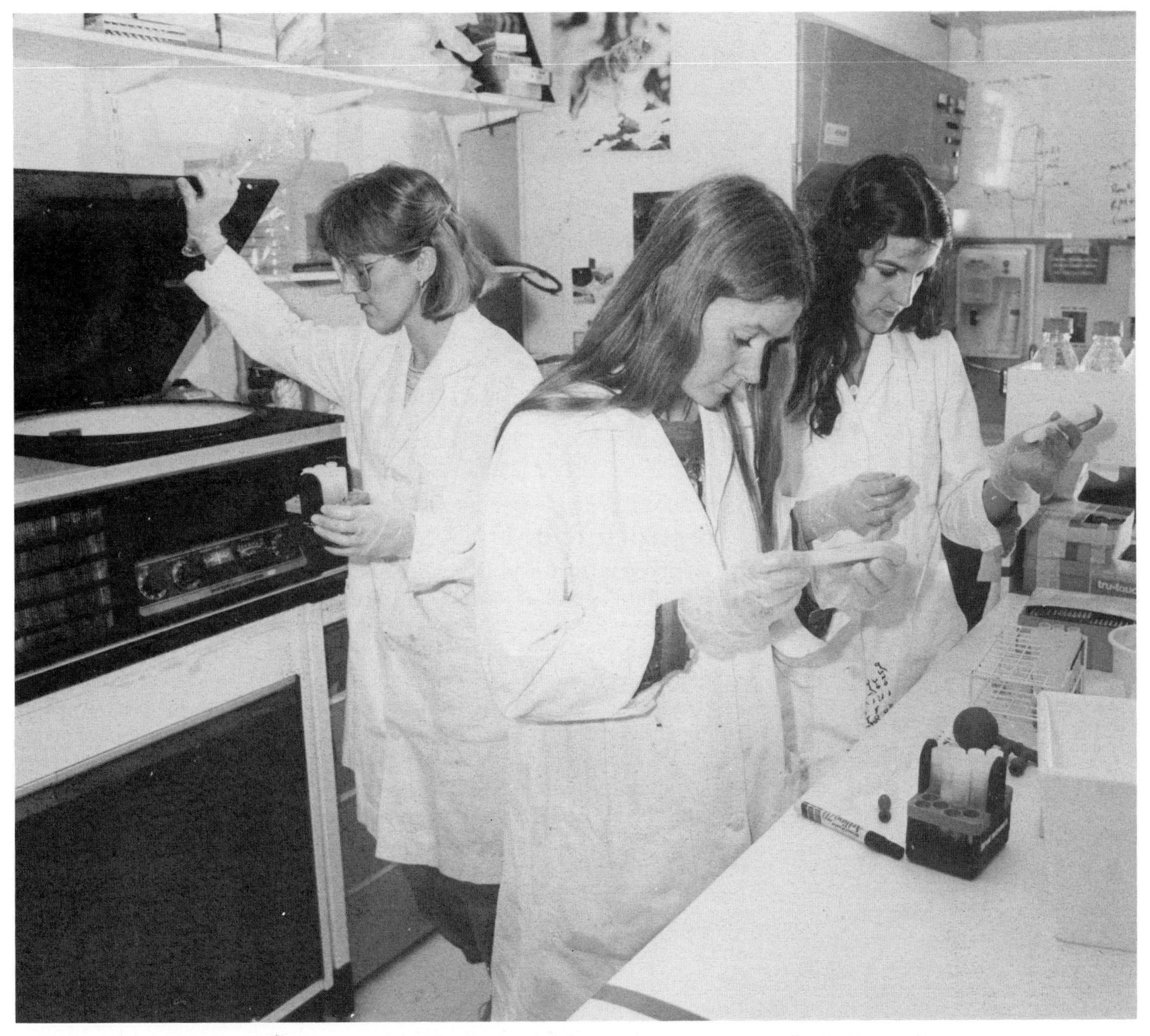

Dr Kay Davies, in the foreground, and her colleagues preparing DNA from blood samples of muscular dystrophy patients prior to analysis by recombinant DNA techniques.

Chapter 11

GENETIC ENGINEERING

To understand and eventually find cures for inherited diseases like Duchenne muscular dystrophy, cystic fibrosis and Huntington's chorea we need to understand how genes work at the molecular level. However, molecular geneticists are faced with a seemingly insurmountable problem: the sheer amount of DNA contained in a cell. For example, the genome of a human cell contains about 3 thousand million base pairs of DNA encoding some 100 000 genes. How do you isolate a gene from so much DNA, let alone work out how it works and what it does? This is where recombinant DNA technology and genetic engineering come in. Genetic engineering allows geneticists to isolate, purify and study individual genes. In this chapter we will look at the basic toolkit of the genetic engineer and how this toolkit can be used to isolate (clone) genes. In the next chapter we will investigate how these techniques can be used by molecular geneticists in industry, agriculture and medicine. First, though, you will need to revise the structure of DNA and RNA.

LEARNING OBJECTIVES

After completing the work in this chapter you will be able to:

1. describe the basic tools of the genetic engineer: restriction enzymes; DNA ligase; vectors; host systems and radioactive DNA probes;
2. explain how to clone and isolate a gene using shotgunning and colony hybridisation;
3. outline some alternative approaches to gene cloning;
4. explain how λ Charon phages can be used to make a gene library.

11.1 SOME BASIC BIOCHEMISTRY

The structure of DNA

Deoxyribose nucleic acid (DNA) is an extraordinary molecule, since it can make faithful copies of itself. The process of complementary base pairing, which forms the basis of this copying process, also lies at the heart of recombinant DNA technology. DNA is a polymer made up of the four subunits shown in Fig 11.1. These molecules, called nucleotides, are made up of a sugar molecule (deoxyribose) to which are attached a phosphate group and a nitrogenous base. The only major difference between the nucleotides is which nitrogenous base they contain.

The polymer is put together by joining the sugar molecules to each other through the phosphate groups to produce a sugar-phosphate backbone, as shown in Fig 11.2. Notice that the two ends of the sugar-phosphate backbone are different. At one end of the molecule there is a phosphate group; this is called the 5' (5 prime) end. At the other end there is a hydroxyl group; this is called the 3' end. The 5' and 3' refer to a particular carbon atom in the deoxyribose molecule. You can add further nucleotides at either the 5' or the 3' end to produce extremely long DNA molecules.

The molecule shown in Fig 11.2 only consists of one strand of DNA, but you probably know that the DNA in your cells actually exists as a double strand. Part of such a double strand is shown in Fig 11.3. Here we have two sugar-phosphate backbones (shown in black) being held together by hydrogen bonds (the red dotted lines) between the bases (shown in grey). The bases always pair up in the way shown in Fig 11.3: adenine (A) – thymine (T); cytosine (C) – guanine (G). This specific pairing process, called **complementary base pairing**, is the key to how DNA copies itself and to the processes involved in recombinant DNA technology.

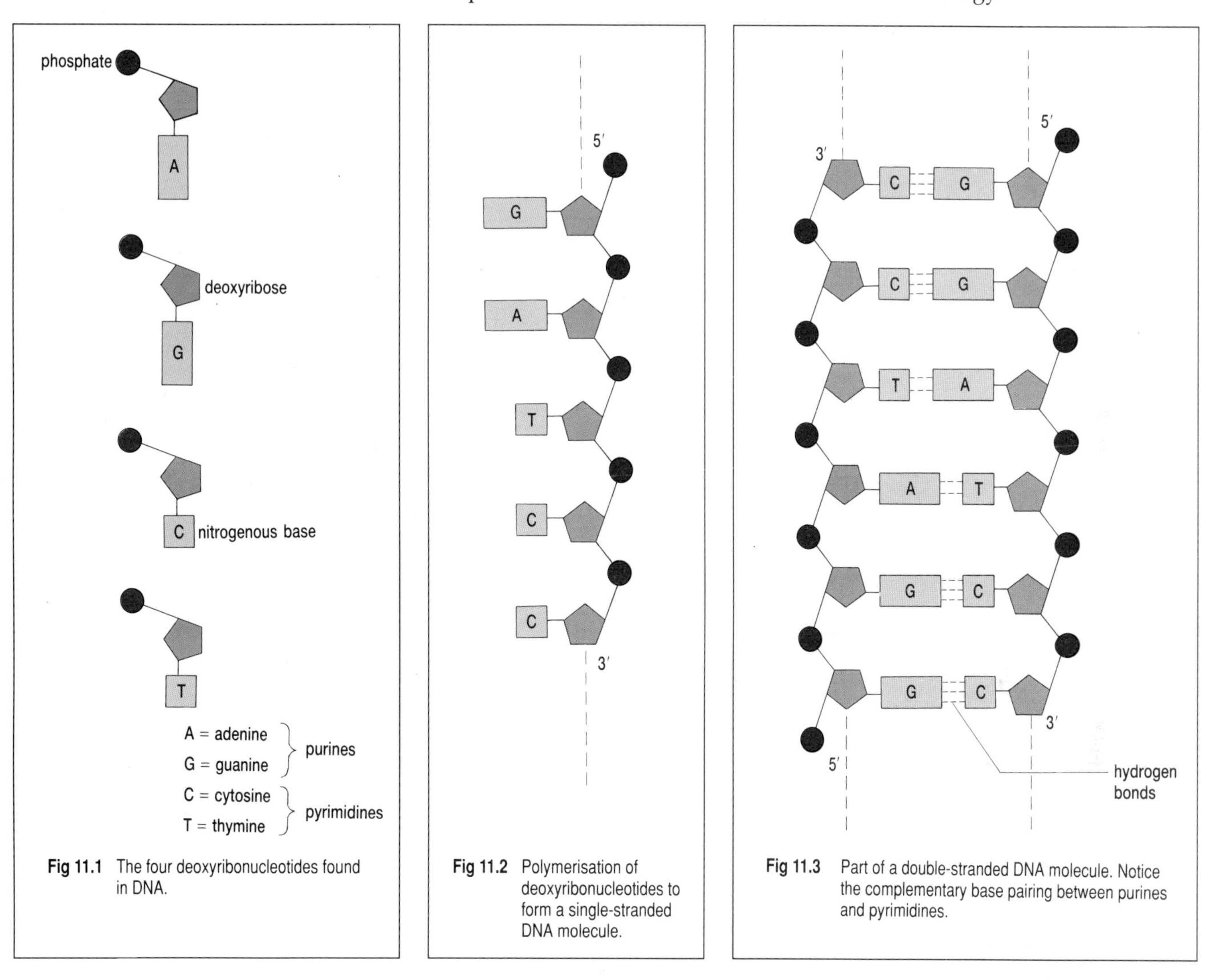

Fig 11.1 The four deoxyribonucleotides found in DNA.

Fig 11.2 Polymerisation of deoxyribonucleotides to form a single-stranded DNA molecule.

Fig 11.3 Part of a double-stranded DNA molecule. Notice the complementary base pairing between purines and pyrimidines.

One last thing about the molecule shown in Fig 11.3. Look at the top of the molecule. Can you see that the ends are different? The left-hand strand has the 3' end at the top but the right-hand strand has the 5' end at the top. The two strands of the DNA molecule shown in Fig 11.3 are running in opposite directions; they are said to be **antiparallel**.

Ribonucleic acid – RNA

RNA is also a polymer made up by joining together nucleotides. The sugar part of the molecule, ribose, is very similar to deoxyribose and the sugar-phosphate backbone of RNA is put together in exactly the same way as the sugar-phosphate backbone of DNA. Once again an RNA molecule has a 5' end and a 3' end. RNA also contains the same bases as DNA, with one exception – thymine is replaced by uracil. RNA also differs from DNA in that it exists as single strands not as double strands. RNA is produced by the process of transcription from the information contained in the sequence of bases on the coding strand of DNA (Fig 11.4). This process depends upon complementary base pairing, and the RNA produced is said to be complementary to the DNA sequence used to produce it.

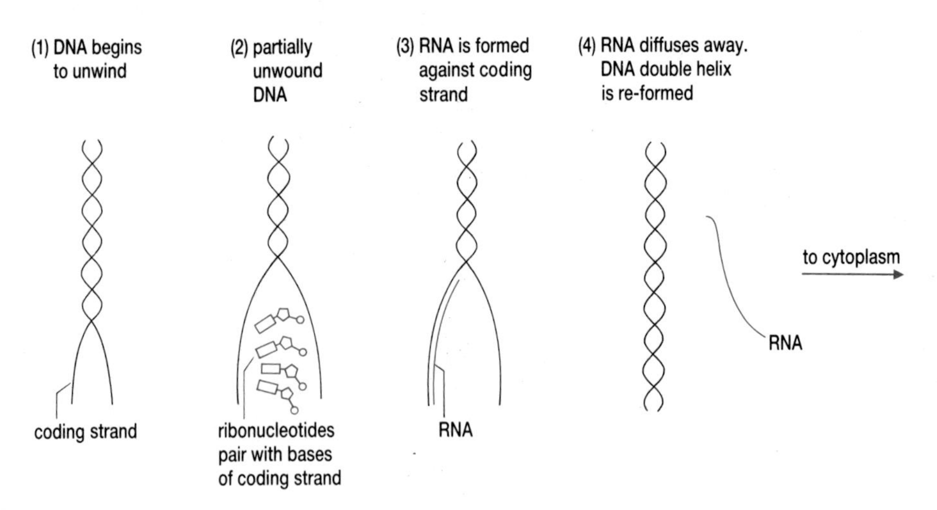

Fig 11.4 The four steps of transcription.

QUESTIONS

11.1 Which of the following descriptions, (i) – (vi), apply to
(a) both RNA and DNA
(b) RNA but not to DNA
(c) DNA but not to RNA?

Descriptions:
(i) linked nucleotides
(ii) typically found in the nucleus but not in the cytoplasm
(iii) typically found as single strands
(iv) contain(s) as many purine bases as pyrimidine bases
(v) contain(s) nucleotides in which the sugar is ribose
(vi) contain(s) nucleotides, some of which contain uracil.

(From an exercise in S102 - *A Science Foundation Course*, Open University, 1988.)

11.2 The following string of bases CGACGGCTACCA is found on the coding strand of a DNA molecule. What are the complementary sequences on
(a) the non-coding strand of the DNA
(b) an mRNA molecule transcribed from the coding strand?

The relationship between DNA, messenger RNA (mRNA) and protein

The central dogma of molecular biology is summarised as follows:

DNA $\xrightarrow{\textbf{transcription}}$ **RNA** $\xrightarrow{\textbf{translation}}$ **protein**

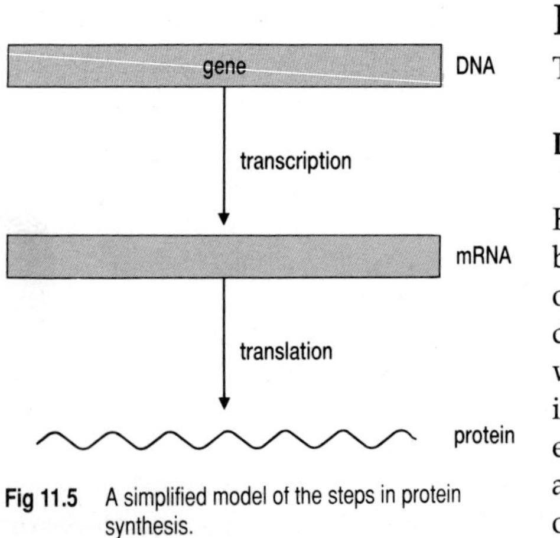

Fig 11.5 A simplified model of the steps in protein synthesis.

For many years it was assumed that the sequence of codons (triplets of bases) in the DNA of a gene encoding a given protein was in the form of one continuous strand (Fig 11.5). In 1977 this view was shattered by the discovery of several eukaryotic genes which were not constructed in this way. These newly discovered **split genes** have the general structure shown in Fig 11.6. The **exons** contain the genetic information which is ultimately expressed in the protein whilst the intervening **introns** do not code for anything. For example, the gene which encodes egg white protein, ovalbumin, has eight exons and seven intervening introns.

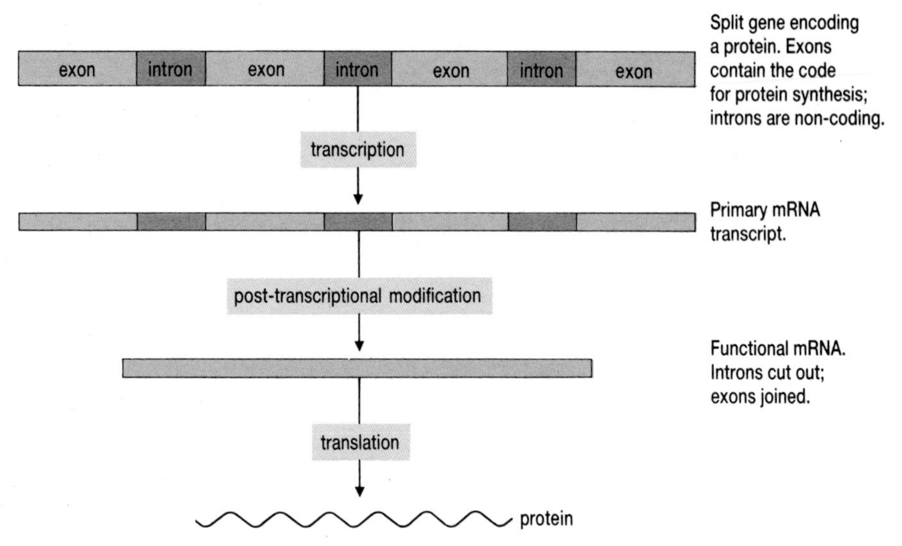

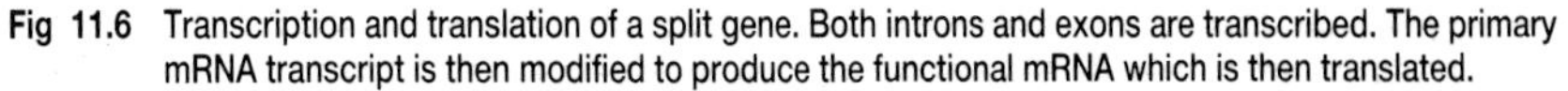

Fig 11.6 Transcription and translation of a split gene. Both introns and exons are transcribed. The primary mRNA transcript is then modified to produce the functional mRNA which is then translated.

The mRNA molecule first produced by transcription contains the RNA equivalents of both the introns and exons. To produce the functional mRNA molecule, the unwanted introns must be removed (tailored out) and the exons joined (spliced) together in the process called post-transcriptional modification (Fig 11.6). These processes are unique to eukaryotes. Neither split genes nor the enzymes required for tailoring and splicing mRNA are found in prokaryotes such as bacteria.

To further complicate the story, only about 50 per cent of the DNA in your cells actually codes for the production of mRNA and hence protein. Some of the rest encodes ribosomal and transfer RNA, but as much as 40 per cent of human DNA has no known coding function. These apparently non-functional sequences are sometimes referred to as 'junk' DNA.

Recent research has also highlighted differences in the way mRNA synthesis is controlled in prokaryotic and eukaryotic cells. In both types of cell RNA is synthesised by the enzyme RNA polymerase II. However in prokaryotes the genes often occur clustered in **operons** (Fig 11.7) under the control of a single **promoter** region. The promoter is where the RNA polymerase binds before it starts transcribing the DNA. The RNA polymerase copies the DNA of all the genes of an operon to yield a single mRNA which, when translated, yields several types of protein, one corresponding to each gene in the operon. Eukaryotes do not have operons. Rather, each gene is separate and under the control of its own promoter (Fig 11.7).

These fundamental differences in the way protein synthesis is directed in eukaryotic and prokaryotic cells have important implications for the genetic engineer.

QUESTION

11.3 List the ways in which prokaryotic and eukaryotic genes differ.

11.2 A BASIC TOOLKIT FOR GENETIC ENGINEERS

To isolate a gene the genetic engineer needs four basic tools. These are described in detail below.

Restriction enzymes

In the early 1950s it was noted that some strains of bacteria seemed to be immune to attack by phages. In 1970 it was discovered that this defence

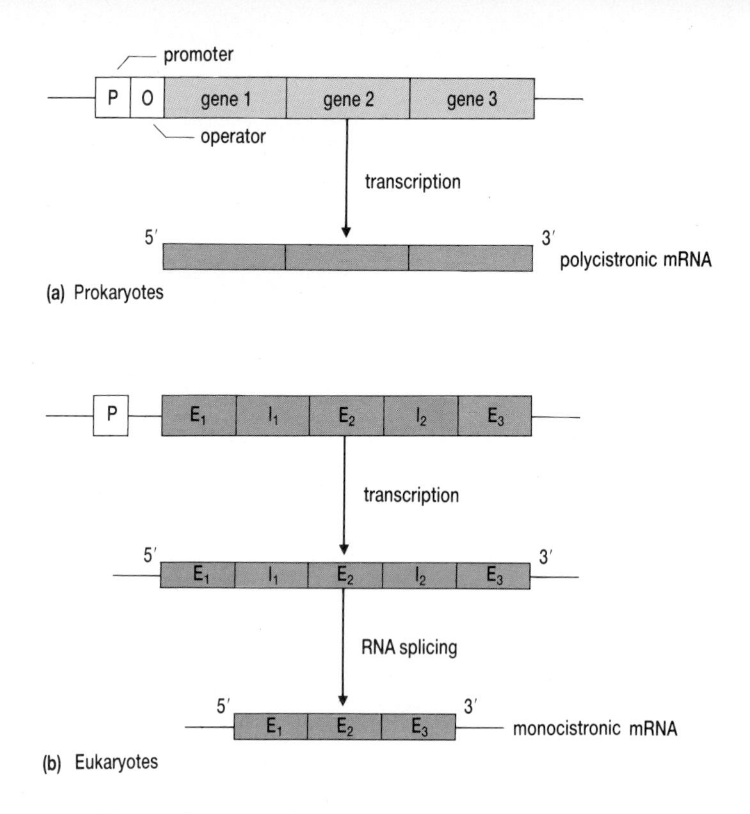

Fig 11.7 The process of transcription in (a) prokaryotes (b) eukaryotes.

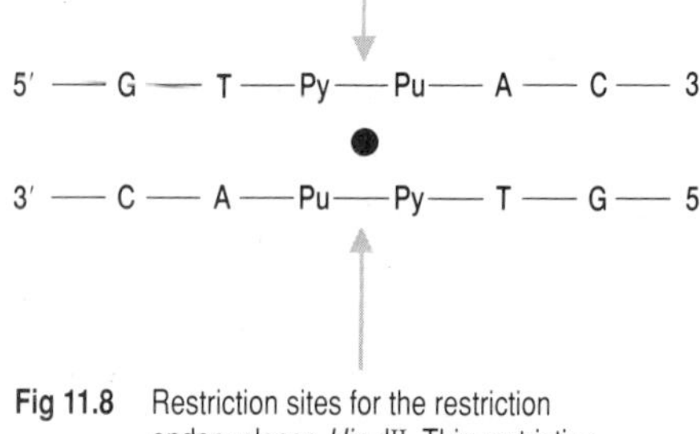

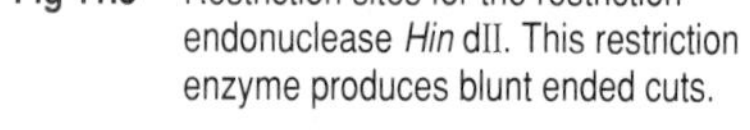

Fig 11.8 Restriction sites for the restriction endonuclease *Hin* dII. This restriction enzyme produces blunt ended cuts.

mechanism was due to a group of enzymes. Since these enzymes restricted the growth of foreign DNA in the bacterial host, they became known as **restriction enzymes** (technically, restriction endonucleases). These enzymes act like molecular scissors, chopping up DNA molecules into **restriction fragments**. However, unlike other DNA digesting enzymes, for example the deoxyribonuclease produced by your pancreas, restriction enzymes will only cut DNA molecules at certain specific sites, called **restriction sites**. These sites are determined by the sequence of bases in the DNA.

The bacterium *Haemophilus influenzae* produces a restriction enzyme called *Hin* dII (pronounced hindee 2) which always cuts the DNA from a phage called T7 into exactly 40 pieces. Only these fragments are produced because the enzyme cuts the DNA only at the points indicated by the arrows in the specific sequence shown in Fig 11.8. Py represents a pyrimidine, that is, thymine or cytosine; Pu represents a purine, that is, adenine or guanine. Notice the symmetry of this sequence: a rotation of 180 degrees around the dot in the centre leaves the sequence unchanged. *Hin* dII is very specific, hence the reason why it always cuts T7 phage DNA into exactly the same 40 pieces.

Investigations of other restriction enzymes have produced similar results. However some restriction enzymes, unlike *Hin* dII, produce staggered rather than blunt ended cuts in DNA. For example, consider the enzyme *Eco* RI which is encoded by a gene on an R plasmid in *E. coli*. This enzyme will only cut DNA where it finds the bases GAATTCC, in that order. Using the complementary base pairing rule, the complementary sequences at the *Eco* RI cleavage site can be worked out and are shown in Fig 11.9. The arrows indicate where the enzyme cuts the DNA. Compare the sequence of bases in the two DNA strands at the *Eco* RI restriction site. Remember, the strands are running in opposite directions so you have to read them in opposite directions. Can you see that the sequences are the same? A sentence or word which reads the same forwards or backwards (for example rotor) is called a palindrome and DNA sequences, like the one at the *Eco* RI site, are described as being palindromic.

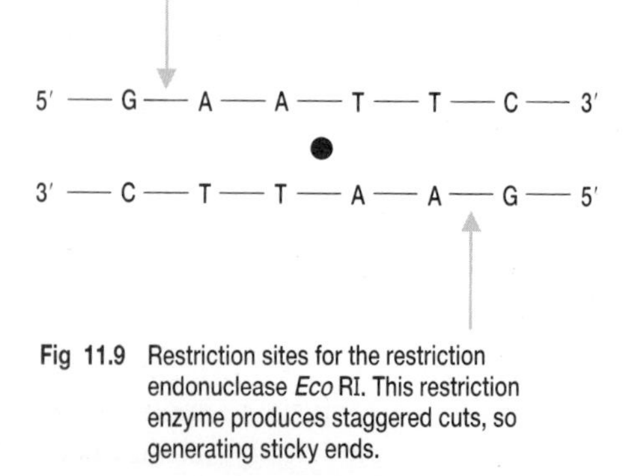

Fig 11.9 Restriction sites for the restriction endonuclease *Eco* RI. This restriction enzyme produces staggered cuts, so generating sticky ends.

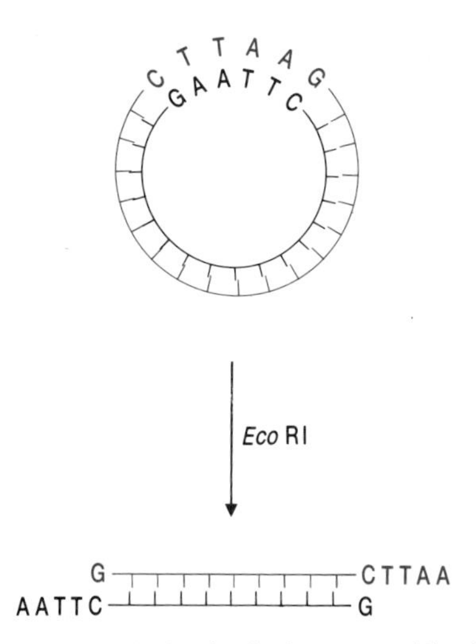

Fig 11.10 Cutting the circular genome of the virus SV40 with *Eco* RI produces a linear DNA molecule with sticky ends.

The important thing to notice about the cuts made by *Eco* RI is that they are staggered. The significance of this is shown in Fig 11.10. Here *Eco* RI has been used to cut open the circular DNA of SV40 (a virus). Since SV40 DNA only has one restriction site for *Eco* RI, the ring is opened out to form a linear molecule. Now look at the ends of the linear molecule. Can you see that each end has a series of unpaired bases, AATT on the left and TTAA on the right? This string of unpaired bases has been produced because *Eco* RI makes staggered cuts.

Such a string of bases will 'want' to pair up with a string of complementary bases. This 'desire' to pair up makes these unpaired base sequences 'sticky'. Unpaired base sequences at the ends of DNA molecules produced by restriction enzymes like *Eco* RI are called 'sticky ends'. About 300 restriction enzymes have now been identified, each of which has its own specific restriction site. Some further examples are given in Table 11.1. To summarise, restriction enzymes provide a way of cutting DNA from any organism at a specific sequence, so producing a collection of DNA fragments of variable length which may have sticky ends.

Table 11.1 Some examples of restriction enzymes

Enzyme	Source	Restriction site
Eco RII	*E. coli*	5' - G ↓ C - C - T - G - G - C - C - G - G - A - C - C ↑ G - 5'
Pst I	*Providentia stuartii*	5' - C - T - G - C - A ↓ G - - G ↑ A - C - G - T - C - 5'
Bam HI	*Bacillus amyloliquefaciens*	5' - G ↓ G - A - T - C - C - - C - C - T - A - G ↑ G - 5'

QUESTION

11.4 *Hin* dIII is a restriction enzyme which cuts DNA where the base sequence is AAGCTT. *Hin* dIII cuts between the AA on each strand. Work out the sticky ends produced when you cut a DNA molecule with *Hin* dIII.

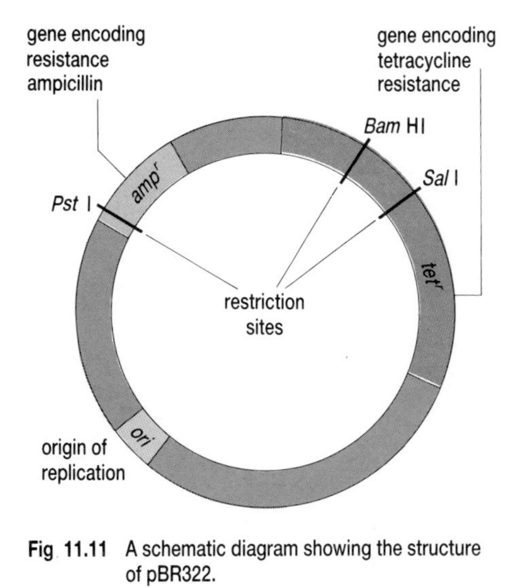

Fig 11.11 A schematic diagram showing the structure of pBR322.

Vectors – molecular lorries

If we digest all the DNA in a human cell with a restriction enzyme, we would end up with a mixture containing several hundred thousand different fragments of DNA. One of those fragments will, hopefully, contain the DNA sequence, the gene, we are interested in. However, we cannot simply pick out the fragment we wish to purify. Instead we have to rely on bacteria to do this for us. So we need to induce the bacteria to take up the DNA fragments from the mixture we have created. We could get the bacteria to take up the DNA fragments as they stand, but these are linear DNA fragments and they will not survive long in a bacterial cell unless they are recombined into a circular molecule of DNA. So the genetic engineer's strategy is to recombine the DNA fragments produced by digestion with a restriction enzyme with a circular DNA molecule, and then get the bacteria to take up this **recombinant DNA molecule**. These 'molecular lorries', which will carry the DNA fragments into the bacterial cell, are called **vectors**.

A vector is a DNA molecule, derived from a plasmid or a bacteriophage, into which fragments of DNA may be inserted. Vectors have the following characteristics in common.

- They are small molecules with a known structure.
- They all contain an origin of replication so that the plasmid can replicate itself and the inserted fragment inside the host cell.
- They may contain genetic markers, often antibiotic-resistance genes, which confer some well-defined phenotype on the host organism which can be selected for.

Charon is the name of the ferryman who, in Greek mythology, carried the spirits of the dead across the River Styx.

Two common vectors are a bacterial plasmid, pBR322 (Fig 11.11), which was tailor-made specifically for use in genetic engineering, and λ phage. Geneticists have engineered this phage to produce specialised vectors called λ Charon phages.

QUESTION

11.5 What phenotypic characteristics would pBR322 confer on a bacterial cell which acquired it?

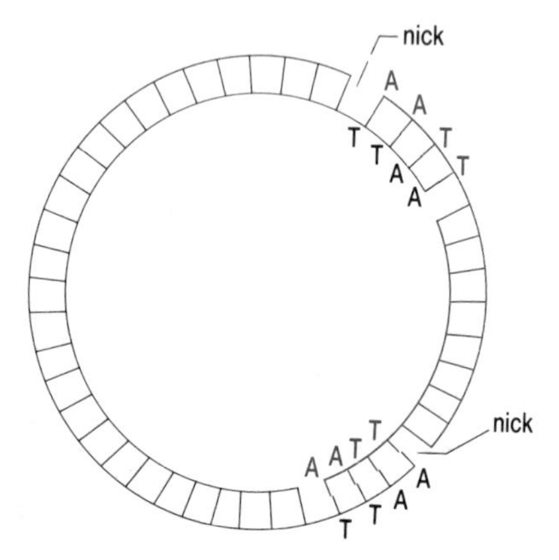

Fig. 11.12 A recombinant DNA molecule.

A recombinant DNA molecule which contains sequences from more than one organism is called a chimeric molecule. It is named after the Chimera, a mythical beast, which had the body of a goat, the head of a lion and the tail of a serpent.

DNA ligase

A vector molecule and DNA from a target organism are cut with the same restriction enzyme, say *Eco* RI. The sticky ends produced on the vector and the restriction fragments will be complementary, as shown in Fig 11.12. Such molecules can therefore pair up, a process called annealing, to produce a recombinant DNA molecule. Look carefully at the recombinant DNA molecule in Fig 11.12. Can you see that, whilst the hydrogen bonds between the bases on the complementary DNA strands have re-formed, the covalent bonds which hold the sugar-phosphate backbone together have not? These breaks are labelled as 'nicks' in Fig 11.12. To produce a stable molecule these nicks must be sealed, and this is the job of another viral enzyme, **DNA ligase**. This process is called **ligation**.

So using both restriction enzymes and DNA ligase genetic engineers are able to cut and then recombine *in vitro* (in a test tube) any two DNA molecules, providing they have complementary sticky ends. This procedure is the basis of genetic engineering.

A host

Vector molecules cannot replicate in a test tube. However the vector can be taken up by a cell inside which it can then replicate. *E. coli* is the most widely used host organism. Its biochemical and genetic characteristics have been studied extensively and it reproduces rapidly, doubling every half hour under favourable conditions. For some purposes, for example producing vaccines, other host systems are preferable. These include yeast and mammalian cells grown in tissue culture.

QUESTION

11.6 You have been given a set of four test tubes containing:
(a) the restriction enzyme *Bam* HI
(b) some purified mammalian DNA
(c) a large quantity of the vector pBR322 and
(d) a solution of DNA ligase.
How would you produce a chimeric DNA molecule using these ingredients? Using simple diagrams, explain what is happening at each stage of your procedure.

11.3 HOW TO ISOLATE A GENE

In order to illustrate the general principles of genetic engineering we will follow an imaginary procedure designed to isolate the gene which encodes human growth hormone (HGH). This process, called gene cloning, involves the production of many copies of an identical gene.

There are a variety of gene cloning strategies but all of them have the following four stages in common:

1. a means of producing DNA fragments which carry the HGH gene;
2. a method for inserting these DNA fragments into a suitable vector to produce recombinant DNA molecules;
3. inducing host cells to take up the recombinant DNA molecules;
4. finding the cell which contains the required DNA fragment.

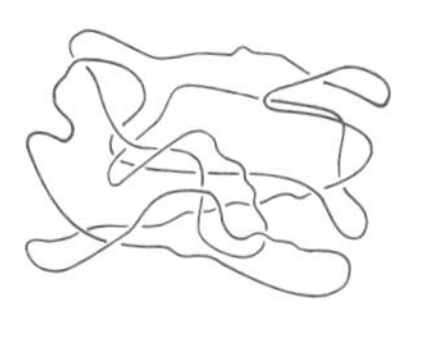

Pure sample of DNA.

Digest with restriction enzyme.

Insert restriction fragments into a suitable vector, e.g. pBR322, to produce recombinant DNA molecules.

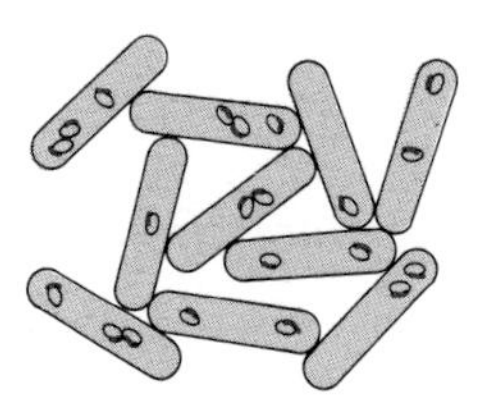

Transform bacteria with the recombinant DNA molecules. The bacteria multiply and increase the number of each recombinant plasmid.

Fig 11.13 An outline of a shotgunning experiment.

To begin with, we will look at a gene cloning procedure called shotgunning, the details of which are summarised in Fig 11.13. The starting point for our experiment is a solution of purified human DNA extracted from millions of cells, which contains many copies of the HGH gene plus all the other genes in the genome. In addition, we have available:

- a test tube containing the vector molecules, in this case pBR322;
- a solution of the restriction enzyme *Bam* HI;
- a solution of DNA ligase.

Our host system will be a strain of *E. coli* which is not resistant to either ampicillin or tetracycline. The procedure can be broken into five stages.

Step 1 – digestion

Half of the restriction enzyme solution is mixed with the human DNA. The restriction enzyme cuts the human DNA into short pieces, so producing a large number of target DNA fragments (several hundred thousand) with sticky ends (Fig 11.14).

The remainder of the restriction enzyme solution is added to the test tube containing the vector molecules. This produces several hundred thousand linear pBR322 molecules each of which has sticky ends complementary to those of the target DNA fragments.

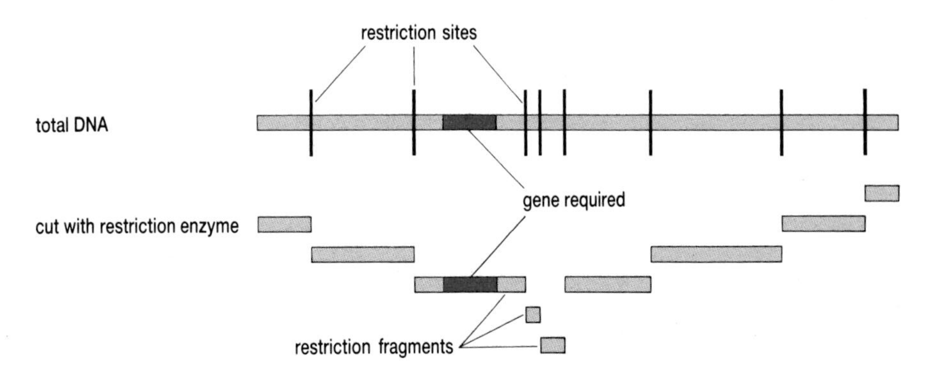

Fig 11.14 The first stage of a shotgunning experiment. Cutting the total DNA content of a cell with a restriction enzyme produces a mixture of restriction fragments, one of which contains the desired gene. In reality, the DNA from thousands of cells would be cut, so the gene required will be present many times over in the mixture of restriction fragments.

QUESTION

11.7 Produce a drawing to show a pBR322 molecule before and after it has been cut by *Bam* HI. Include in your diagram the relative position of the *amp*^{r} and *tet*^{r} genes.

Step 2 – recombination

The solutions containing the target DNA fragments and the cut pBR322 DNA are now mixed. ATP and DNA ligase are added. A whole range of reactions can take place, but by carefully controlling the conditions three main classes of products will be formed (Fig 11.15).

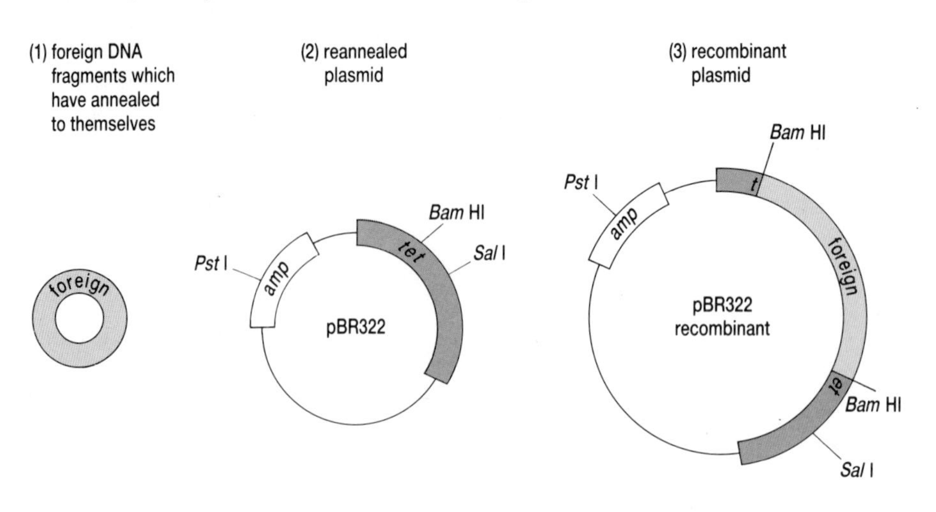

Fig 11.15 The three major products after recombination has occurred.

Step 3 – transformation

The mixture of DNA molecules produced in step 2 is now added to a flask containing the growing culture of *E. coli*. The bacteria are made competent, that is, able to take up DNA from the solution, by the addition of calcium ions (usually in the form of calcium chloride). Of the millions of bacteria in the flask, only about 1 per cent will take up DNA from the solution, and only a small proportion of these transformed cells will contain the required recombinant plasmids. The next step is to identify and isolate these cells.

Step 4 – identification

The bacteria are now plated onto nutrient agar plates which contain the antibiotic ampicillin, resistance to which is encoded by pBR322. Since the strain of *E. coli* we are using is killed by ampicillin, only those cells which have acquired the plasmid, and hence the ampicillin-resistance gene carried on it, in step 3 will be able to grow. How do we distinguish between those bacteria which have taken up a recombinant plasmid and those which contain a non-recombinant plasmid (3 and 2 in Fig 11.15 respectively)?

Look carefully at the difference between the recombinant and non-recombinant plasmids. The site into which we have inserted the target DNA fragment – the *Bam* HI site – is in the middle of the gene encoding resistance to tetracycline. This has the effect of inactivating the tetracycline-resistance gene. Consequently, bacterial cells carrying recombinant plasmids will be resistant to ampicillin but sensitive to tetracycline. By contrast, bacteria carrying non-recombinant plasmids will be resistant to both antibiotics. This phenotypic difference is the key to distinguishing between bacterial cells which are carrying the two different types of plasmid.

The bacteria growing on the ampicillin plate are now replica plated onto a tetracycline plate (Fig 11.16). The only colonies which grow will consist of bacteria containing non-recombinant plasmids. By comparing the tetracycline replica plate with the ampicillin plate, we can pick out those colonies which contain recombinant plasmids. These can be picked off the ampicillin plate and spread onto a new nutrient agar plate which now becomes our master plate.

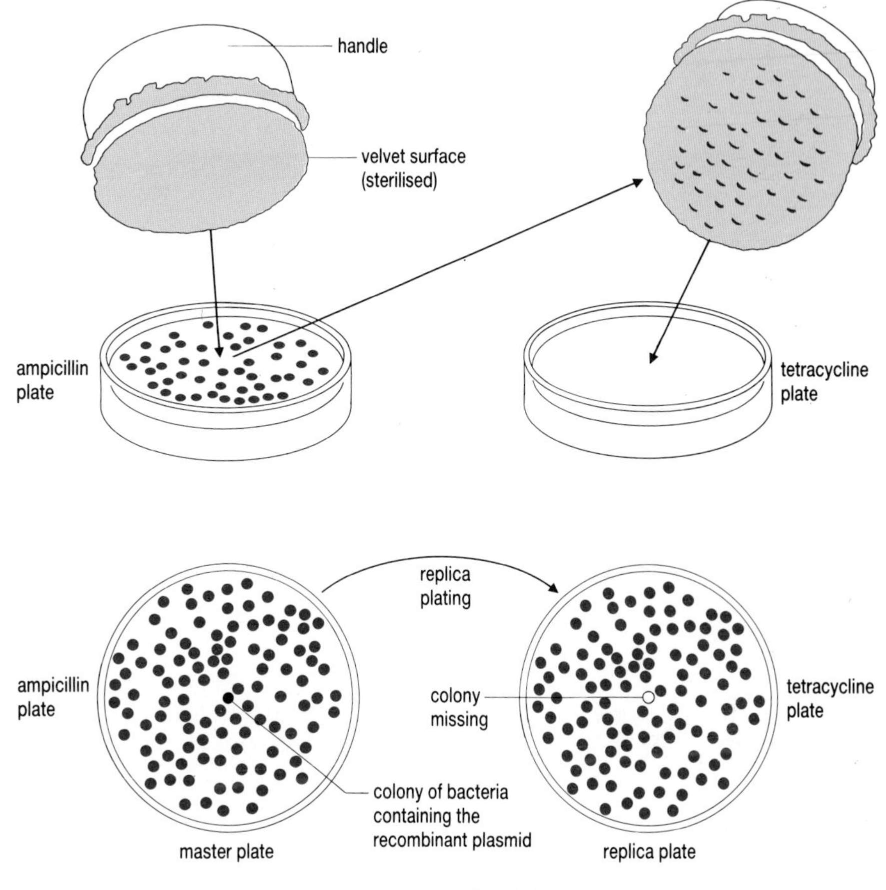

Fig 11.16 Replica plating.

At this point we have probably identified several thousand colonies consisting of bacteria carrying recombinant plasmids. How can we determine which of these colonies contain the DNA sequence which encodes the human growth hormone?

Step 5 – colony hybridisation

A nitrocellulose membrane is laid on the surface of a nutrient agar plate and the bacterial colonies on the master plate are replica plated onto the surface of this filter (Fig 11.17). The nutrients in the agar diffuse through the filter and nourish the bacteria which grow on top of the filter.

After a suitable incubation period (1–2 days) the filter, carrying the bacterial colonies, is stripped away from the agar plate and laid on blotting paper soaked in alkali (sodium hydroxide). This diffuses through the filter and bursts the bacterial cells. It also denatures the DNA they contain by breaking the hydrogen bonds between the strands and so rendering the DNA single stranded. This denatured DNA sticks to the filter.

The filter is incubated with a protein-digesting enzyme for a few minutes, washed with chloroform and saline to remove the debris of the burst bacterial cells and baked in an oven at 80 °C for a few minutes. So we now have a filter replica of the bacterial colonies growing on the master plate, in the form of filter-bound DNA. Remember, this filter-bound DNA is single stranded. This means it will hybridise (pair up) with any nucleic acid molecule which contains a complementary sequence of bases. Our next task is to produce such a molecule, called a **gene probe**. Since the sequence of amino acids in HGH is known, we can actually synthesise a short sequence of nucleotides (15 to 20 will do) arranged in an order which corresponds to the sequence of amino acids in HGH. By using nucleotides which contain ^{32}P we can make our gene probe radioactive.

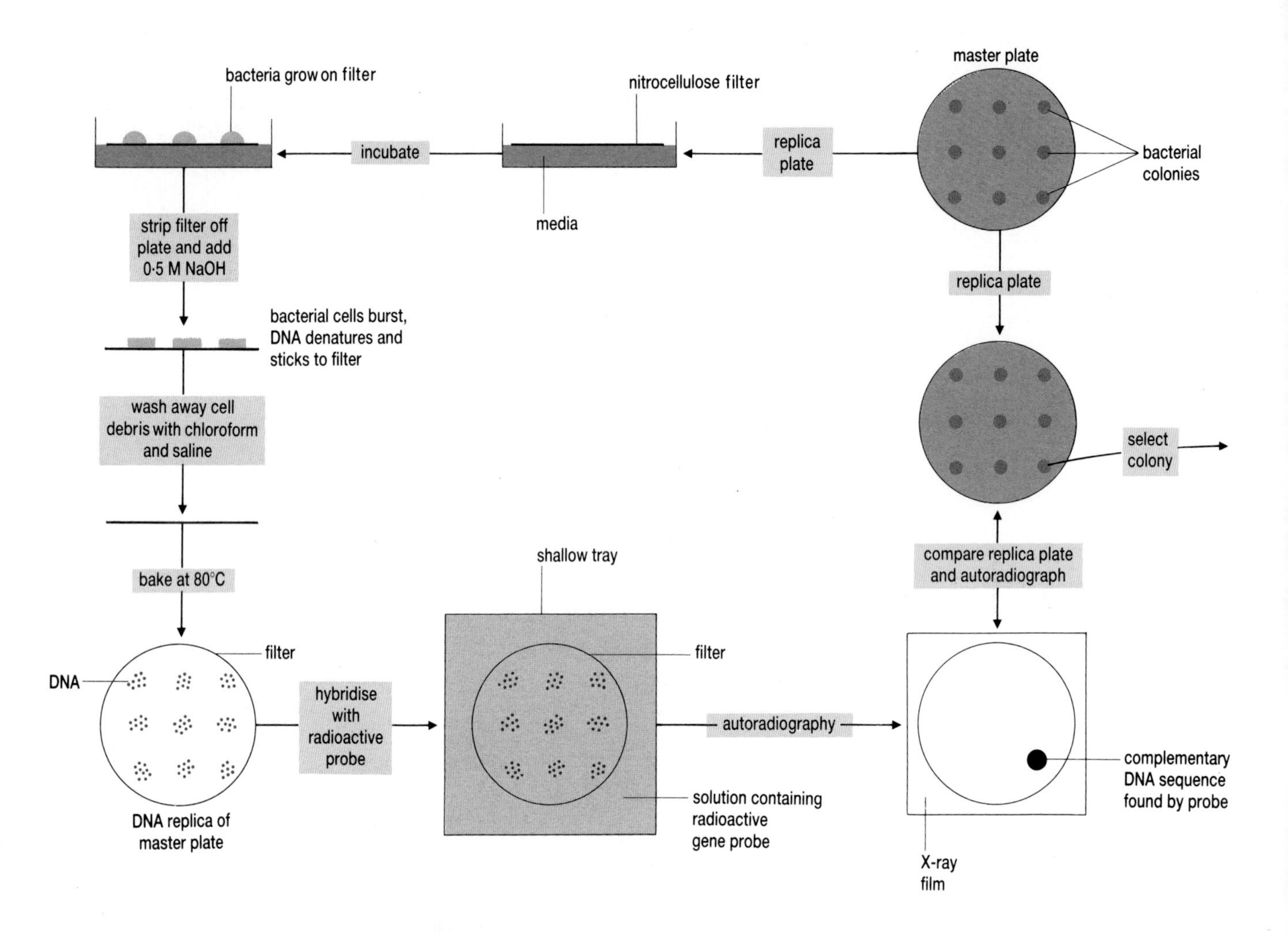

Fig 11.17 Colony hybridisation.

The filter is now washed with a solution which contains the radioactive gene probe. Wherever our gene probe meets a complementary DNA sequence on the nitrocellulose filter it will bind, and a DNA-probe hybrid will be formed. The location of these hybrids, which correspond to the position of the colonies containing the DNA sequence encoding human growth hormone on the master plate, can be identified using autoradiography.

The hybridised filter, which has been washed to remove any excess probe, is placed on an X-ray film and left in the dark for several days. The colonies containing the gene hybridised to the radioactive gene probe will blacken the film and so reveal their position. By comparing the blackened areas on the X-ray film with the original master plate, we can select those colonies which contain the DNA sequence encoding human growth hormone. We can pick those colonies off the master plate and grow up a large quantity of bacteria, each of which contains several copies of the human growth hormone. From these cultures we can, therefore, extract substantial quantities of chemically pure recombinant plasmids carrying the human growth hormone gene. The isolated and chemically purified gene can then be investigated in detail. Alternatively, we might try to get the bacteria carrying the recombinant plasmids to produce large quantities of human growth hormone. However, this is more difficult than it sounds and is discussed further in the next chapter.

QUESTION

11.8 (a) Human growth hormone contains 191 amino acids. What is the minimum number of DNA base pairs that could encode this protein?
(b) Why may the gene actually be longer than this?
(c) Why may the sequence of nucleotides in our synthetic gene probe not match exactly the sequence of nucleotides in the HGH gene trapped on the nitrocellulose filter?

11.4 VARIATIONS ON A THEME

Genetic engineering is not a science. Rather, it is a collection of techniques which allows geneticists to manipulate DNA molecules and move fragments of DNA from one organism to another. Very often it is nothing more than cookery. Just as cooks modify a recipe depending upon the circumstances, so genetic engineers vary the precise details of the techniques which they use. We cannot go into all these variations, but you do need to know about some of the more important ones.

Alternative strategies for producing target DNA fragments

Shotgunning a whole genome to isolate a single gene is a gene cloning strategy with very long odds against success. Only a few bacterial cells out of millions will contain recombinant plasmids, and only a very few of these will contain the DNA sequences we are interested in. We could shorten the odds considerably if we knew the target DNA fragments we were cloning were all the same and all contained the DNA sequences we were interested in. There are two ways of achieving this.

Make your own gene. Somatostatin, a hormone, is a small protein which contains only 14 amino acids. Since the sequence of amino acids is known we can work out the sequence of nucleotides that would direct the synthesis of somatostain. Using this information it is possible to make the gene in the laboratory. The synthetic gene can then be recombined with an appropriate vector and so introduced into a suitable host. The drawback of this technique is that most proteins are simply too big to make the synthesis of the genes which encode them possible. However, the advent of gene machines, which can be programmed with the amino acid sequence of a particular protein and then left to generate the gene automatically, will increase the usefulness of this approach.

Reverse transcriptase and complementary DNA. One group of organisms, the retroviruses, violate the central dogma of molecular biology. They are able to run the first step of the sequence, transcription, in reverse because they contain a special enzyme called reverse transcriptase. The importance of this enzyme is that it allows the genetic engineer to make DNA that complements any mRNA molecule which can be purified. (Two molecules which can join up through the process of complementary base pairing are said to complement each other.) The process is shown in Fig 11.18.

All the cells in a particular animal or plant contain the same genes, but different genes are expressed (that is, transcribed into mRNA) in different tissues (for example, the insulin gene in the pancreas, the globin gene in immature red blood cells). Such cells, whilst only containing two copies of the gene encoding the protein, will contain thousands of copies of the mRNA transcript directing protein synthesis on the ribosomes. Isolation of mRNA from such specialised protein-producing cells is relatively easy. If we now run transcription in reverse, using reverse transcriptase, we can produce DNA molecules (cDNA) complementary to the mRNA isolated from the cells.

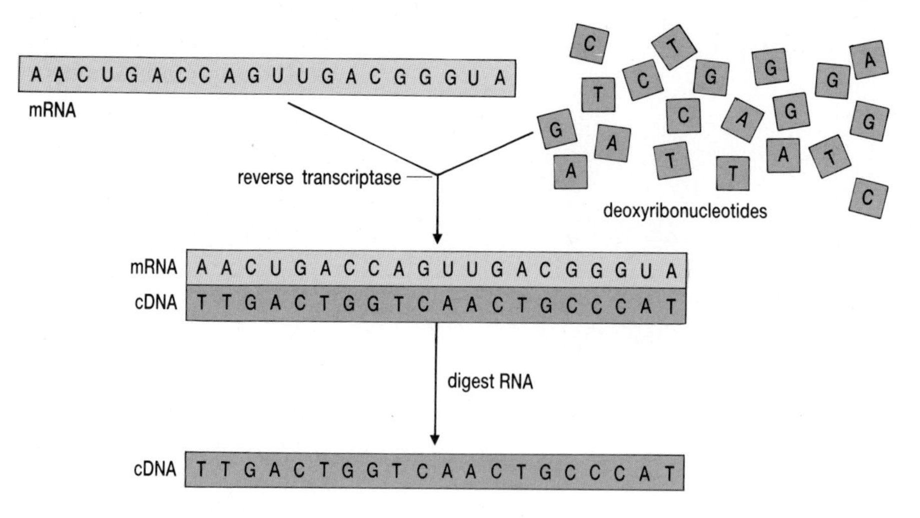

Fig 11.18 Making complementary DNA (cDNA). If the original deoxyribonucleotides contained a radioactive marker then the cDNA will also be radioactive. Such molecules are useful as gene probes.

This cDNA can be made double stranded, recombined into a suitable vector and used to transform a bacterial cell in exactly the same way as before. Alternatively, if we had made the cDNA with nucleotides containing radioactive phosphorous (^{32}P) then it can be used as a DNA probe.

Split genes. A further advantage of using the two techniques described above over shotgunning occurs when the gene we are trying to clone is a split gene. Shotgunning will put the entire gene, exons and introns, into a bacterium. If we want the bacterium to express that gene then we hit a major problem. Bacteria do not contain the necessary enzymes to tailor and splice the primary mRNA transcript to produce the final mRNA transcript which makes sense in terms of protein synthesis. By contrast, cDNA, produced using reverse transcriptase, only contains exons since it has been made using the mature mRNA which directs protein synthesis on the ribosomes. Such a gene is similar to a prokaryotic gene. A bacterium should, therefore, in principle, have no difficulty in producing proteins using such a gene.

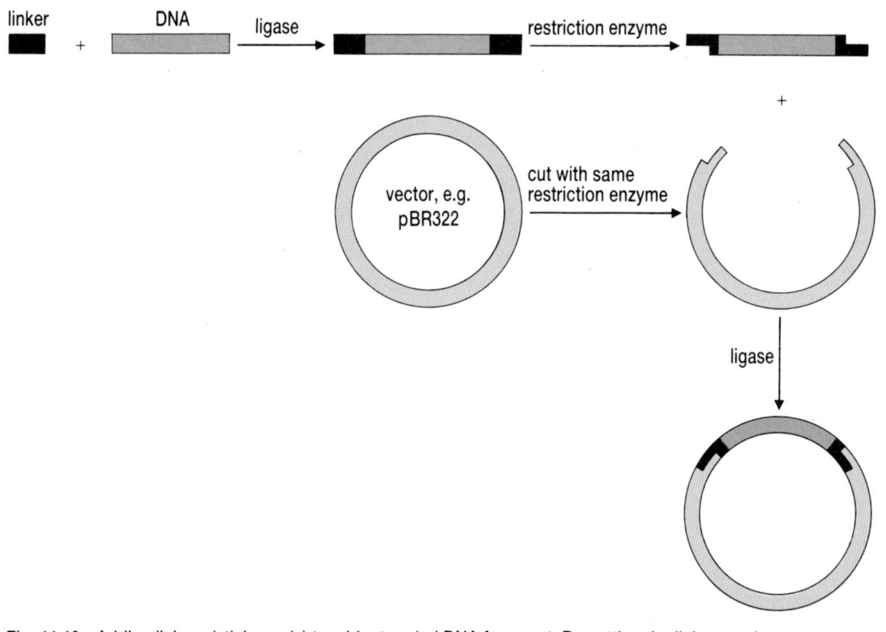

Fig 11.19 Adding linkers (sticky ends) to a blunt ended DNA fragment. By cutting the linkers and a vector molecule with the same restriction enzyme, a recombinant DNA molecule can then be made.

Adding sticky ends. One problem with the DNA molecules produced using the two techniques described in this section is that they do not have sticky ends. However, these can be added as shown in Fig 11.19.

QUESTION

11.9 Explain why using cDNA or a synthetic gene increases the probability of producing a recombinant plasmid containing the desired DNA sequence.

Using λ phage as a vector

λ phage is useful for cloning larger fragments of DNA, between 15 and 20 kb, since plasmids that contain DNA inserts as large as these tend to lose them.

λ phage heads can package DNA molecules upto 45 kb in length; this property can be used to produce chimeric DNA molecules in which the central portion of the phage DNA is replaced by a target DNA fragment, as shown in Fig 11.20. The portion of the phage genome that is replaced controls the recombination of the prophage into the bacterial DNA during lysogenisation. The portion of the phage genome left controls replication of the phage in *E. coli.* These specially engineered λ phages are called λ Charon vectors. The rapid replication of λ phage, coupled with its ability to carry large target DNA fragments, makes it very useful for creating gene libraries by shotgunning. The idea here is to clone enough DNA fragments so that the collection of recombinant phages produced by inserting target DNA fragments into λ Charon vectors should contain at least one copy of every DNA sequence in the target organism's genome.

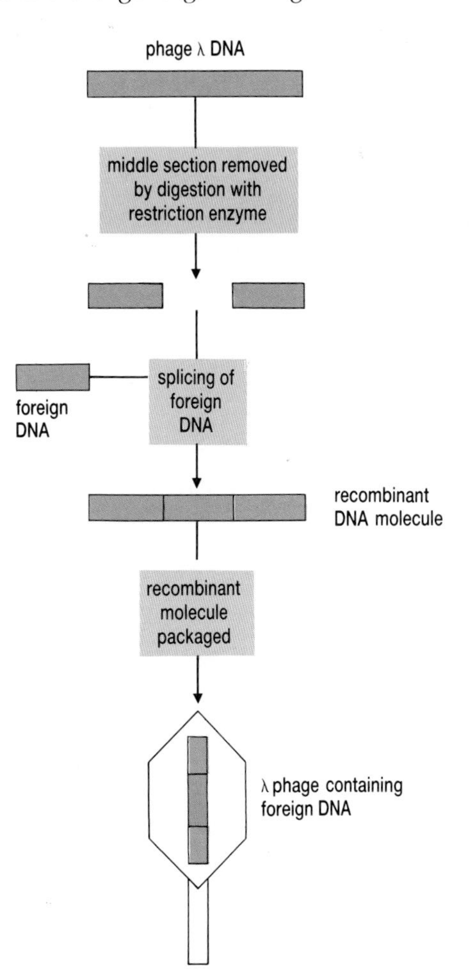

Fig 11.20 Using λ phage as a cloning vehicle.

The procedure for producing a gene library of the human genome is summarised in Fig 11.21. The test tube contains a complete library of the human genome packaged into about 1 million λ Charon phages. The problem with such a library is the small number of copies of each DNA sequence. These can be multiplied by allowing the λ Charon vectors to reproduce on a lawn of *E. coli* growing on a nutrient agar plate. Each 15 cm nutrient agar plate can accommodate about 10 000 λ Charon phages, so the total human DNA library will fit on about 100 plates. The rapid multiplication rate of phages means that after only a few hours the numbers of each phage and the DNA fragments they are carrying will have increased several hundred times. The agar can now be scraped off the plates and mixed with chloroform to burst any cells which still contain the λ Charon phage. This solution is then spun in a centrifuge to remove the agar debris and the supernatant, containing the greatly amplified human gene library can be stored in a bottle in the fridge. Such a library can then be screened, using colony hybridisation, for genes of particular interest. The only difference is that this time we will be screening plaques and not bacterial colonies for the gene we are interested in. What the geneticist can do with such isolated genes is discussed in the next chapter.

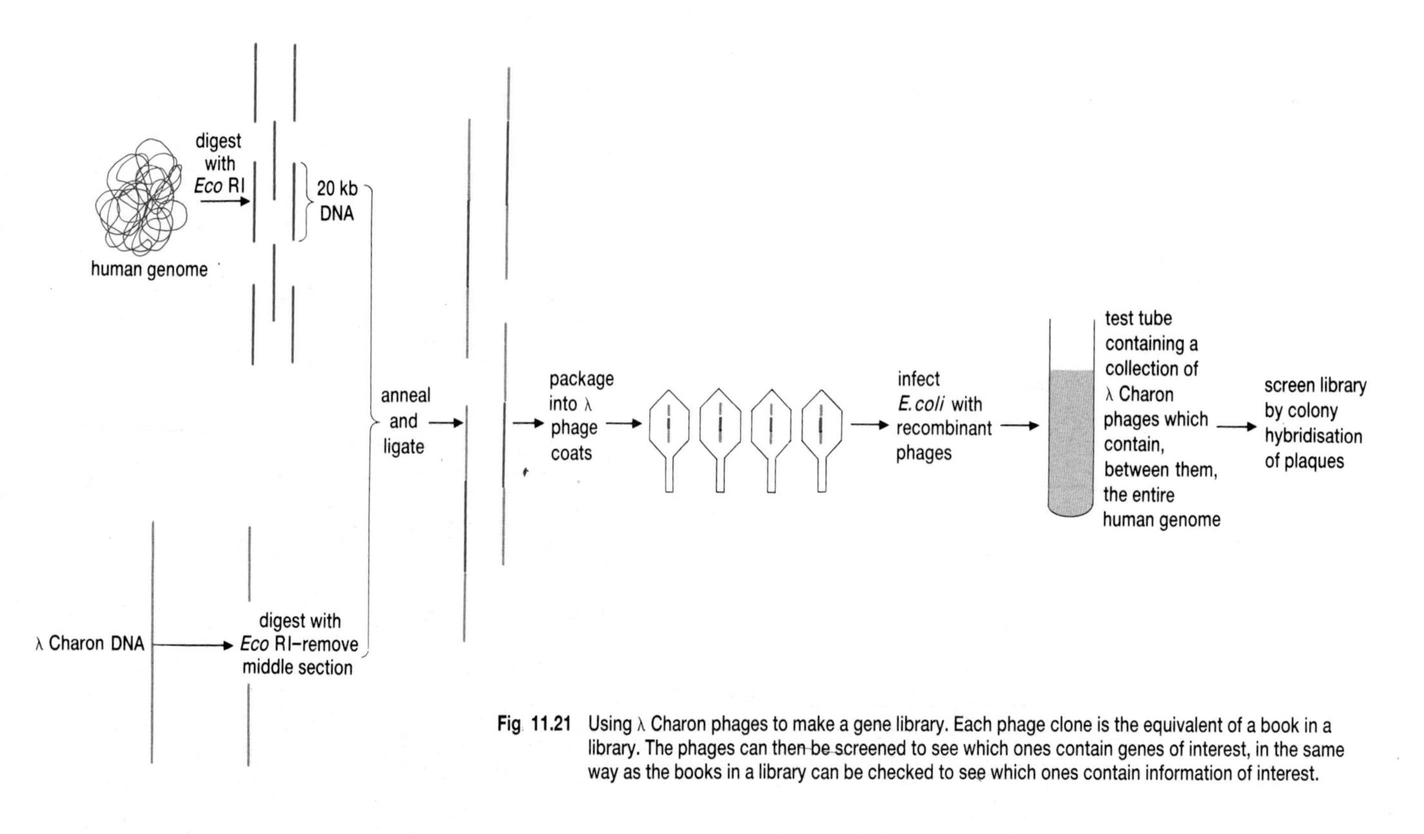

Fig. 11.21 Using λ Charon phages to make a gene library. Each phage clone is the equivalent of a book in a library. The phages can then be screened to see which ones contain genes of interest, in the same way as the books in a library can be checked to see which ones contain information of interest.

QUESTION

11.10 (a) A restriction enzyme like *Bam* HI which has a restriction site containing six bases will, on average, cut a DNA molecule every $4^6 = 4096$ bases, say approximately every 4 kb. How many restriction fragments would digesting the DNA of a human cell with *Bam* HI produce, given that each cell contains 3×10^9 base pairs of DNA? Why would the digestion of an actual sample of purified DNA yield many more DNA fragments than this?

(b) Give two reasons why producing a gene library by shotgunning the human genome into pBR322 would not be feasible.

SUMMARY ASSIGNMENT

1. What is complementary base pairing and why is it essential to the genetic engineer?
2. Produce a table to show the functions of the different enzymes used by genetic engineers.
3. What is a vector? Give two examples of vectors used by genetic engineers.
4. **(a)** Using the diagrams in this chapter, produce a flow chart to illustrate the process of gene isolation by shotgunning using pBR322, *E. coli* and colony hybridisation.

   (b) Now add to your flow chart details of the two additional methods of producing target DNA fragments, that is, gene synthesis and cDNA, described in section 11.4. Don't forget to include the addition of linkers (sticky ends).
5. What is a gene library? How are such libraries produced? Why are they useful?
6. Describe two methods of producing a gene probe.

Chapter 12

THE APPLICATIONS, LIMITATIONS AND IMPLICATIONS OF GENETIC ENGINEERING

The techniques of genetic engineering are very new and the general public's awareness of its applications are extremely hazy. Very often the terms genetic engineering and biotechnology are used as though they are the same thing. This is not the case. Genetic engineering is just one example of biotechnology. Equally important areas include the production of monoclonal antibodies and the industrial use of enzymes. You can read about these in another book in this series called *Microorganisms and Biotechnology*. Remember, genetic engineering is still first and foremost a research tool. In this chapter you will have the opportunity to explore some of the possible applications of genetic engineering. You will develop a better understanding of what the technology can and cannot do. Finally, you will be asked to consider the safety and social aspects of the process.

This chapter also has an additional aim. The field of genetic engineering is moving so fast that written material rapidly becomes out-of-date. To get up-to-date information you must read science journals and extract the relevant information from them. This chapter contains several extracts from relevant articles so you can practise this skill.

LEARNING OBJECTIVES

After completing the work in this chapter you will be able to:

1. give several examples of the applications of genetic engineering;
2. analyse the relationship between genetic engineering and its commercial applications;
3. evaluate the safety implications of recombinant DNA technology.

12.1 RECOMBINANT DNA AND THE PHARMACEUTICAL INDUSTRY

Genetic engineering is a relatively new technology and yet enormous claims are already being made about its industrial applications. Are such claims exaggerated? Do we really stand at the beginning of a biological revolution? Will drugs become cheaper and safer as a result of this new technology? Do we now have the ability to produce limitless supplies of vaccines?

Current technical limits on genetic engineering

From the picture painted so far in this book, you may think that practically any genetic manipulation is possible. Unfortunately, this is not the case as there exist a number of technical limitations to the technique. These include:

- Genetic maps – we simply do not know where most of the genes are on the human chromosomes.
- Vectors are still at an early stage of development, especially for eukaryotic genes.
- Many eukaryotic genes are poorly expressed in prokaryotic cells. More research is needed on genetic engineering using eukaryotic cells, for example yeast.
- How gene expression is controlled is still poorly understood.
- Much research is still needed to identify all the enzymatic steps involved in the synthesis of useful cellular products.
- The number of genes involved in synthesis is a major limitation. Currently, genetic engineering is most successful when only a single, easily identifiable gene is involved. It is more difficult where several genes have to be transferred.
- Many genes remain to be identified. This is the case with many traits of agricultural importance, for example plant height.
- Further research is needed on methods to identify the cells carrying the transferred gene.

In addition to these genetic problems there are also the challenges to be met in scaling up the genetic engineering process from a laboratory flask to an industrial fermenter the size of a small tower block. To keep billions and billions of microorganisms growing under ideal conditions, producing the product you want in such a fermenter poses a new set of engineering, biochemical and economic problems. Add to this the problem of separating the final product from the culture in the fermenter and you will realise that the gene cloning is the easy part of the operation. You can find out how these problems are solved in another book in this series called *Microorganisms and Biotechnology*. Finally, and perhaps most importantly, there are commercial criteria which must be met before genetic technology becomes commercially feasible. These criteria represent major hurdles that any industry must overcome before genetic engineering can play a part in producing a marketable product. They include the need for:

- a useful biochemical product;
- a useful biological fermentation approach to commercial production;
- a useful genetic approach to increase the efficiency of production.

To illustrate these challenges, we will consider how genetically engineered insulin is made.

Case study – manufacturing insulin

Diabetes is a common condition. In Britain there are 600 000 diabetics, many of whom need insulin each day. Worldwide the number of insulin users is in excess of 2 million. Consequently the market for insulin is enormous. In the United States of America alone it is worth more than $200 million per year, and in 1981 the world market was worth $400 million. Insulin therefore has a **mass market** which is both **established** and **growing** as the number of diagnosed cases of insulin-dependent diabetes rises each year.

Traditionally, diabetes has been treated with **bovine** or **porcine insulin**. This is extracted from the pancreases recovered from dead animals from slaughterhouses. Unfortunately, animal insulin is not identical to human insulin. As a result some diabetics produce antibodies against the insulin they need to keep them alive. The effect of this is to destroy the insulin before the body can use it. This problem would be overcome by using human insulin produced by genetically engineered bacteria which contain the human insulin gene. So this would seem to be a product ideally suited to production using a genetically engineered cell.

The problem. In principle, transferring genes from eukaryotes to prokaryotes seems straightforward. Why can't we just shotgun the human genome into a suitable vector and so isolate the insulin gene? The human insulin gene could then be transferred to a bacterium which then grows and divides rapidly, so producing a large number of bacteria containing the gene. Since the genetic code is to all intents and purposes universal, the bacterial cells should be able to transcribe the information contained in the insulin gene into mRNA, which should then be translated to make insulin.

Unfortunately, things are not quite as simple as that. To understand the problems faced by the genetic engineers we need to look at how insulin is made in a pancreatic cell.

How a eukaryotic cell makes insulin. The actual process by which insulin is made in your pancreas is shown in Fig 12.1.

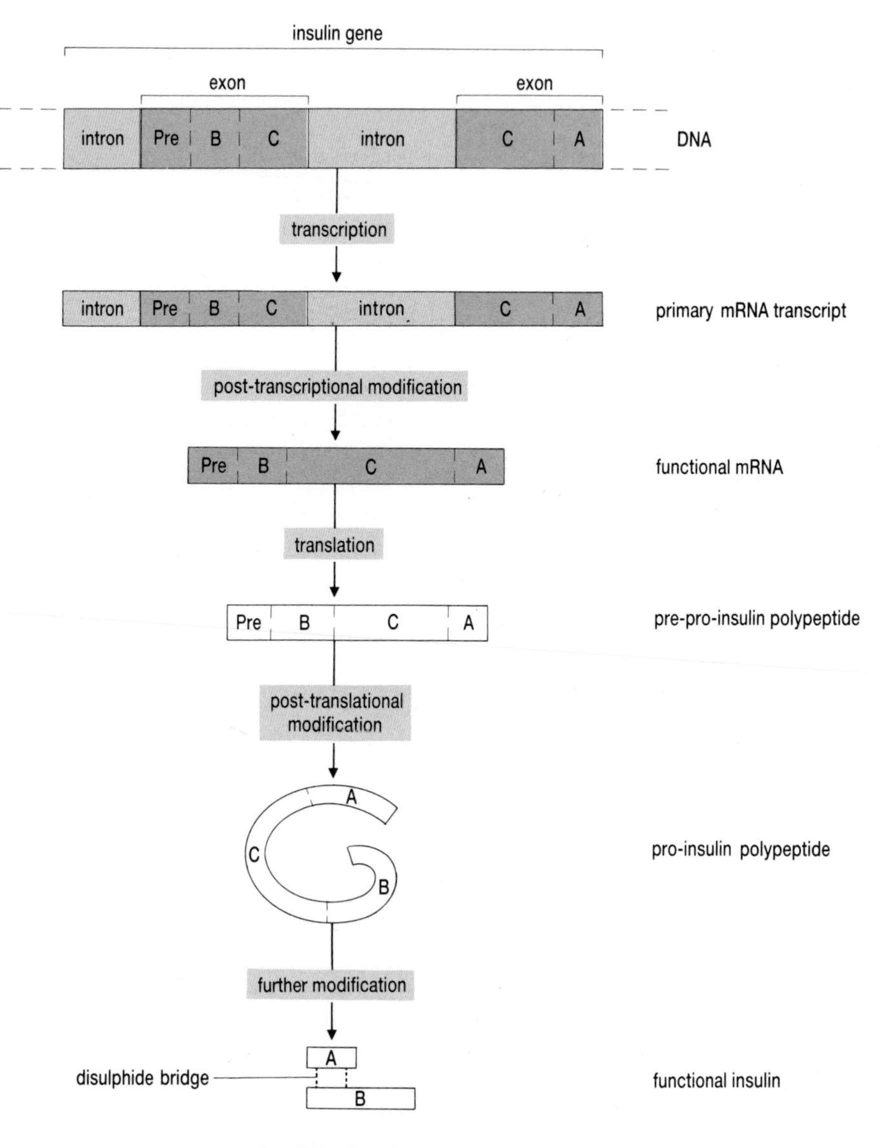

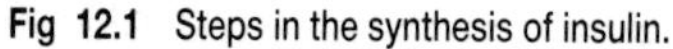

Fig 12.1 Steps in the synthesis of insulin.

QUESTION

12.1 Why could bacteria not make insulin using the entire human insulin gene?

Clearly, then, shotgunning is not going to work. One possible way around this problem would be to make a cDNA copy of the functional mRNA that codes for insulin. This cDNA could then be introduced into a plasmid and so into a bacterium. This gene would then be transcribed and translated into functional insulin.

QUESTION

12.2 Why will this approach not work?

An early solution. This approach was devised by scientists working for the American biotechnology company Genentech in the late 1970s. The answer they came up with was to make the DNA coding for each of the two polypeptides in the functional insulin molecule. These synthetic genes would then be inserted into different bacteria and the two products produced in separate fermenters. The A chain and the B chain would then be chemically united after they had been extracted from the bacteria making them. It took three months to synthesise the gene encoding the A chain and two months to make the gene encoding the B chain.

QUESTION

12.3 What information would be needed to make the two genes?

One final problem remained. How do we turn on the insulin gene after we have inserted it into a bacterium? Remember that in bacteria the genes are arranged in operons with a promoter region which controls the transcription of the genes (Fig 11.7). The promoter is in effect a molecular switch. Under the correct conditions it will allow RNA polymerase to bind and so transcribe the genes in the operon. If we are going to get the bacteria to produce insulin then we will have to tie the insulin gene to a bacterial promoter. If we switch on the promoter then the bacteria should transcribe the insulin gene.

One of the best known operons in *E. coli* is the *lac* operon. This gene encodes an enzyme, beta-galactosidase, involved in the metabolism of the sugar lactose. If *E. coli* are grown on a medium containing lactose then the *lac* operon is switched on and beta-galactosidase is produced. So if we tie our synthetic insulin genes to the *lac* operon, insert this composite gene into a vector and then into *E. coli*, we should be able to get the bacteria to make insulin. The process is shown in Fig 12.3, and it worked.

QUESTION

12.4 Under what conditions will we have to grow the *E. coli* to make them produce insulin?

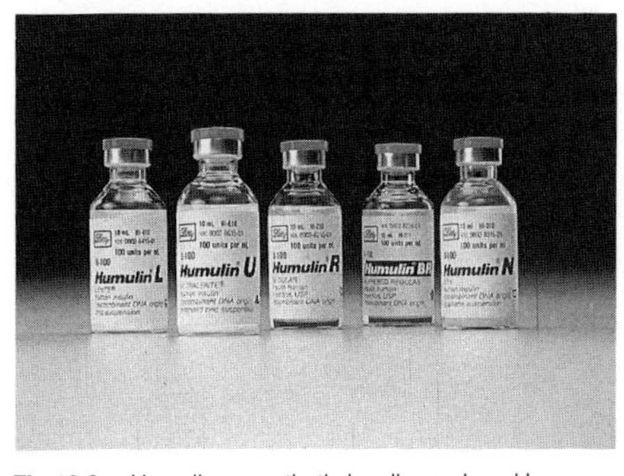

Fig 12.2 Humulin, a synthetic insulin produced by genetic engineering.

The process was not very efficient and synthetic insulin is now made in a different way. A cDNA copy of the functional mRNA, in effect a pre-pro-insulin gene, is used. The bacteria produce pro-insulin which is then extracted. Enzymes are then used to cut out the C region of this molecule and combine the A and B chains in the correct configuration. By 1981 this synthetic insulin (Fig 12.2) was undergoing clinical trials, and in 1983 a licence was granted to the American drug company Eli Lilly to produce genetically engineered insulin for human use.

Clearly this was a major breakthrough for the young genetic engineering industry. Unfortunately for Eli Lilly, in 1980 a rival drug company, Novo Industry of Denmark, developed a process which converts pig insulin into human insulin. This process used existing chemical technology and plant which Novo Industry already possessed. The product is just as good but cheaper than its genetically engineered competitor.

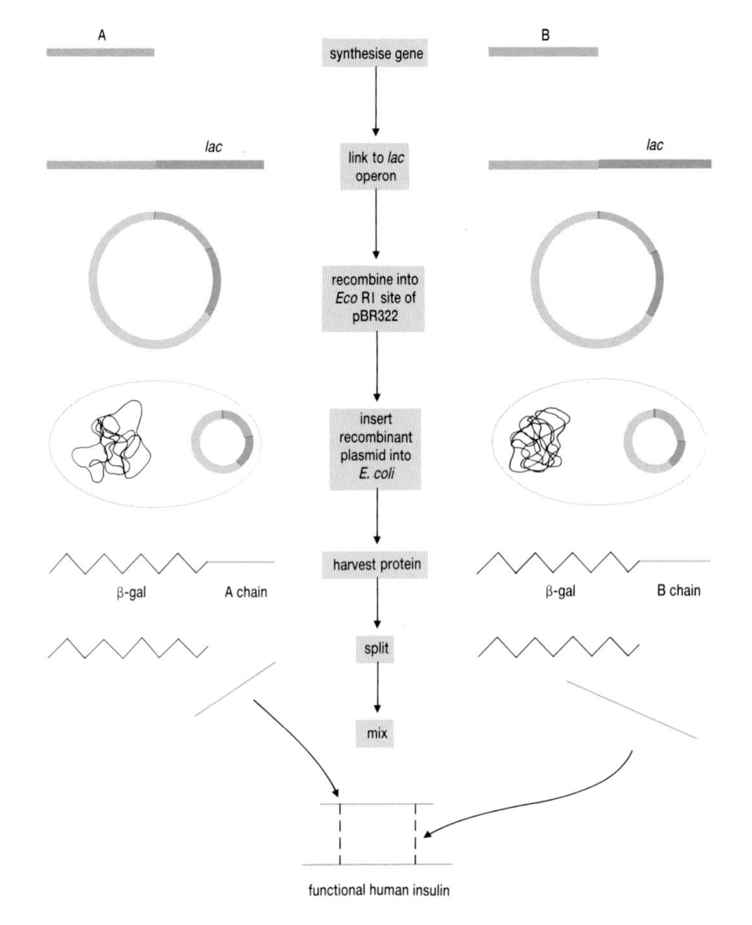

Fig 12.3 The method used by Genentech to produce synthetic human insulin.

Clearly, then, just because a particular process is technically feasible and socially desirable will not make it a commercial success. Indeed, one British newspaper, *The Independent*, concluded in August 1988 that only one genetically engineered pharmaceutical product, TPA (tissue plasminogen activator – an anticoagulant) was actually making a profit. This drug, which is used to treat people who have suffered heart attacks, was itself in danger of being dropped for an alternative therapy based on two other, cheaper, drugs, streptokinase and aspirin.

QUESTIONS

12.5 Using a library, find four more examples of the industrial application of genetic engineering. For each example, explain:
(a) how the gene was isolated;
(b) the cloning procedure used;
(c) the commercial feasibility of the project.

12.6 Malaria kills about 1 million people per year, mainly in poor tropical countries. How would you convince a pharmaceutical company to fund your research into developing a genetically engineered vaccine against the disease? Bear in mind the commercial constraints listed earlier.

A cautionary note

You have convinced the pharmaceutical company to fund your research. You now have a grant of several million pounds and you probably think that your research will save thousands of lives each year. Think again. To distribute your vaccine, poor countries will have to invest millions of pounds in building new roads, improving their refrigeration technology (vaccines must be kept cold) and training staff to inject your wonder drug. In truth, that money would probably be better spent on providing better sanitation, clean water and more food.

Genetic engineering is very glamorous and a lot of scientists want to be involved in it. It is an important research tool and it may, eventually, become a commercial way of producing new chemicals. It will, however, not solve the problems of the Third World as it is a Western technology. But scientists are only human. Putting in sewerage systems in a poor tropical country is not as newsworthy as developing a new wonder cure!

12.2 GENETIC ENGINEERING IN PLANTS

So far we have only considered genetic engineering using prokaryotic cells, for example bacteria. However, genetic engineering techniques which use eukaryotic cells are becoming increasingly common. For example, genetically engineered factor VIII, needed by haemophiliacs, will be produced using mammalian cells in tissue culture.

The first clinical trial in Britain using genetically engineered factor VIII commenced at the Royal Free Hospital in London on October 11 1988.

The best examples of gene cloning in eukaryotes to date come from the plant world. A basic technique is shown in Fig 12.4. An example of such a successful gene transfer using this technique involves BASTA resistance. BASTA is the active ingredient in the herbicide phosphotricin. It acts by

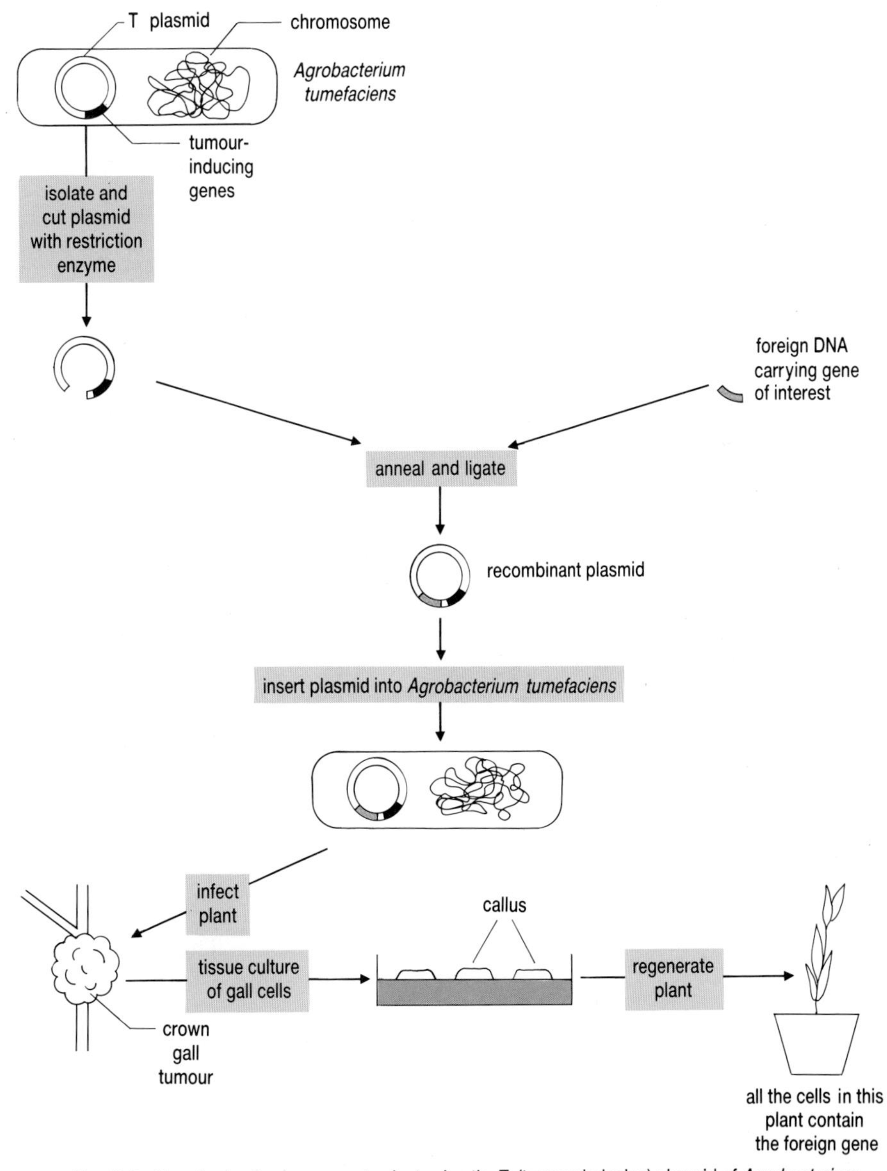

Fig 12.4 Transferring foreign genes to plant using the T_i (tumour-inducing) plasmid of *Agrobacterium tumefaciens.* This plasmid produces crown gall tumours in infected plants.

blocking the action of an enzyme called glutamine synthetase which is essential to the plant's physiology. However, some bacteria belonging to a group called the streptomycetes are resistant to the effects of phosphotricin because they possess an enzyme which deactivates BASTA. So somewhere in the genome of this bacterium there must be a gene which codes for the production of this BASTA deactivating enzyme. This gene was isolated and inserted into tomato, tobacco and potato plants using the technique shown in Fig 12.4. The result was plants resistant to BASTA. These crops could then be sprayed with phosphotricin which will kill the competing weeds but would have no effect on the BASTA-resistant plants. One slight problem is that the tomatoes apparently taste disgusting.

Again, much research is going on into suitable vectors for transferring genes into plants. In addition to the T plasmid of *Agrobacterium tumefaciens,* the cauliflower mosaic virus (CMV) is widely used. Other current genetic engineering ideas in plants include transferring the genes which control nitrogen fixation in *Nitrobacter* to cereal crops and moving pest resistance genes from one plant to another. Again, if you look in a few science magazines you are bound to find lots of other examples.

QUESTION

12.7 Why would genetic engineers want to transfer genes for nitrogen fixation and pest resistance into plants? What economic and environmental advantages would such genetically engineered plants offer?

12.3 THE USE OF RECOMBINANT DNA IN MEDICINE

Two examples are discussed.

The detection of disease-causing organisms

Infectious disease caused by viruses (for example, AIDS) and bacteria (for example, Legionnaire's disease) can be diagnosed rapidly using recombinant DNA techniques. A radioactive probe, specific to the infectious agent, can hybridise with foreign nucleic acids present in infected tissue or body fluids. The presence of the hybridised probe can then be detected using autoradiography. Such diagnoses are very sensitive and can be completed within ten to twelve hours.

The diagnosis of inherited disease

Recombinant DNA technology, combined with classical pedigree analysis (Chapter 15), is proving to be very useful in the early diagnosis of genetic diseases such as Duchenne muscular dystrophy, Huntington's chorea and cystic fibrosis, and in detecting those individuals who are carriers (individuals who are carrying one copy of a deleterious recessive allele). To illustrate the principles involved, we will examine how recombinant DNA can be used to diagnose sickle-cell anaemia. The procedure may seem complicated, but to understand it you only need to remember two facts, one from the last chapter the other from classical genetics.

1. Restriction enzymes always cut DNA at a specific sequence of nucleotides. This sequence is often quite rare in any given chromosome.

2. Genes (that is, stretches of DNA) on a chromosome are linked to one another during inheritance.

If you are not familiar with this second idea, consider the following, rather absurd, example. Imagine that being a great geneticist and having green eyes are controlled by two genes which lie next to each other on a chromosome. Because they are so close together the chance of them being separated by crossing over during meiosis is very small. This means that every person with green eyes will be a great geneticist. We could look at a baby and, just by noting the colour of their eyes, we could predict whether they are going to be a great geneticist. In other words, the genes are linked. By looking at one gene we can predict the presence of the other.

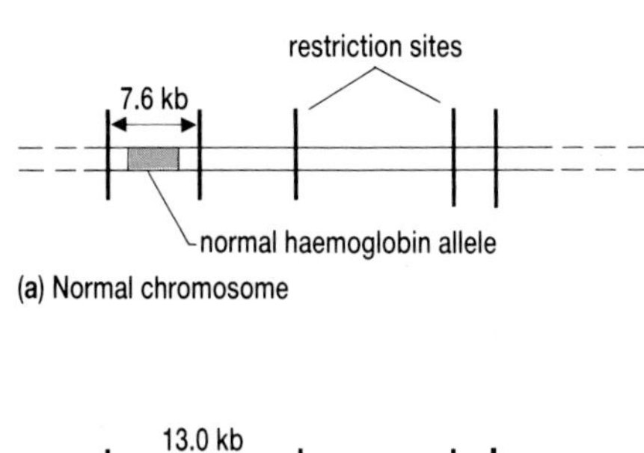

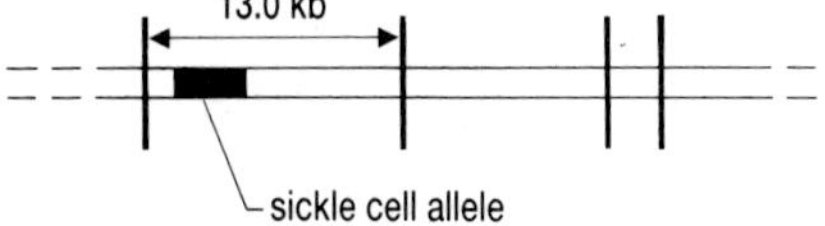

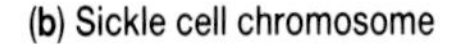

Fig 12.5 Restriction fragment length polymorphism (RFLP) of human chromosomes which contain (a) the normal haemoglobin allele; (b) the sickle-cell allele.

Restriction fragment length polymorphism. Now back to diagnosing sickle-cell anaemia. Look at Fig 12.5(a). This shows the restriction sites for the enzyme *Hpa* I on a chromosome which contains the normal haemoglobin allele. *Hpa* I will cut the DNA at these points, producing pieces of DNA with a characteristic size. In particular, notice that the normal haemoglobin allele is contained in a piece of DNA which is 7.6 kb long.

Now look at Fig 12.5(b). This shows the restriction sites for a chromosome which contains the abnormal sickle-cell allele. Can you see the difference in the pattern of restriction sites? For some unknown reason virtually every chromosome that bears the sickle-cell allele also has another rare abnormality at the *Hpa* I recognition site, just to the right of the sickle haemoglobin allele. In other words, the sickle allele and this abnormality which produces a change in the pattern of *Hpa* I restriction sites are linked. If we cut the chromosome with *Hpa* I, we end up with the sickle-cell allele in a fragment which is 13.0 kb long rather than 7.6 kb long. This variation in the size of restriction fragments is called restriction fragment length polymorphism (RFLP). Analysis of the lengths of the fragments when DNA is digested with *Hpa* I forms the basis of the system for detecting the sickle-cell allele.

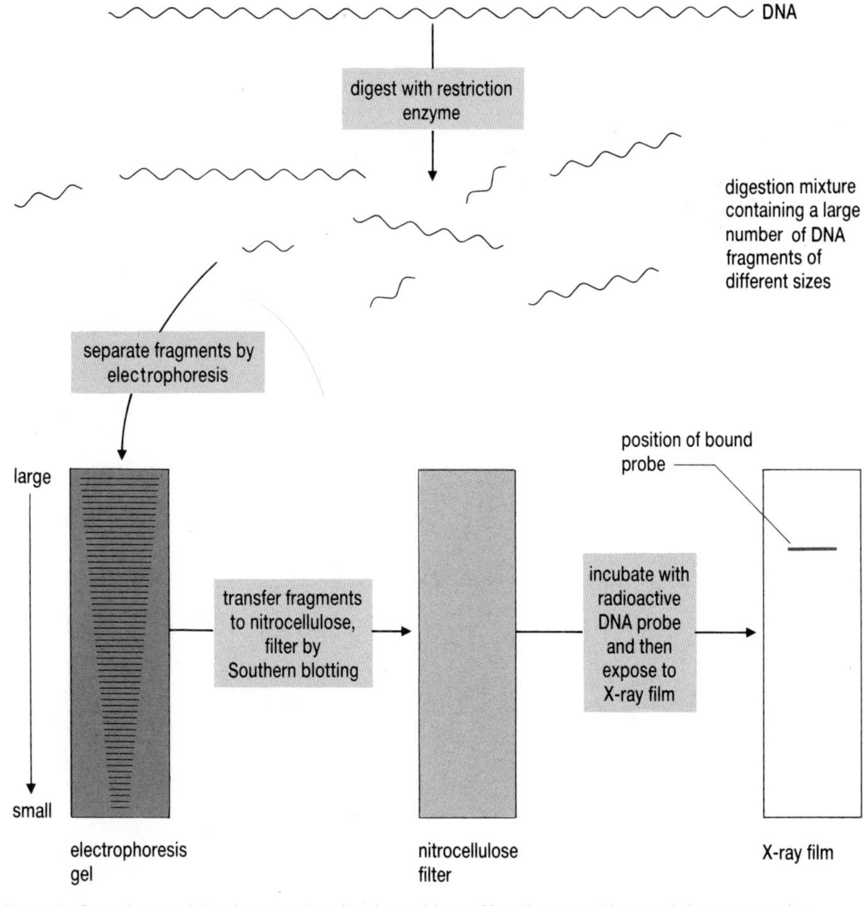

Fig 12.6 Detecting restriction fragment length polymorphisms. After digestion with a restriction enzyme, the fragments are separated, on the basis of size, by agarose gel electrophoresis. The fragments are transferred to a nitrocellulose filter by Southern blotting (see Fig 12.7). Particular restriction fragments are then located by hybridisation with a radioactive gene probe followed by autoradiography.

The technique. This is shown in Fig 12.6. Purified DNA, obtained from either fetal cells (either by amniocentesis or chorionic villus sampling – see Chapter 15) or the blood cells of a suspected carrier, is digested with a restriction enzyme, in this case *Hpa* I. The resulting mixture of restriction fragments is then separated by electrophoresis. Notice that the fragments are separated on the basis of size not charge; small fragments move further than large fragments. The separated fragments are now transferred to a nitrocellulose filter by Southern blotting (Fig 12.7).

Notice that although the DNA fragments are drawn in Fig 12.6, they are actually not visible unless they are labelled in some way. This is the next job. A radioactive gene probe with a nucleotide sequence complementary to the β-globin gene is synthesised. The nitrocellulose filter is washed with a solution containing the gene probe.

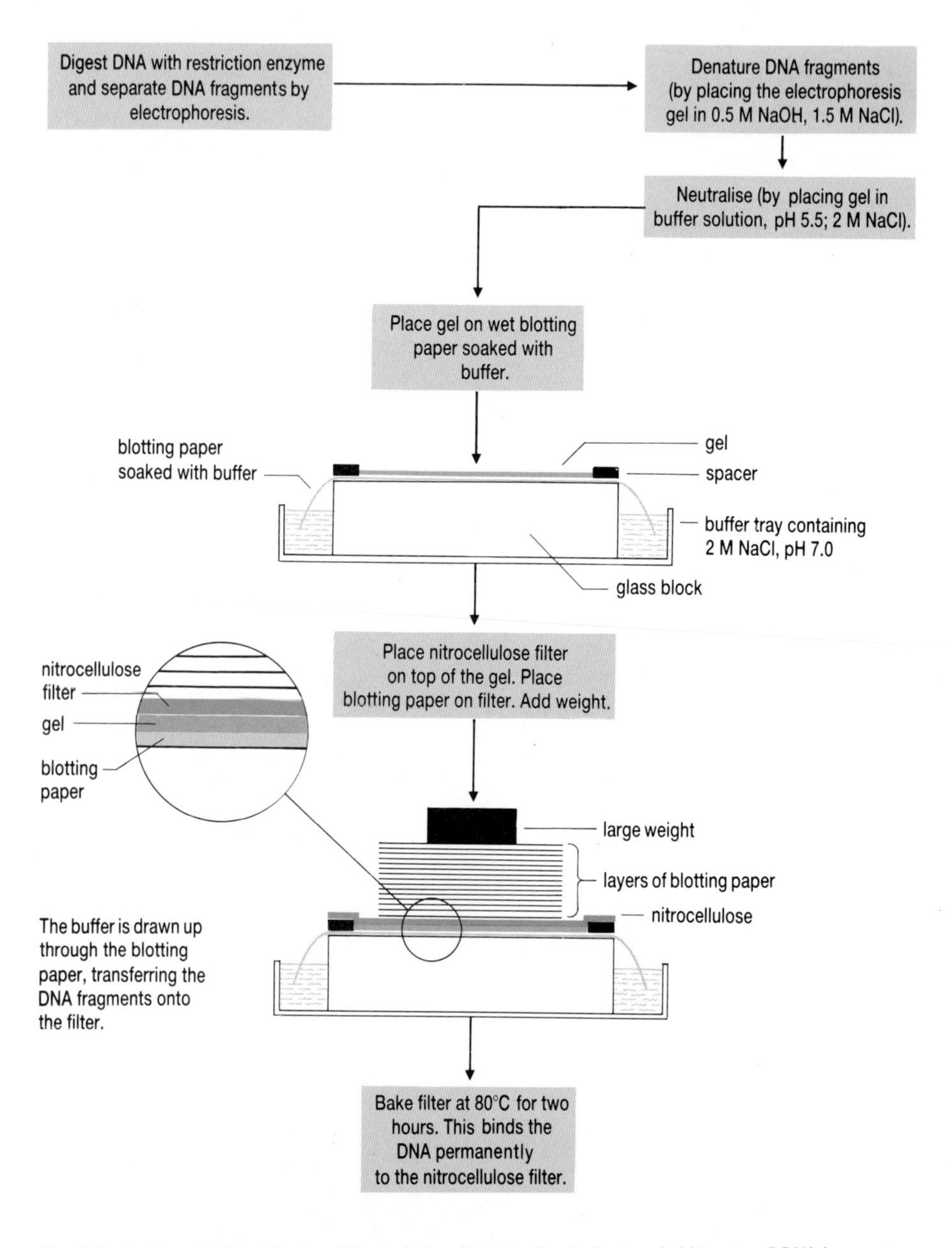

Fig 12.7 Southern blotting. The aim of this technique is to transfer single-stranded (denatured) DNA fragments from the agarose gel used for electrophoresis to a nitrocellulose filter. The technique is named after its inventor, Dr E M Southern.

QUESTIONS

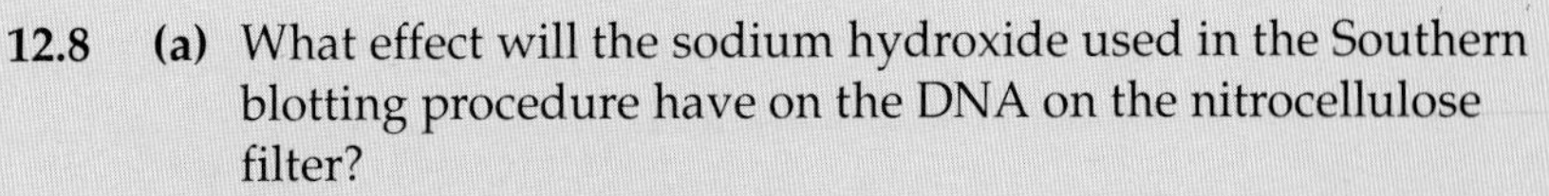

12.8 **(a)** What effect will the sodium hydroxide used in the Southern blotting procedure have on the DNA on the nitrocellulose filter?

(b) What will happen to the gene probe when it is washed on to the nitrocellulose filter?

(c) How would you detect where it binds on the nitrocellulose filter?

12.9 You are now going to use the technique outlined above to predict the genotypes of three people. Remember, the normal haemoglobin allele is on an *Hpa* I restriction fragment which is 7.6 kb long whilst the sickle allele will be on a restriction fragment which is 13 kb long. This means that a restriction fragment which contains the normal allele will move further up the electrophoresis gel than the one which contains the sickle-cell allele. The DNA probe will anneal to both the normal allele and the sickle-cell allele, since there is only a very small difference between the two alleles. So we can get three patterns on our electrophoresis gels which correspond to homozygous normal, heterozygous and homozygous sickle. These are shown in Fig 12.8. Your job is to work out which gel corresponds to which genotype. Explain how you reach your conclusion.

direction of electrophoresis

X-ray film

(a) (b) (c)

Fig 12.8 Autoradiographs from three patients using the technique shown in Fig 12.6.

12.10 Analysis of RFLP is extremely useful in detecting individuals who are carrying one copy of a recessive allele. Can you suggest why?

12.11 Huntington's chorea is caused by a dominant allele, the phenotypic effects of which often do not become apparent until an individual is in their 40s or 50s. Why is it important for an individual who has a history of Huntington's disease in their family to be screened for the allele before they have children?

Towards gene therapy

In addition to providing an important diagnostic tool, RFLP and other recombinant DNA techniques will also allow geneticists to actually find where the defective alleles are in the human genome and what is wrong with them. This raises the interesting possibility that we may then be able to use recombinant DNA techniques to actually correct the inherited metabolic disorders.

The two articles below summarise the state of the art in terms of gene therapy in 1988. This is such a fast moving area of science that by the time you come to read this book the situation may have altered quite radically. You should try to find out if anything new has happened before you answer the next set of questions.

QUESTIONS

12.12 Using your knowledge of retroviruses and the information contained in the articles, explain how gene therapy might work.

12.13 Why is 'somatic gene therapy' more likely to be successful than 'germline therapy' ?

12.14 Discuss some of the medical and moral problems that could be associated with gene therapy.

Europe's researchers design code for gene therapy

The treatment of some hereditary diseases by gene therapy is likely to be "clinically justified" in the near future, according to the medical research councils of Europe. In a joint statement published in *The Lancet*, the councils give guidelines for research into the genetic manipulation of humans.

The research councils approve the notion of inserting genes into certain body cells, such as bone marrow cells, in an attempt to cure genetic diseases such as thalassemia. People with this severe anaemia have inherited defective genes from both parents. Such a treatment — yet to be successfully attempted in humans — is known as "somatic gene therapy". Genetic changes in the cells of a particular tissue would not be passed on to children. The guidelines say that "germline therapy" — the insertion of foreign genes into a fertilised egg to create heritable changes — is "not acceptable".

Few researchers would disagree, if only because germline gene therapy has little chance of success. Experiments with animal embryos, where researchers inject foreign genes to create "transgenic" animals, usually fail. The embryos die or do not take up the foreign genes.

An easier approach in humans would be to screen embryos for a defective gene, then doctors could transfer only normal embryos into a woman's uterus. Embryologists and molecular geneticists are now joining forces to perfect such "pre-implantation diagnosis".

The European research councils also stress the issue of the safety of gene therapy. Most researchers use modified retroviruses to carry foreign genes into cells, but this approach has problems. As it integrates into the genetic material of human cells, the virus can cause mutations which might lead to cancer.

There is also a small chance that the retrovirus carrier, or viral vector, might combine with a contaminating virus to spread to other tissues or even to other people. "Much further work is required in the development of safe species-specific and tissue-specific retrovirus vectors," the guidelines say.

To regulate such research, say the councils, "an expert national body" and local ethical committees should approve proposals for therapy. A "central body" should also assess the outcome of early trials of human gene therapy.

From New Scientist June 9 1988

The Route to Gene Therapy

Viruses are ideally suited to deliver genes into cells, because they have evolved to deliver their own genes into cells during infection. Modified viruses are among the most effective "vector" systems in routine use. Most serve simply to manufacture proteins. But many researchers have taken on the bolder task of trying to "design" viruses that will give human cells the genes they need to compensate for faulty genes in serious genetic diseases, such as thalassaemia - a sometimes fatal form of anaemia - or rare disorders of the immune system.

Research towards gene therapy focuses largely on the so-called somatic cell approach. Somatic cells are all the cells apart from the reproductive "germ" cells. A more adventurous idea is to manipulate the genes within the germ cells - those that give rise to sperm and egg. This has the potential advantage of correcting a genetic disease forever, because the correction could be passed on to the patient's children. The difficulty of this approach is that the genetic tinkering must be done in a very young embryo so that every cell in the embryo contains a copy of the new gene.

The troublesome class of viruses known as retroviruses, responsible for AIDS and some cancers, is particularly suited to the task of ferrying genes into cells. As part of their natural life cycle, retroviruses insert a DNA copy of their genetic material into the chromosomes of an infected cell. So if a researcher wants to make some new gene a permanent part of a cell's DNA, the first step is to insert the gene into the genetic material of an appropriate retrovirus, and then let the retrovirus insert the gene into the chromosomes of the cell.

Richard Mulligan and his colleagues at the Massachusetts Institute of Technology have taken this approach in trying to tackle thalassaemia. People with thalassaemia lack a proper gene coding for one of the globin proteins that forms haemoglobin, which carries oxygen around the body. Mulligan's team inserted a human globin gene into the genetic material of a retrovirus and then allowed the virus to infect some bone marrow cells isolated from a mouse. After a while they returned the infected cells to the mouse to resume the task of making blood cells. The new red blood cells contained significant amounts of the human globin protein. Mulligan's results show that retroviruses can do the job of delivering a gene into the genetic material of blood-forming cells, and that the new gene then works properly, generating a good supply of the protein it codes for. Eventually, Mulligan's retroviruses could carry good globin genes into the bone marrow cells of people with thalassaemia. Mulligan and other researchers are also trying to develop similar cures for other genetic diseases. It will probably be years before retroviruses are saving lives rather than taking them, but the research is an inspiring "turning of the tables" for virologists to aim at.

From New Scientist May 19, 1988

12.4 GENETIC FINGERPRINTS

This application was developed by Dr Alec Jeffreys of the University of Leicester, in England. Whilst working on the globin gene, he discovered several regions of non-coding DNA which had a very odd structure. They consisted of short DNA sequences which were repeated many times. Jeffreys was able to isolate these satellite DNA sequences and use them to develop a probe to look for similar sequences elsewhere in the human genome. Again, human DNA is digested with a restriction enzyme and the fragments are separated on the basis of size by electrophoresis. The fragments are transferred to nitrocellulose by Southern blotting and the filters are washed with the probe. The resulting autoradiograph (Fig 12.9) looks rather like the bar codes you can see on groceries in the supermarket. Clearly, these satellite regions occur throughout the human genome. The exciting thing about them is that they are unique. Your 'bar code' genetic fingerprint will look quite different from mine.

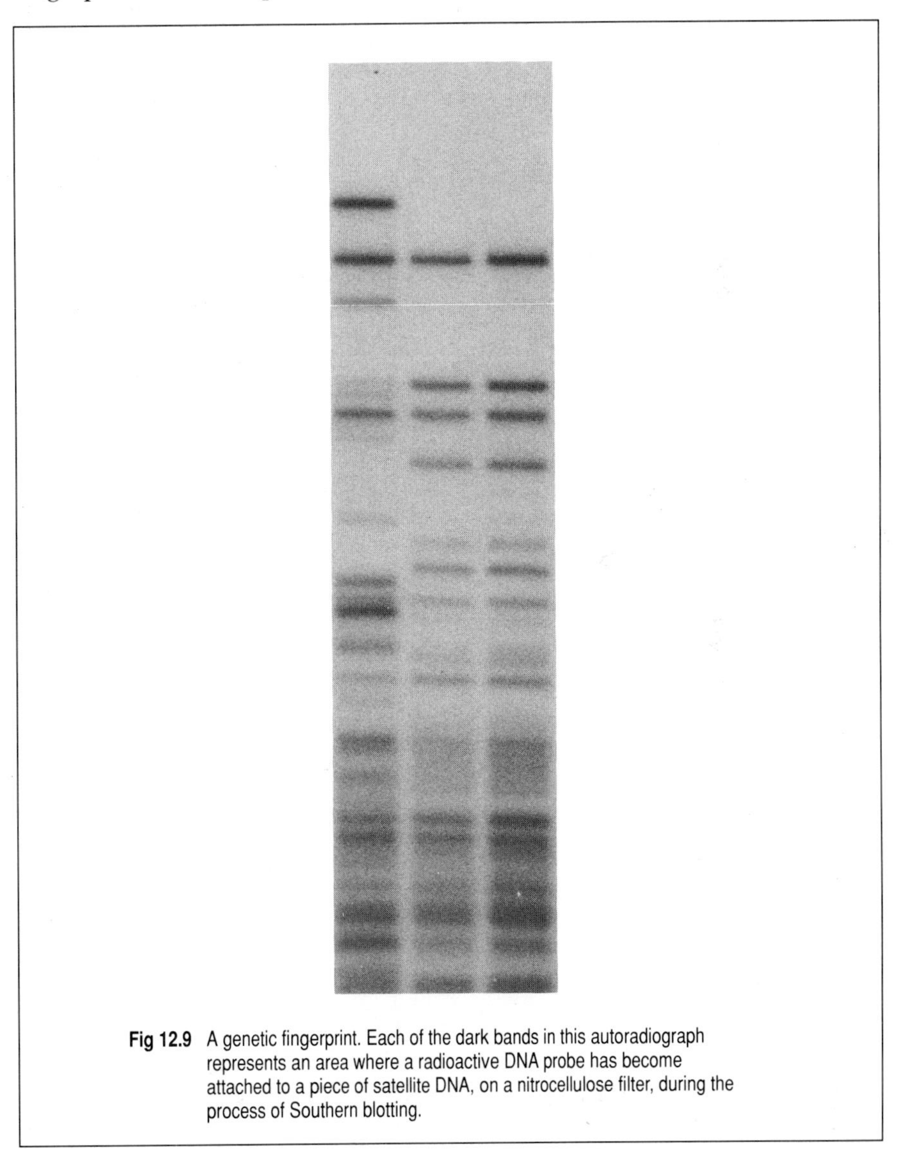

Fig 12.9 A genetic fingerprint. Each of the dark bands in this autoradiograph represents an area where a radioactive DNA probe has become attached to a piece of satellite DNA, on a nitrocellulose filter, during the process of Southern blotting.

How can we use this procedure to catch, say, a rapist? Sperm collected from the rape victim will contain sufficient DNA to produce a genetic fingerprint of the rapist. Blood taken from a suspect will provide enough DNA to produce his genetic fingerprint. If the two match then you have got your man, because the chances of the match occurring by chance are vanishingly small. Clearly the process can work in reverse. A man who was accused of raping a woman was later shown to be innocent when his genetic fingerprint did not match that of the rapist.

Interestingly, the bar code patterns of the genetic fingerprints are inherited in a very simple way. This has been used to settle the issue in a number of paternity/maternity cases like the one shown in the article below.

The technique may also have other wider applications, for example in analysing the population biology of whales.

How Jeffreys secured a Ghanaian boy's right to enter Britain

One of the first real tests of genetic fingerprinting involved a boy who was born in Britain but emigrated to Ghana to join his father. When he returned to Britain to live with his mother, the immigration authorities claimed that he was not the woman's real son.

Alec Jeffreys analysed blood samples from the boy, the alleged mother and the alleged brother and two sisters of the boy, who lived in Britain. The father in Ghana was not available for testing. Jeffreys extracted DNA from the white blood cells (red blood cells do not have nuclei and so have very little DNA), and subjected it to enzymes that chopped it into tiny pieces. The pieces were separated in an electric field and stuck to a nylon membrane to retain their relative positions; the smaller pieces of DNA move faster and further in the electric field than larger DNA strands.

The next step was to subject the DNA to his probes, which identified and labelled certain strands of DNA containing hypervariable regions. These appeared as dark bands on the nylon membrane. The patterns that result are unique to an individual, who inherits half from the biological mother and half from the biological father.

In the case of the Ghanaian boy, the woman shared half her DNA pattern with the boy, and half with the other children. The alleged brother and two sisters of the boy also shared the similar paternal bands in the fingerprint, showing that they all had the same father.

Presented with the evidence, the Home Office had to admit the boy into the country in order to join his mother.

Below: An imaginary case study. Bands in the children's prints that do not come from their mother can be seen to derive from their father. Thus the status of the parent is confirmed.

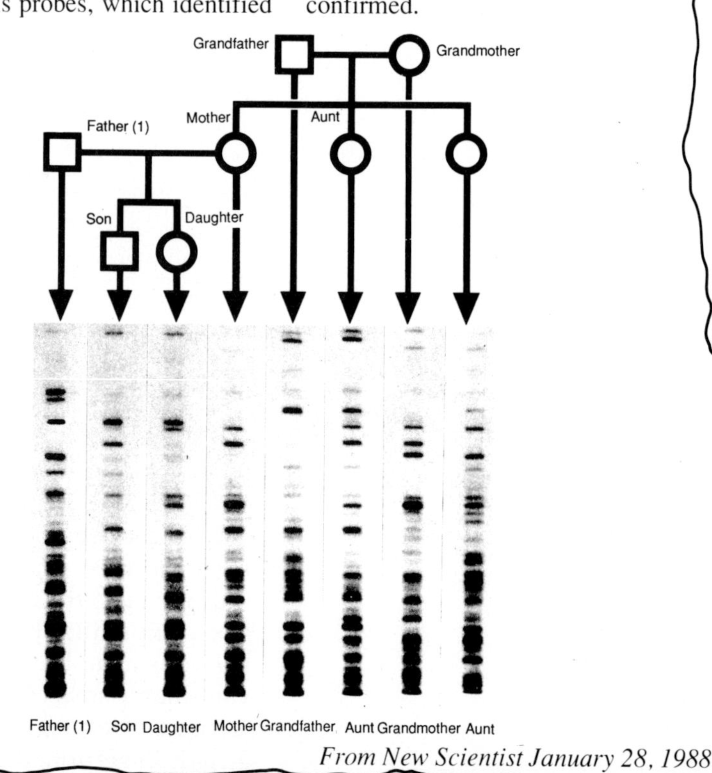

From New Scientist January 28, 1988

Fingerprinting undermines the case for killing whales

There is now less reason to kill whales as part of scientific research. Workers at the University of Cambridge have developed a way to monitor the size of whale populations and the age and genetic status of individual animals using the technique of genetic fingerprinting.

At the end of last year, the International Whaling Commission (IWC) agreed reluctantly to issue Japan with a permit to harvest 825 Minke whales and 50 sperm whales for research.

Some members of the commission's scientific committee questioned the value of this research, aware that it may be used as an excuse to circumvent the IWC's three-year moratorium on commercial whaling.

The proposed Japanese catch was reduced, in December, to 300 Minke whales. Now, Rus Hoelzel and William Amos of the Department of Genetics at Cambridge have shown that DNA fingerprinting, developed by Alec Jeffreys of Leicester University for use in humans, is equally applicable to whales. In humans, the technique confirms identity through analysis of samples of body tissue or fluids.

Hoelzel and Amos obtain DNA samples from whales with a small dart that abstracts skin. They fire the dart from a crossbow or a rifle into the hide of a whale swimming underwater. The dart has at its tip a hollow metal cylinder which abstracts a pellet of skin weighing somewhere around 0.5 milligrams.

The researchers say that the operation is painless to the animal. Amos says that the sample yields far more information than is available traditionally through radio-tracking, tagging or counting.

They report in *Nature* (vol. 333, p305) that the test can identify individuals, patterns of paternity and in some cases, maternity, with greater certainty. From these data, the researchers believe that they can gain a better understanding of the geographical dispersal and reproductive biology of whales (see also *New Scientist*, 29 January 1987, p44).

Also, by examining mitochondrial DNA, they can assess the amount of genetic variation within individual populations of whales. The genetic diversity within a population affects markedly its prospects for survival.

Amos says that the case for killing whales to collect physiological data is now without foundation. He says that this information is already available from studies carried out during heavy whaling in the 1930s and 1940's.

From New Scientist June 2, 1988

QUESTION

12.15 Two of the genetic fingerprints shown in Fig 12.9 belong to identical twins. Can you identify them? Explain the reasoning behind your choice.

12.5 IS GENETIC ENGINEERING SAFE ?

There is understandable concern about genetic engineering. After all, using the technique, geneticists can effectively create organisms which have never existed before. What if such an organism should escape from the laboratory? Such concern caused a group of American geneticists to write the letter reproduced below.

Potential Biohazards of Recombinant DNA Molecules

Recent advances in techniques for the isolation and rejoining of segments of DNA now permit construction of biologically active recombinant DNA molecules in vitro. For example, DNA restriction endonucleases, which generate DNA fragments containing cohesive ends especially suitable for rejoining, have been used to create new types of biologically functional bacterial plasmids carrying antibiotic resistance markers and to link ***Xenopus laevis*** *DNA to DNA from a bacterial plasmid. This latter recombinant plasmid has been shown to replicate stably in* ***Escherichia coli*** *where it synthesises RNA that is complementary to* ***X. laevis*** *ribosomal DNA. Similarly, segments of* ***Drosophila*** *chromosomal DNAs have been incorporated into both plasmid and bacteriophage DNAs to yield hybrid molecules that can infect and replicate in* ***E. coli.***

Several groups of scientists are now planning to use this technology to create recombinant DNAs from a variety of other viral, animal and bacterial sources. Although such experiments are likely to facilitate the solution of important theoretical and practical biological problems, they would also result in the creation of novel types of infectious DNA elements whose biological properties cannot be completely predicted in advance.

There is serious concern that some of these artificial recombinant DNA molecules could prove biologically hazardous. One potential hazard in current experiments derives from the need to use a bacterium like ***E. coli*** *to clone the recombinant DNA molecules and to amplify their number. Strains of* ***E. coli*** *commonly reside in the human intestinal tract, and they are capable of exchanging genetic information with other types of bacteria, some of which are pathogenic to man. Thus, new DNA elements introduced into* ***E. coli*** *might possibly become widely disseminated among human, bacterial, plant or animal populations with unpredictable effects.*

Concern for these merging capabilities was raised by scientists attending the 1973 Gordon Research Conference on Nucleic Acids, who requested that the National Academy of Sciences give consideration to these matters. The undersigned members of a committee, acting on behalf of and with the endorsement of the Assembly of Life Sciences of the National Research Council on this matter, propose the following recommendations.

First, and the most important, that until the potential hazards of such recombinant DNA molecules have been better evaluated or until adequate methods are developed for preventing their spread, scientists throughout the world join with the members of this committee in voluntarily deferring the following types of experiments:

** Type 1: Construction of new, autonomously replicating bacterial plasmids that might result in the introduction of genetic determinants for antibiotic resistance or bacterial toxin formation into bacterial strains that do not at present carry such determinants; or construction of new bacterial plasmids containing combinations of resistance to clinically useful antibiotics unless plasmids containing such combination of antibiotic resistance determinants already exist in nature.*

** Type 2: Linkage of all segments of the DNAs from oncogenic (cancer-inducing) or other animal viruses to autonomously replicating DNA elements such as bacterial plasmids or other viral DNAs. Such recombinant DNA molecules might be more easily disseminated to bacterial populations in humans and other species, and thus possibly increase the incidence of cancer or other diseases.*

Second, plans to link fragments of animal DNAs to bacterial plasmid DNA or bacteriophage DNA should be carefully weighed in light of the fact that many types of animal cell DNAs contain sequences common to RNA tumour viruses. Since joining of any foreign DNA to a DNA replication system creates new recombinant DNA molecules whose biological properties cannot be predicted with certainty, such experiments should not be undertaken lightly.

Third, the director of the National Institutes of Health is requested to give immediate consideration to establishing an advisory committee charged with (i) overseeing an experimental programme to evaluate the potential biological and ecological hazards of the above types of recombinant DNA molecules; (ii) developing procedures which will minimise the spread of such molecules within human and other populations; and (iii) devising guidelines to be followed by investigators working with potentially hazardous recombinant DNA molecules.

Fourth, an international meeting of involved scientists from all over the world should be convened early in the coming year to review scientific progress in this area and to further discuss appropriate ways to deal with the potential biohazards of recombinant DNA molecules.

The above recommendations are made with the realisation (i) that our concern is based on judgments of potential rather than demonstrated risk since there are few available experimental data on the hazards of such DNA molecules, and (ii) that adherence to our major recommendations will entail postponement or possible abandonment of certain types of scientifically worthwhile experiments. Moreover, we are aware of the many theoretical and practical difficulties involved in evaluating the human hazards of such recombinant DNA molecules. Nonetheless, our concern for the possible unfortunate consequences of indiscriminate application of these techniques motivates us to urge all scientists working in this area to join us in agreeing not to initiate experiments of types 1 and 2 above until attempts have been made to evaluate the hazards and some resolution of the outstanding questions has been achieved.

Paul Berg, Chairman
David Baltimore
Herbert W. Boyer
Ronald W. Davis
Daniel Nathans
Richard Roblin
James D. Watson
Stanley N. Cohen
Donald S. Hogness
Sherman Weissman
Norton D. Zinder

Committee on Recombinant DNA
Molecules Assembly of Life Sciences
National Research Council
National Academy of Sciences
Washington DC 20418

As the result of such fears, regulatory bodies, for example the Genetic Manipulation Advisory Group in Great Britain, were set up to produce guidelines which genetic engineers should follow in order to prevent accidents. In particular, they recommended that:

- genetic engineering experiments could only be carried out in specialised laboratories which have the physical containment facilities used for handling pathogenic organisms;
- that the strains of bacteria used as hosts of recombinant DNA molecules should be disabled so that they can only grow under special culture conditions *not* found in the general environment;
- finally, the vector molecules themselves should not contain the genes needed to bring about conjugation.

These safeguards, based on physical and biological containment, have worked well and the rules have now been relaxed a little. However, scientists must still get approval from the appropriate committee before they can carry out a research programme using recombinant DNA molecules.

A potentially bigger challenge for society is the problem of deliberately releasing genetically engineered organisms into the environment. For example, a programme to assess the use of genetically engineered viruses to control the moth caterpillars which defoliate pine trees is currently underway in Scotland. The viruses involved are natural pathogens on the caterpillars and there is no health risk to people. The virus has simply been engineered to make it more effective at killing caterpillars. The problem in this case is not, therefore, whether the virus constitutes a health hazard but whether it constitutes an ecological hazard. What will the impact be on the pine forest ecosystem if the virus is released in large numbers and is successful in destroying large numbers of moth caterpillars? What, for example, will happen to the birds which rely on the caterpillar for food?

SUMMARY ASSIGNMENT

Using the information contained in this chapter and from other sources, for example *New Scientist, Scientific American* and other magazines, analyse the benefits and possible dangers of genetic engineering. Then prepare a report for a parliamentary subcommittee, which is considering legislation to control genetic engineering experiments, entitled:

'Do the benefits of genetic engineering outweigh any potential hazards?'

Discuss your report with your tutor and friends. Remember, you want to convince the committee that your point of view is correct, so you will need to support your arguments with plenty of factual information. Remember these are Members of Parliament so they will not know much about the techniques of genetic engineering, therefore you will need to explain these as clearly as you can.

Theme **5**

GENETICS AND PEOPLE

Genetics is a fundamental biological science which affects us all. In this theme we will examine some of the ways in which genetics can be used to help people directly, including transplant surgery, prenatal screening and genetic counselling.

PREREQUISITES

To complete the work in this theme successfully, you will need to have completed Themes 1, 3 and 4.

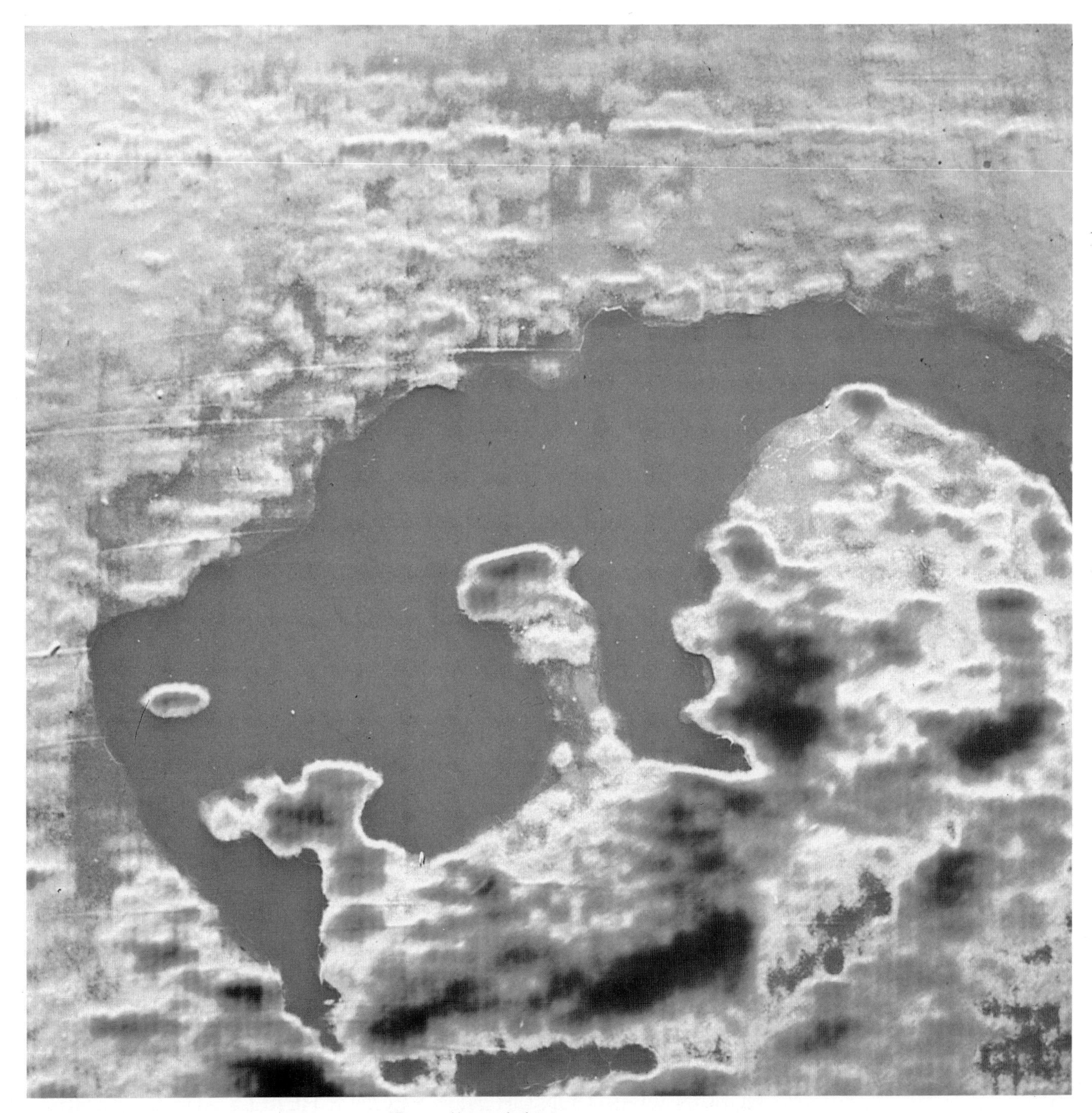

Ultrasound image of a fetus.

Chapter 13

HUMAN VARIATION OR WHY PEOPLE REALLY ARE DIFFERENT

Human variation depends both on the genes we contain and the environments in which we develop. The most obvious differences between us are features on the body surface, and the size and shape of our bodies. We each have a unique face and absolutely distinct fingerprints. We can be tall or short, black or white, fat or thin and so on. At the biochemical level, the amount of variation in human populations is enormous.

For example, each of us belongs to one of the more than 60 different blood groups; we can have one or more of several alternative forms of the same enzyme; and there are molecules on the surfaces of our cells (antigens) that provide a cellular recognition system that is essentially unique. To illustrate the amount of genetic variation in human populations we will look at a biochemical system: proteins. Then we shall consider two of the implications of the genetic diversity of human populations: the challenges of transplant surgery and the contentious issue of the so-called human races.

LEARNING OBJECTIVES

After completing the work in this chapter, you will be able to:

1. calculate heterozygosity from data on human protein variation;
2. explain the importance and genetics of the HLA system;
3. evaluate whether human races are really genetically different.

13.1 THE POLYMORPHIC NATURE OF HUMAN PROTEINS

Protein molecules consist of long chains of amino acids. The sequence in which the amino acids are joined together during protein synthesis is ultimately determined by the sequence of nucleotide bases in the gene encoding the protein. (Genes which encode proteins are called **structural genes**.) So by looking at variation in the structure of proteins, we are essentially looking at variation in the structural genes themselves.

For example, imagine we are examining the sequence of amino acids in a blood protein. We take a blood sample from 1000 people and isolate the protein we are interested in. We then determine its amino acid sequence, that is, we sequence the protein, using an amino acid analyser. If we find that the amino acid sequence of the protein is the same in every single individual, in other words, the structure of the protein does not vary, then we can conclude that the structural gene which encodes the protein also does not vary. However, if we find that the structure of the protein does vary then we know that the gene which encodes it also varies. One person has got one form of the gene but another person may have a different form of the gene. Furthermore, we can measure how much the gene varies in a population by looking at the number and frequency of the different forms of the protein which it encodes.

QUESTION

13.1 What are these alternative forms of the same gene called?

Only a small fraction of the thousands of proteins encoded by structural genes have had their exact sequence determined. Of the 100 or so human enzymes which have been sequenced, about 25 per cent have been found to exist in different forms, that is, they are polymorphic. Table 13.1 gives the frequencies of two of the most highly polymorphic enzymes in the British population.

Table 13.1 Frequencies of polymorphic variants of two enzymes in the British population

	Form					
Enzyme	1	2	3	$\frac{1}{2}$	$\frac{2}{3}$	$\frac{1}{3}$
red cell acid phosphate	0.13	0.36	0.0	0.43	0.05	0.03
placental alkaline phosphatase	0.41	0.007	0.1	0.35	0.05	0.12

QUESTION

13.2 **(a)** How many forms of each enzyme could an individual have?
(b) What percentage of people have each form of red cell acid phosphatase?

Measuring genetic variation

Can we use this sort of information to estimate the amount of genetic variation in a population? The answer is yes, but first of all we have to think about some statistical implications of the problem of measuring genetic variation.

To actually measure the genetic variation in a population we would have to look at every single gene in every single individual in that population. For example, suppose we wanted to measure the amount of genetic variation in the human population of London, England. There are about 10 million people who live in London, and each of those individuals has about 100 000 genes. Since each gene is represented by two copies in each cell, and these copies can be different, that means we would have to screen $2 \times 100\,000 \times 10\,000\,000 = 2 \times 10^{11}$ genes. Obviously this is just not practical. Furthermore, we do not know exactly how many genes there are in the human cell and we do not know what the majority of genes actually do.

Clearly, then, we cannot look at every single gene locus in every person living in London. Instead, we have to take a sample of gene loci and use this sample to estimate the amount of genetic variation in the human population of London. For such a sample to be unbiased and therefore representative, it must be a random sample of all the gene loci in the population under study. This is analogous to trying to count the numbers of a particular plant on a lawn. You cannot count every single plant so you try to estimate the number using a quadrat. However, to get a representative set of samples you must put your quadrat down at random. Similarly, to estimate the amount of genetic variation in a population we must use a set of genes chosen at random, or rather a set of proteins chosen at random. We can do this by choosing to study proteins without knowing whether or not they are variable in a population. Such a set of proteins will therefore represent an unbiased sample of all the structural genes in an organism.

Measuring variation in protein structure

One way of measuring the amount of genetic variation in a population would be to select, say, 20 proteins without knowing whether they were variable or not. Samples of each protein could be obtained from, say, 100 people chosen at random from our population, giving us a total of 2000 protein samples. Each of the proteins in these samples would now have to be sequenced and the amount of variation, if any, determined for each protein. The problem with this approach is that it takes several months, perhaps even years, to sequence each protein. Sequencing 2000 proteins is not really a practical proposition. However, **electrophoresis** will allow us to screen rapidly the sample proteins for structural variation. We can use the data from electrophoretic studies to estimate the genetic variation of our population.

QUESTION

13.3 Explain why electrophoresis can be used to screen proteins for structural variation.

Heterozygosity

This is the measure of genetic variation preferred by most geneticists. It is calculated by measuring the frequency of heterozygous individuals at each locus you are interested in and then averaging these frequencies across all loci. An example will make the procedure clear.

Assume we are studying five gene loci in a population, each of which encodes a protein. Using electrophoresis we determine the number of individuals in our random sample of 100 people who are heterozygous at each locus. The results are given in Table 13.2.

Table 13.2 Calculation of the average heterozygosity at five loci

	Number of individuals		
Locus	Heterozygotes	Total	Heterozygosity
1	31	100	31/100 = 0.31
2	17	100	17/100 = 0.17
3	6	100	6/100 = 0.06
4	14	100	14/100 = 0.14
5	0	100	0/100 = 0

$$\text{The heterozygosity of this population} = \frac{0.31 + 0.17 + 0.06 + 0.14 + 0}{5}$$

$$= 0.14$$

$$= 14\%$$

This figure is an estimate of the probability that two alleles drawn at random from a population will be different.

QUESTION

13.4 Table 13.3 lists 20 variable loci out of 71 loci sampled in a population of Europeans. The data are expressed as the number of individuals heterozygous at a particular locus per 100 people sampled.

(a) Calculate the frequency of heterozygotes at each of the 20 variable gene loci.

(b) Calculate the heterozygosity of the population based on these 71 gene loci.

Table 13.3

Gene locus	Enzyme encoded	Number of individuals heterozygous
ACP1	acid phosphatase	52
PGM1	phosphoglucomutase-1	36
PGM2	phosphoglucomutase-2	38
AK	adenylate kinase	9
PEPA	peptidase-A	37
PEPC	peptidase-C	2
PEPD	peptidase-D	2
ADA	adenosine deaminase	11
PGD	phosphogluconate dehydrogenase	5
ACP2	alkaline phosphatase (placental)	53
AMY2	α-Amylase (pancreatic)	9
GPT	glutamate-pyruvate transaminase	50
GOT	glutamate-oxaloacatate transaminase	3
GALT	galactose-1-phosphate uridyltransferase	11
ADH2	alcohol dehydrogenase-2	7
ADH3	alcohol dehydrogenase-3	48
PG	pepsinogen	47
ACE	acetylcholinesterase	23
ME	malic enzyme	30
HK	hexokinase (white cell)	5

You may feel that the figure you have calculated in question 13.4 is not very large and that this indicates that this population is not, genetically, very variable. However, to realise the full implications of this figure we have to do just a little more arithmetic.

Assume that there are 30 000 structural genes in a person (this is probably an underestimate). You should have calculated that the average heterozygosity of the population is 0.067. This means that an individual would therefore be heterozygous at $30\,000 \times 0.067 = 2010$ loci. If an individual is heterozygous at just one locus, two different types of gamete can be produced. If an individual is heterozygous at n loci (where n = any number) then 2^n different gametes are possible. So an individual heterozygous at 2010 loci can, in theory, produce $2^{2010} = 10^{605}$ (10 followed by 605 zeros!) different kinds of gametes. By comparison, the total number of sperm ever produced by all human males who have ever lived is between 10^{23} and 10^{24}.

Whilst all possible combinations of genes are not equally likely (some genes will be linked, for example) the above calculation does indicate that no two individual human gametes are likely to be identical. What this means is that the combination of alleles that you received from your parents was a one off event – **you are unique**!

QUESTION

13.5 (a) Given that no two human gametes are identical, why do identical twins have the same genotype? What type of reproduction has produced these two individuals? Will identical twins necessarily have the same phenotype?

(b) Evolution is said to occur more rapidly in populations of sexually reproducing organisms. Given the calculations outlined above and your answer to part (a) of this question, can you suggest why?

Mapping the human genome

At the time of writing (1988) the prospect of sequencing the human genome was being actively considered. Given that there are 3×10^9 nucleotide pairs in the haploid genome of a human egg or sperm which carry the information representing an unknown number of genes (estimates vary from 100 000 to 2.5 million), this is a daunting task. Furthermore, a lot of the DNA is 'junk' – highly repetitious sequences of bases that apparently have little function, and given that the coding sequences of the genes are embedded in sequences which regulate the action of the genes or act as linkers between different parts of the genes, the challenge of interpreting the DNA sequences is a daunting one.

INVESTIGATION 13.1

Collect information from popular science magazines like *Scientific American* and *New Scientist* on how the human genome mapping project is going. A good starting place is the *New Scientist* edition of 5 March 1987 (vol. 113, no. 1550, p. 35). Having collected your data, summarise how the project is going and comment on whether you think the effort is worthwhile.

13.2 TRANSPLANT SURGERY

If skin from one mouse (the donor) is grafted onto another mouse (the recipient) then the recipient mouse will mount an **immune response** against the grafted tissue and ultimately destroy it. This is the process known as **tissue rejection**. Such an effect is also observed if we try to graft skin from one human to another or try to transplant a kidney. This effect suggests that the body has an ability to identify self and non-self. This recognition system depends upon the presence of particular molecules, usually proteins, which are present on the surface of cells. Such molecules are called antigens since they can produce an antibody response. To illustrate the general principles of tissue incompatibility we will first look at the ABO blood group system and then at the genes of the **major histocompatibility complex (MHC)** which, in people, is sometimes called the **human leucocyte antigen (HLA)** system.

In addition to the ABO blood group system, there are at least a dozen other human blood group systems. In cattle, in excess of 200 different blood groups have been identified.

The ABO blood group system

If you have ever had an operation one of the things that happens before surgery is that you will have a sample of blood taken. This sample is used to match your blood to that of potential donors from the blood bank. If this procedure were not adopted, there could be disastrous results. For example, if you are blood group A then you cannot receive group B blood. The reason for this is that the surfaces of your red blood cells carry antigens called **agglutinogens**. If you only have agglutinogen A then you are blood group A.

QUESTION

13.6 Which blood group type would you have if your red blood cells carried both agglutinogen A and agglutinogen B?

Your blood plasma also contains genetically determined antibodies which are called **agglutinins**. If blood serum containing the agglutinin anti-A is mixed with red blood cells which carry agglutinogen A then the red blood cells agglutinate (clump).

QUESTIONS

13.7 (a) Table 13.4 shows the agglutinins present in the blood serum from people with different blood groups. Copy and complete the table to show what will happen when red blood cells from the groups listed across the top of the table are added to the serum from the groups listed at the left of the table. (Serum is blood plasma minus the proteins responsible for blood clotting.) Record your results as either A (the cells agglutinate) or N (the cells do not agglutinate).

(b) Blood group O is sometimes called the universal donor whilst blood group AB is sometimes called the universal recipient. Explain these descriptions.

Table 13.4

Serum from blood group	Agglutinins present in serum	Reaction when red blood cells from groups listed below are added to serum from groups listed at left			
		O	A	B	AB
O	Anti-A Anti-B				
A	Anti-B				
B	Anti-A				
AB	——				

13.8 A man who is blood group A marries a woman who is blood group B and they have three children. One is blood group AB, one is blood group O and the other is blood group B. Explain these results. Note the ABO blood group genotypes are given in Table 1.2.

The major histocompatibility complex

In the 1950s it was discovered that white blood cells also carried antigens on their surfaces. Later it became apparent that all the cells in our body, with the exception of red blood cells, carry these antigens. These antigens are encoded by a group of four genes located on one region of chromosome number 6. These four genes constitute the major histocompatibility complex (MHC) and the histocompatibility antigens they encode are called MHC antigens. Since these antigens were first discovered in humans on the surface of the white blood cells called leucocytes, these genes are sometimes called the human leucocyte antigen (HLA) system. The term HLA is used in the rest of this book because we will only be dealing with humans.

It is the difference in HLA antigens between people which is mainly responsible for the failure of tissue transplant operations. There are, however, only four genes in the HLA system, so how can we account for the fact that practically everybody has a different set of HLA antigens? The answer is that the genes in this system, called HLA-A, HLA-B, HLA-C, and HLA-D, are the most polymorphic known in humans. Each of the HLA alleles so far identified has been given a number. For example, HLA-B27 is the twenty-seventh allele of the HLA-B locus. The HLA antigens which an individual is carrying, and hence their HLA genotype, can be determined by testing white blood cells.

QUESTION

Table 13.5 The number of alleles at each HLA locus

Locus	Number of alleles
HLA-A	20
HLA-B	40
HLA-C	8
HLA-D	12

13.9 Table 13.5 shows the number of known alleles for each of the HLA genes.

(a) How many different HLA alleles can an individual carry?

(b) Given that all the HLA alleles are codominant, how many HLA antigens could an individual have on the surface of (i) their liver cells; (ii) their red blood cells?

(c) For one locus we can generate $\frac{n\,(n+1)}{2}$ genotypes, where n is the number of alleles at the locus. Calculate the number of genotypes that can be produced for each gene locus in the HLA system.

(d) Multiply these four numbers together. This will give you the total number of different genotypes that these four genes can produce.

(e) Given your answer to (d), what will the probability be of finding a suitable kidney donor drawn at random from the general population?

Fortunately for our kidney transplant patient, things are not quite as bad as your answers to question 13.9 might suggest. There are three reasons for this.

1. Some of the HLA alleles are more common than others.
2. Some of the HLA antigens produce stronger rejection reactions than others. Providing we ensure these HLA antigens are the same in both donor and recipient then a transplant stands a good chance of success.
3. The four genes are very tightly linked, which means that they cannot assort independently at meiosis. This block of genes, called a haplotype, tend to be inherited together because crossing over between the genes within the haplotype occurs very rarely.

QUESTIONS

13.10 How many haplotypes will you have in each of your liver cells?

13.11 Fig 13.1 shows the inheritance of HLA haplotypes in a family.

(a) What is the probability that the second child born to this couple will have (i) the same HLA antigens as the first; (ii) four of the HLA antigens the same; (iii) less than four the same?

(b) Explain why a bone marrow donor is usually sought within the family of the intended recipient.

(c) In addition to matching the HLA antigens, what else would have to be matched before our bone marrow transplant could proceed?

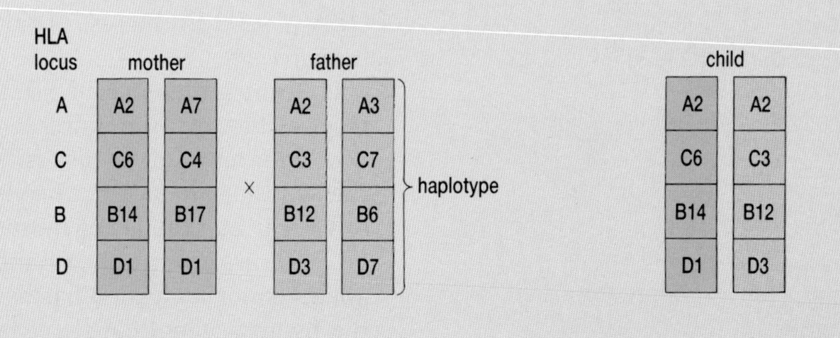

Fig 13.1 The inheritance of HLA haplotypes. Notice that each haplotype is inherited as a block of four genes.

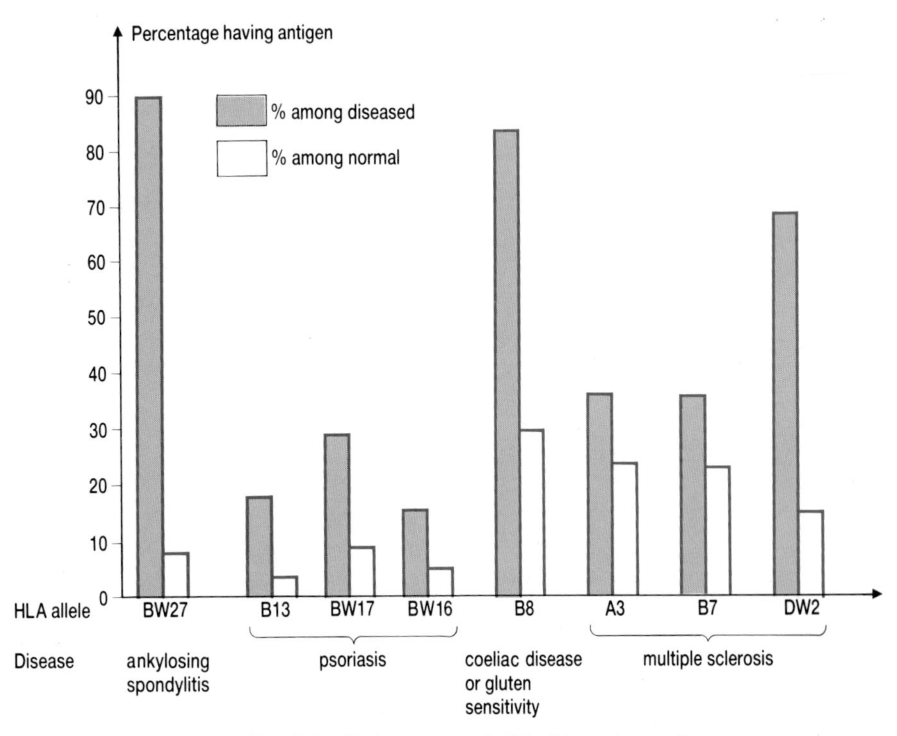

Fig 13.2 The relationship between certain HLA alleles and some diseases.

Interestingly, certain HLA alleles seem to be associated with particular diseases, as shown in Fig 13.2. The strongest association is between an allele called BW27 at the B locus (the W stands for Workshop) and ankylosing spondylitis. This is a disease primarily of young men characterised by rigidity of the back.

Care needs to be exercised in interpreting such associations. Obviously, having a particular haplotype does not mean you are going to develop the particular disease. Nor do such associations mean that the allele causes the disease. Nonetheless, such relationships may be useful in diagnosing a disease or spotting those people who may be at risk. The biological basis of these relationships still remains something of a mystery, but an interesting feature of the diseases associated with HLA is that they appear to be connected with abnormal immune responses in which the body literally attacks itself. This category of diseases is called autoimmune disorders.

Transplant surgery also emphasises that just because something is encoded in our genes does not mean that it is inevitable. We have been able to overcome our genetic defence system using immunosupressant drugs like cyclosporin. So by altering the environment inside the human body using drugs, transplant surgery is much more feasible, providing you have ABO compatibility, a fresh kidney and a good surgeon.

13.3 HUMAN RACES

Given that there is so much variation in human populations, we should not find it surprising that certain alleles are commoner in some parts of the world compared with other parts. This is the basis on which a species can be divided into races. The concept of race, particularly when applied to people, is greatly misunderstood, even abused, and consequently needs consideration. The diversity of people was recognised by Carl Linnaeus who identified four human races: African, American, Asiatic and European. The five 'colour' races were established by Johann Blumenbach in 1775: yellow or Mongolian; black or Ethiopian; white or Caucasian; red or American; and brown or Malayan. But is the classification of races on the basis of skin colour biologically justified? The answer is a resounding **no**.

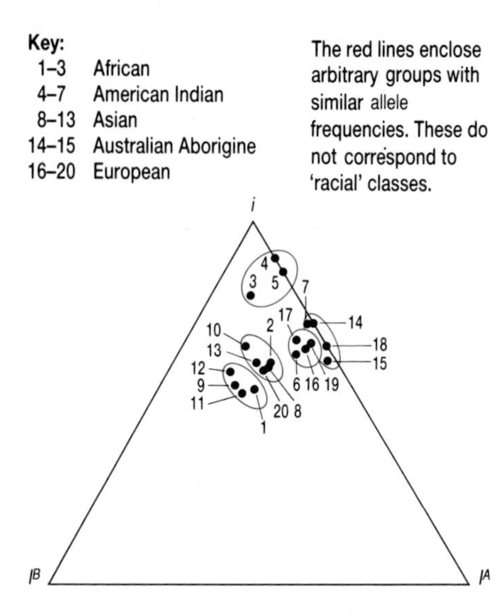

Fig 13.3 A triallelic diagram showing the relative frequencies of ABO alleles in 20 different human populations.

Geneticists define race as simply a group of individuals who share certain genes with higher frequency than other groups. People who resemble each other in skin colour differ markedly with respect to other traits. To define a group of people as a race on the basis of just one characteristic is biological nonsense. To illustrate this point, consider a classification of human race which was based on ABO blood groups rather than skin colour (Table 13.6). If races can be discerned on the basis of one trait alone, it is as valid to group the Chinese, Japanese, African pygmies and Hottentots as one race based on the frequencies of the ABO alleles in these populations as it is to separate them into different races on the basis of the colour of their skin.

Table 13.6 The frequency of *i*, I^A and I^B alleles in nine different populations.

Population	Number tested	i	I^A	I^B
Japanese	29 799	0.55	0.28	0.17
Chinese (Huang Ho)	2 127	0.59	0.22	0.20
African pygmies	1 032	0.55	0.23	0.22
Hottentots	506	0.59	0.20	0.19
Americans (white)	20 000	0.67	0.26	0.07
Swiss	275 644	0.65	0.29	0.06
Eskimos	484	0.64	0.33	0.03
Navajo Indians	359	0.87	0.13	0.00
Blackfoot Indians	115	0.49	0.51	0.00

This becomes even more obvious if we plot the frequencies of these alleles on a triallelic diagram as shown in Fig 13.3. Each point represents the allelic composition of a population. The diagram shows that all human populations are bunched together in the region of high *i*, intermediate I^A and low I^B frequencies. Furthermore, neighbouring points (enclosed by red lines) do not correspond to geographical races, so geographical races are not distinguished from each other by characteristic allelic frequencies for this gene.

QUESTION

13.12 Table 13.7 gives allelic frequencies at seven polymorphic loci for a European and an African population. On the basis of this large data set, can you differentiate between the African and the European population?

Table 13.7

	Europeans			Africans		
Locus	Allele 1	Allele 2	Allele 3	Allele 1	Allele 2	Allele 3
red cell acid phosphatase	0.36	0.60	0.04	0.17	0.83	0.00
phosphoglucomutase-1	0.77	0.23	0.00	0.79	0.21	0.00
phosphoglucomutase-3	0.74	0.26	0.00	0.37	0.63	0.00
adenylate kinase	0.95	0.05	0.00	1.00	0.00	0.00
peptidase A	0.76	0.00	0.24	0.90	0.10	0.00
peptidase D	0.99	0.01	0.00	0.95	0.03	0.02
adenosine deaminase	0.94	0.06	0.00	0.97	0.03	0.00

Source: R.C. Lewontin, The Genetic Basis of evolutionary change, Columbia University Press, 1974. Adapted from H. Harris, The Principles of Human Biochemical Genetics. North Holland, Amsterdam and London, 1970.

How genetically different, then, are the so-called human races? To answer this question we once again need to take an unbiased sample of gene loci. So let us go back to the analysis of polymorphic proteins. Consider two populations of people who we think might belong to separate races. We calculate the average heterozygosity for each group using data obtained from electrophoretic studies of protein polymorphism. Let us say we get a figure of 12 per cent in one population and 14 per cent in the other. Now, if the two populations really do belong to separate races then we would expect each group, by definition, to have certain alleles which were peculiar to it, or at least occurred at a higher frequency than in the other group. So when we amalgamate the data on protein polymorphism for the two groups and then calculate the average heterozygosity we would expect there to be a large increase inthe value for heterozygosity. In fact when we do this calculation for a real set of data for, say, an African and a European population, the increase in average heterozygosity is quite small. In other words, the genetic differences between groups are tiny in comparison to the genetic differences within groups. **The genetic differences between 'human races' are small.** This is best expressed by the American geneticist Richard Lewontin who points out that if every member of the human race were to die now except for one tribe inhabiting a remote part of Africa, that group of people would still contain 85 per cent of the genetic variability of the human species.

From the analysis of a large number of loci, six major human racial groups can be identified: Caucasian, African, Asian, North American Eskimos and Indians, South American Indians and Australian aboriginal groups. These are not 'pure' races – they do not share a uniform genetic identity. Remember 94 per cent of the genetic variability among humans comes from differences between individuals within the same racial group, and only six per cent comes from differences between racial groups.

This raises an interesting question. If the genetic differences between races are small, why is it that we find differences between people so conspicuous? The answer is that our everyday impressions of racial differences do not come from genes but from visible traits, some of which are extremely conspicuous, for example skin colour. How can we explain these differences?

Skin colour certainly varies in response to the environment. All people, including black people, tan when exposed to the sun. However, the differences in skin colour are too great to be explained by environmental effects. Clearly skin colour has a large genetic component. If we look at the distribution of skin colour in the world, we find that skin colour is usually darker in the tropics compared with more temperate regions. Apparently, then, skin colour is correlated with climate. One possible explanation of this is that a dark skin protects against sunburn and hence skin cancer in tropical regions. But the incidence of skin cancer is so low, even amongst light skinned individuals, near the equator that it is hard to believe that this is the only selective agent acting.

An alternative hypothesis is concerned with the production of vitamin D which is involved in the metabolism of calcium in the body. A deficiency of vitamin D leads to rickets, whilst too much vitamin D causes an excess of calcium in the bones, making them brittle. Vitamin D is made by cells just under the surface of the skin using ultraviolet light from the sun. The pigmentation of the skin determines the amount of sunlight which is absorbed. It seems quite reasonable, therefore, to suppose that as people moved out of Africa and started to colonise more northern latitudes, the need for vitamin D gave a selective advantage to lighter skinned individuals at high latitudes where the sun is very weak. The lightest skins are possessed by the northern Caucasians.

Two pieces of evidence support this hypothesis. Firstly, when the Industrial Revolution was in full swing in nineteenth century England, the skies above industrial cities became grossly polluted, cutting out sunlight. At the same time rickets reached epidemic proportions. Secondly, the Eskimos and Lapps have relatively dark skins even though they live at extremely high latitudes. However, unlike those Caucasians whose diets have been based on cereals for the last 5 to 8 thousand years, the Eskimos and Lapps have diets based on meat which are therefore much richer in vitamin D. Consequently, they have much less need to produce the vitamin with the help of the sun.

Like skin colour, other body surface traits often show great differences between people from different parts of the world for example in eye colour, hair texture and body size. It is not really surprising that adaptations to climate produce readily observable differences in body surface characteristics. We tend to be unduly impressed by these superficial differences when evaluating differences between human races. Superficial differences which are the result of major climatic adaptations represent a biased sample of loci which over-emphasises the differences between races. When we look at a larger sample of gene loci, we find that the differences between races are very small. The truth of the matter is that we are all gloriously different from each other. Each person has a unique genotype – it is different from that of every other person, whether or not they belong to the same race.

SUMMARY ASSIGNMENT

1. **(a)** What is polymorphism?
   (b) How does polymorphism originate in a population?
2. An individual is waiting for a kidney transplant. Explain why a kidney donated by a member of the general public chosen at random is likely to be rejected, whilst one from a person's close family might prove satisfactory.
3. 'Race is a cultural or social entity because the concept has little biological basis' (Stewart, 1987). Discuss.

Chapter 14

GENETIC DISEASE

We should not, indeed cannot, ignore the social consequences of science. Scientists, after all, are people just like you and me and their ideas are shaped by the society in which they live. Just as scientists' ideas are shaped by society, so their discoveries and interpretations of their experimental results affect all of us. Nowhere is this more true than in the application of genetics to people, especially in the area of so-called inherited or genetic diseases. Fortunately, such diseases are very rare (Table 14.1) but we are too often offered a picture of some incurable hereditary curse which is handed down from generation to generation in some predetermined manner. By the time you have finished the work in this and the next chapter, you will be able to convince other people that this view is wrong, and that we should not treat people like light bulbs which either work or do not.

Table 14.1 The percentage of children born who suffer from diseases caused by gene mutation and to chromosome mutation

Type of mutation	Percentage of live births
gene mutation:	
autosomal dominant	0.90
autosomal recessive	0.25
X-linked	0.05
chromosomal mutation:	
autosomal trisomy (mainly Down's syndrome)	0.14
sex chromosomes -	
XYY, XXY and other male	0.17
XO, XXX and other female	0.50

LEARNING OBJECTIVES

After completing the work in this chapter you will be able to:

1. critically discuss the concepts of genetic normality and abnormality;
2. describe some genetic disorders caused by gene and chromosome mutation;
3. suggest why rare, deleterious alleles are not eliminated from gene pools by natural selection.

14.1 WHAT IS AN INHERITED DISEASE?

Towards a definition

Some infectious diseases, like measles and syphilis, are said to be non-hereditary since all people, more or less, can catch them, and it depends primarily on the environment whether one catches them or not. For example, the chances of you catching malaria in Aberdeen, Scotland are zero because the *Anopheles* mosquito which carries the malarial parasite does not occur in Northern Europe. Similarly, syphilis was unknown

Fig 14.1 This young man is an albino. Notice the lack of pigmentation in his hair. Because the eyes of people suffering from albinism are particularly sensitive to light, they often have to wear tinted glasses. Otherwise they live perfectly normal lives.

in Polynesia until it was introduced by Europeans. However, these diseases only develop in humans and a few other animals. Only possessors of human genotypes develop the symptoms of malaria and, as you will see, not all human genotypes are equally susceptible.

Consider now two rather rare human traits, albinism (Fig 14.1) and xeroderma pigmentosum (Fig 14.2). Albinos characteristically have little or no melanin (a pigment) in their skin, hair or the irises of their eyes. This places them at risk from severe sunburn if they are exposed to strong sunlight. Xeroderma pigmentosum is a trait which is characterised by the development of freckles which may become cancerous when the skin is exposed to sunlight. Both of these traits are inherited – albinism is due to the presence of a double dose of a recessive allele, whilst xeroderma is caused by a dominant allele. However, if people who have these traits protect themselves from sunlight they can live normal lives. Obviously they retain their inherited potential to develop the trait, but the 'disease' as such is no longer there. Health and disease are properties of the phenotype, and the phenotype is a product of both the genotype and the environment.

Once again we are dealing with a spectrum of effects in which there is no clear-cut distinction between hereditary and non-hereditary diseases. The best definition is no more than a rule of thumb: **Hereditary disease is due to a more or less rare genotype that reacts to produce ill-adapted phenotypes in environments in which more common genotypes give rise to healthy phenotypes.**

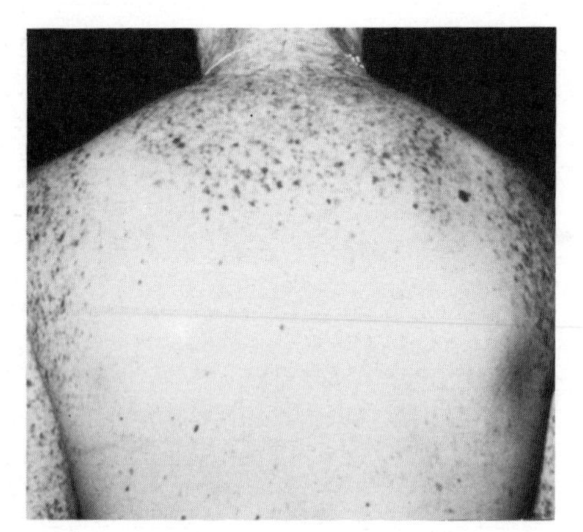

Fig 14.2 A person showing the characteristic freckling of the skin associated with xeroderma pigmentosum.

'Hereditary diseases are incurable – aren't they?'

I suffer from two 'inherited diseases'. I have an optical astigmatism which means my eyes are rugby ball shaped rather than spherical and I am red-green colour blind. By wearing glasses I can cure the first and I only notice the second when I am trying to convince my wife that the brown shirt I wore yesterday is now dark green. The curability of a disease does not depend upon whether it has a hereditary component or not, it depends on the environment. If an appropriate environment can be provided, any disease, in theory, is curable. Obviously there are at present some incurable hereditary diseases like Duchenne muscular dystrophy, but there are also a large number of incurable non-hereditary diseases like AIDS. The reason they are incurable is that we have not yet determined the correct environment needed to cure them. If we can provide appropriate changes in life style or administer certain drugs then life may be prolonged or, unfortunately, the disease may be incurable and the patient will die.

Phenocopies

To cure a hereditary disease we need to obtain the best phenotype from the existing genotype. Ideally we would like to know how the genotype would respond to every single environment we could possibly expose it to, the so-called **norm of reaction**. This is unobtainable in practice for people, but the norms of reaction in experimental animals like *Drosophila* (fruit flies) do suggest a way forward.

The normal body colour of *Drosophila melanogaster* is light brown with black markings on the abdomen. There exists a mutant which has a yellow body colour. However, if you raise the larvae of normal flies on food containing silver salts, the adults have a yellow body colour. So a genotypically normal fly has the potential to develop a yellow phenotype provided it is raised in the correct environment. The yellow bodied but genotypically normal fly is a **phenocopy** of the yellow mutant. Similarly, someone who has dark hair but bleaches it is a phenocopy of a person who has blonde hair.

Fig 14.3 Gary Mabbutt, the Tottenham Hotspur footballer, is a diabetic, but is also a superb athlete.

We can now see clearly just what is needed to 'cure' a hereditary disease: an environment must be found which makes an abnormal genotype produce a phenocopy of a normal healthy phenotype. Such environments may be rare but they can be found. To convince you that this is possible, consider diabetes mellitus. Some forms of this disease have an inherited component. As you probably know the disease involves a loss of the ability to control the amount of sugar in the blood. People who suffered from this disease usually died at an early age until the physiological basis of the disorder was discovered. Once it became known that diabetes was caused by a lack of insulin it was a logical step to supply diabetics with insulin. Diabetics are cured as long as they receive the correct amount of insulin, and they can lead perfectly normal lives (Fig 14.3). Similarly, as long as I wear my glasses my astigmatism is cured. Such a cure is, of course, not like recovering from a bout of flu, since the cure only lasts as long as the correct environment is supplied. If I take my glasses off I cannot see properly, but since I only take them off to go to sleep that does not really matter.

Clearly the terms 'normal' and 'abnormal', as applied to human genotypes, are relative. The phenotype which develops from a particular genotype depends upon the environment. We should not, therefore, classify people on the basis of their genotypes into distinct phenotypic categories of diseased and normal. Please bear this in mind as we proceed through this chapter, looking at human disorders which have a large hereditary component. These shall be referred to as genetic disorders but this is really just a form of shorthand.

The origin of genetic disorders

If a disorder has a large hereditary component then this suggests that an alteration has occurred somewhere in the patient's DNA. Sometimes the change in the DNA affects a whole chromosome, for example Down's syndrome, but often the changes that have taken place are too small to see. Consequently, whilst we can track the pattern of inheritance of a particular disorder, the site or sites of mutation remain a mystery. Even more mysterious are those disorders which seem to have an inherited component but which do not show simple patterns of inheritance, for example high blood pressure, heart disease and some forms of cancer. We can see associations, but so far it is impossible to specify cause and effect. The science of medical genetics is far from exact and very often the answer to the question 'what causes X?' is quite simply 'we do not know'. The remainder of this chapter will give you some basic information about a number of different genetic disorders, their patterns of inheritance and why the alleles which cause the disorders are not eliminated from a population by natural selection. In the next chapter we will go into details of prenatal diagnosis, genetic screening and genetic counselling.

14.2 GENETIC DISORDERS DUE TO SINGLE GENE DEFECTS

About 2300 genetic diseases are known or thought to be determined each by a single mutant gene or a pair of mutant alleles. We can divide these into three groups on the basis of their pattern of inheritance – autosomal dominant, autosomal recessive and sex-linked.

Autosomal dominant inheritance

Unlike recessives, where the effect is usually due to a change in the structure of an enzyme (see section 2.2), the biochemical basis of dominant defects is poorly understood. Dominant disorders may be associated with changes in regulatory mechanisms. Homozygotes for most genes determining dominant defects are almost never seen. This is partly due to the fact that the parents would both have to be heterozygous for the same rare allele (think about it), and also that homozygosity for genes that determine a dominant defect would probably be a lethal condition. Over 1700 traits showing a dominant pattern of inheritance are known in humans. One example is given here.

Huntington's chorea.
Cause: Autosomal dominant gene located on the tip of the short arm of chromosome number 4.
Frequency: 5 per 100 000 of the population. Sufferers can be heterozygous or homozygous for the defective gene.
Symptoms, treatment and detection: A neuropsychiatric disorder resulting in involuntary muscle movement and progressive mental deterioration. Variable age of onset, but usually in the 40s and 50s so people can have children before they realise they have the disorder. No known treatment. Detection of carriers of the gene is now possible using DNA probes and restriction fragment length polymorphism (RFLP), as explained in Chapter 12.

Autosomal recessive inheritance

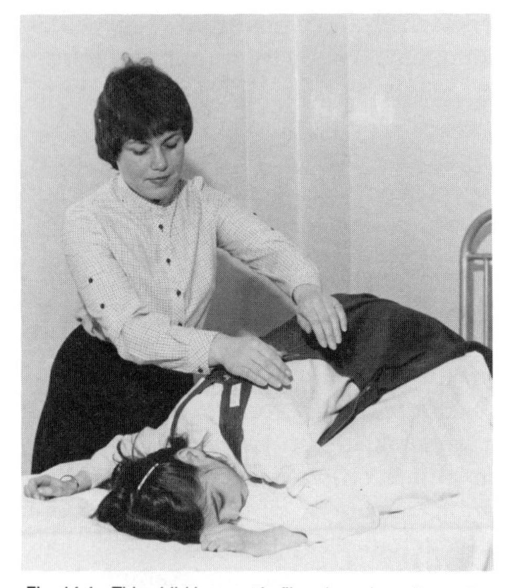

Fig 14.4 This child has cystic fibrosis and needs regular physiotherapy to clear the mucus from her lungs.

Recessive traits, whether autosomal or X-linked, are likely to be due to a mutation producing an allele which encodes a defective enzyme that has a very low level of activity. For this reason they are sometimes called inborn errors of metabolism. Heterozygotes, though they have only half as much of the enzyme, usually metabolise the substrate almost as efficiently as the normal homozygote. Over 1300 autosomal recessive traits are now known in humans.

Cystic fibrosis.
Cause: Autosomal recessive gene probably located on the long arm of chromosome 7.
Frequency: A very common disorder among Europeans. In Britain, 1 in 1600 live births.
Symptoms, treatment and detection: This is a generalised disorder of mucus-secreting glands, particularly those in the pancreas, intestine and lungs. The mucus becomes very thick and sticky so that ducts can become blocked leading to intestinal blockages and pneumonia. The food is not digested and absorbed properly due to blockage of the pancreatic duct. Treatment consists of regular physiotherapy (Fig 14.4), antibiotic therapy and pancreatic extract is sometimes given by mouth. The outlook for most patients is not good, many dying at an early age. Detection of sufferers prenatally (before birth) may become possible using DNA probes and linkage analysis linked to chorionic villus sampling. Detection of the heterozygote carriers on the basis of sodium content of their sweat has been attempted but is not a reliable technique. The use of DNA probes may also help with the detection of carriers in the future.

Phenylketonuria (PKU).
Cause: Autosomal recessive affecting the enzyme phenylalanine hydroxylase which converts phenylalanine into tyrosine (Fig 2.4). Absence of the enzyme causes an increase in the level of phenylalanine in the blood and tissues.
Frequency: 1 in 15 000 births.
Symptoms, treatment and detection: Increased levels of phenylalanine create a wide range of effects including mental retardation, restlessness, muscle stiffness and epilepsy. All children in Great Britain are compulsorily tested at birth (Guthrie test) and those affected are put on a diet low in phenylalanine until the onset of puberty. Expectant mothers who are PKU sufferers are also sometimes put back on the special diet. With the correct treatment, people with the condition lead a perfectly normal life, though apparently the diet is absolutely disgusting.

Sickle-cell anaemia.
Cause: Autosomal recessive. A classic example of a point mutation causing the replacement of the glutamic acid residue at position 6 in the amino acid chain of the β-chain of haemoglobin by valine.
Frequency: Responsible for as many as 100 000 deaths per year. The frequency of the sickle allele reaches 10 per cent in some populations. In some parts of Africa the incidence can be as high as 30 per cent. Under these circumstances, 1 in 50 children may have the disease.
Symptoms and detection: Cells containing the abnormal haemoglobin S (Hb^S) sickle when subjected to low oxygen tensions (Fig 9.5). As a result, blood vessels can become blocked, leading to poor circulation and intense pain. The pleiotropic effects of the disease are shown in Fig 14.5. Successful prenatal diagnosis is based on DNA probes and RFLP. Heterozygotes can be detected using electrophoresis of haemoglobin.

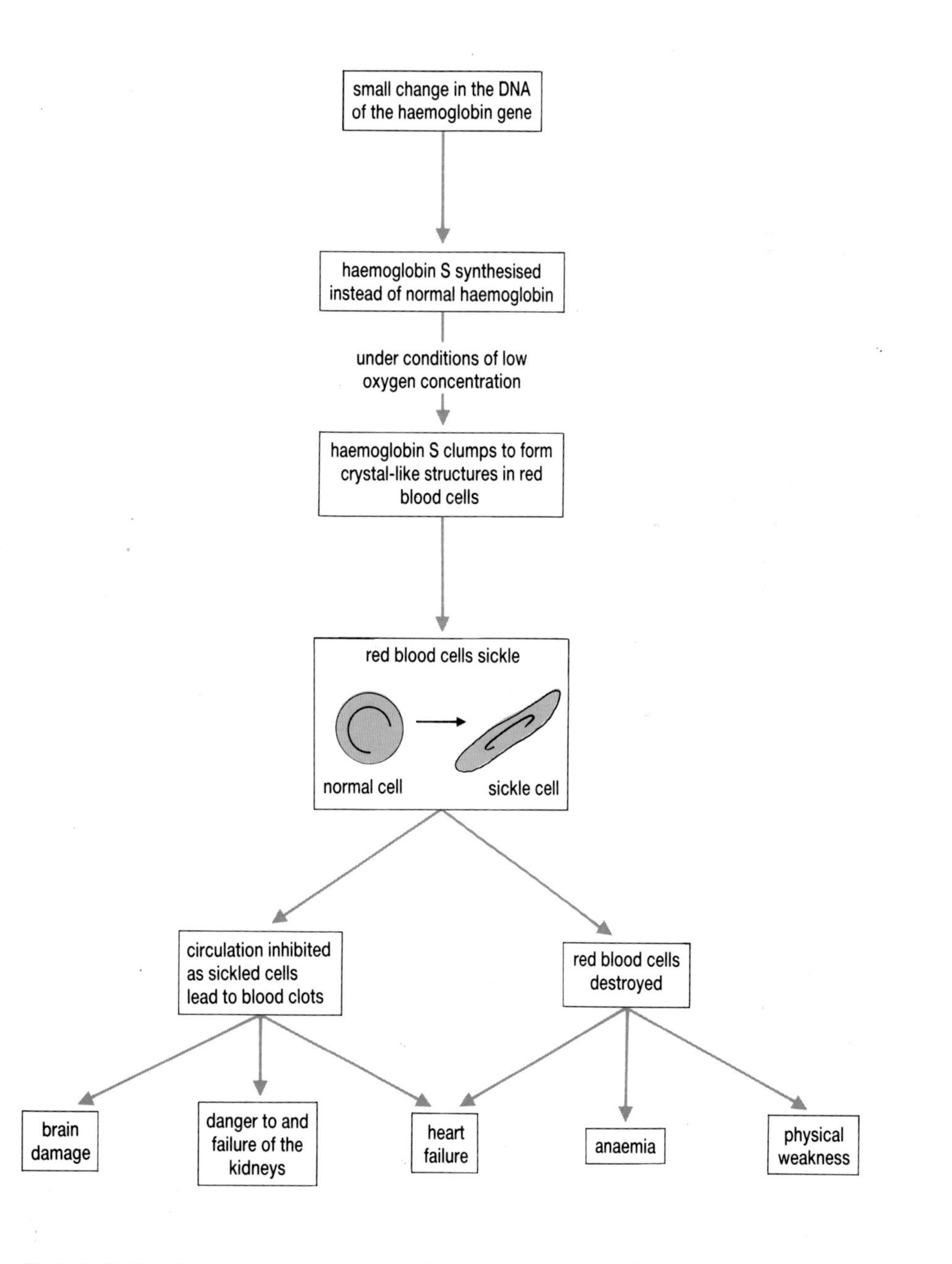

Fig 14.5 Sickle-cell anaemia – how a minor change in a single gene can cause a large number of phenotypic effects.

Galactosaemia.
Cause: Autosomal recessive, possibly on chromosome 9. There may be several types of galactosaemia. The best known involves the absence of a transferase enzyme, galactose-1-phosphate uridyltransferase, so that galactose in milk is not converted into glucose, leading to a build up of galactose in the body.
Frequency: Extremely rare about 1 in 200 000 live births.
Symptoms, treatment and detection: If the build up of galactose goes undetected this may cause severe malnutrition, brain damage and mental retardation. The disorder can be detected prenatally by assaying amniotic fluid (obtained by amniocentesis) for the presence of the enzyme. Heterozygotes can be detected by assaying the level of enzyme activity in their red blood cells.

Sex (X)-linked inheritance

Most known X-linked traits are recessive. One of the exceptions is hypophosphatemic (vitamin D resistant) rickets. The presence of the mutation on the X chromosome means that most sufferers are male since the probability of a female getting a double dose of a rare allele is very small. A provisional map of the X chromosome showing the relationship between some of the genes involved in X-linked traits is shown in Fig 14.6. Note this is only a small proportion of the 250 possible X-linked traits.

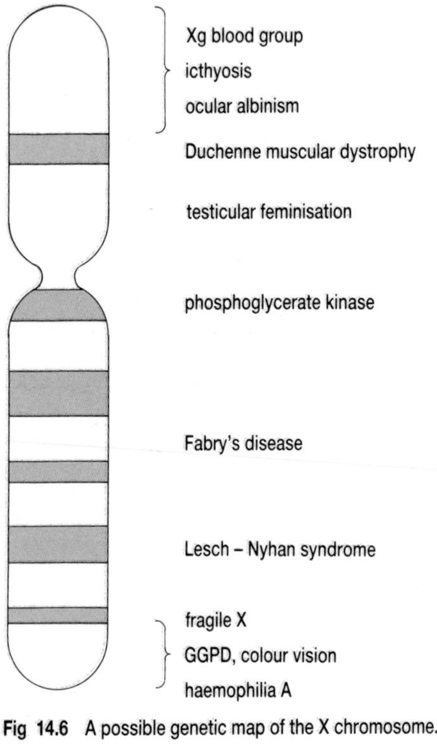

Xg blood group
icthyosis
ocular albinism
Duchenne muscular dystrophy
testicular feminisation
phosphoglycerate kinase
Fabry's disease
Lesch – Nyhan syndrome
fragile X
GGPD, colour vision
haemophilia A

Fig 14.6 A possible genetic map of the X chromosome.

Haemophilia.
Cause: X-linked recessive. A group of disorders characterised by the lack of blood clotting factors, for example factor VIII, leading to excessive bleeding.
Frequency: About 1 in 10 000 live births of males. (Haemophilia A and B combined.)
Symptoms, treatment and detection: Sufferers characteristically suffer excessive bleeding both internally and externally, usually following injury. The bleeding takes the form of a persistent, slow oozing which if unchecked can lead to anaemia. Bleeding into the joints can cause severe pain. Some forms of haemophilia can now be treated using regular injections of factor VIII. Contaminated supplies of this blood product from the United States of America has probably caused the high rate of AIDS among haemophiliacs. All blood products are now pasteurised to kill the virus before they are given to haemophiliacs. The production of genetically engineered factor VIII will overcome the infection problem completely. Rare female sufferers have been recorded who have survived childbirth and borne haemophiliac sons. About 85 per cent of the heterozygous female carriers can be detected by assaying the level of factor VIII in their blood plasma.

Duchenne muscular dystrophy.
Cause: X-linked recessive.
Frequency: 1 in 5000 male live births.
Symptoms, treatment and detection: A disorder found in young boys which starts with a waddling gait and increased difficulty in climbing the stairs. Confined to wheelchairs by 10 years of age, most patients die before their twentieth birthday. No cure is currently available. Some attempt has

been made to detect carriers by using raised serum creatinine kinase levels in the blood but this approach has many pitfalls. Now the use of DNA probes and RFLP can detect carriers in about 95 per cent of families. Prenatal screening using similar techniques may soon be available.

An interesting problem

Obviously, the alleles which we have considered in the previous section are deleterious. They tend to reduce an individual's fitness, that is, they die or they leave fewer children than average. So why, then, has natural selection not weeded these alleles out? There are three possibilities.

Possibility one

Mutation feeds the alleles into the gene pool as fast as natural selection wipes them out. If you imagine a gene pool as a bath, then the new alleles being introduced into the population by mutation will be the water coming out of the taps and the water going down the plug hole will be the alleles being lost from the gene pool. The plug hole is, of course, a selective one, only allowing certain alleles through. If the rate at which the water comes into your bath is the same as the rate at which it leaves then the bath will always have the same amount of water in it. In terms of a gene pool, if mutation supplies deleterious alleles as fast as natural selection destroys them then there will always be a certain number of deleterious alleles in the population, albeit at a very low frequency. The problem with this model is that it is very difficult to test since it is very hard to measure mutation rates in human populations. Nonetheless, genetic disorders due to dominant alleles must largely be maintained by this mechanism, and probably only about 40 of the 2300 genetic variations which are regarded as causing disease have allele frequencies above the mutation rate. This observation has far-reaching implications for genetic screening, discussed in the next chapter.

Possibility two

Natural selection acts on phenotypes. So recessive alleles can hide from the effects of selection provided they are in the heterozygous condition. Look at Table 14.2 which gives details about a number of recessive disorders. In all cases the number of sufferers, the homozygous recessives, is very low. But look at the number of heterozygotes. The number of carriers is much higher. To give you some idea of the implications of this, imagine a crowd of people at a large sporting event, say a football match or the Olympic Games. Let us say there are 100 000 people in the crowd. From Table 14.2 we can say that, on average, about 50 (that is, 100 000 divided by 2000) of

Table 14.2 The frequency of affected and carrier individuals for some genetic diseases showing an autosomal recessive pattern of inheritance

Disease	Frequency of affected individuals	Frequency of carriers
cystic fibrosis	about 1 in 2000	1 in 22
albinism	about 1 in 20 000	1 in 71.9
phenylketonuria	about 1 in 25 000	1 in 80
alkaptonuria	about 1 in 1 000 000	1 in 502.5

those people will have cystic fibrosis, but of the remainder, 4543 will, on average, be carrying the allele. Since each sufferer is homozygous, the number of cystic fibrosis alleles in that part of the population is $50 \times 2 = 100$. Each carrier is heterozygous, so the number of cystic fibrosis alleles in that part of the population is $4543 \times 1 = 4543$. So in our sports stadium only 2 per cent of all the cystic fibrosis alleles present in our little population can be 'seen' by natural selection and eliminated.

Possibility three

A third possibility is that the heterozygotes actually enjoy some selective advantage over both homozygotes. This certainly provides an explanation for the very high incidence of the sickle-cell allele in certain parts of the world. Fig 14.7 shows the distribution of the haemoglobin S allele (Hb^S) in the Old World, whilst Fig 14.8 shows the distribution of malaria.

Notice the relationship. The frequency of the Hb^S allele is highest where malaria is common. Now from the argument given above, most of the people carrying the Hb^S allele will be heterozygous, and it appears that the heterozygotes are more resistant to malaria than either of the homozygotes. The carriers of the allele (the heterozygotes) enjoy what is called heterozygote advantage. I will leave you to suggest why heterozygote advantage will maintain the deleterious Hb^S allele in the population. You will need to think about the three possible genotypes, the two homozygotes and the heterozygote, and their relative chances of surviving to reproduce in an area with a high incidence of malaria.

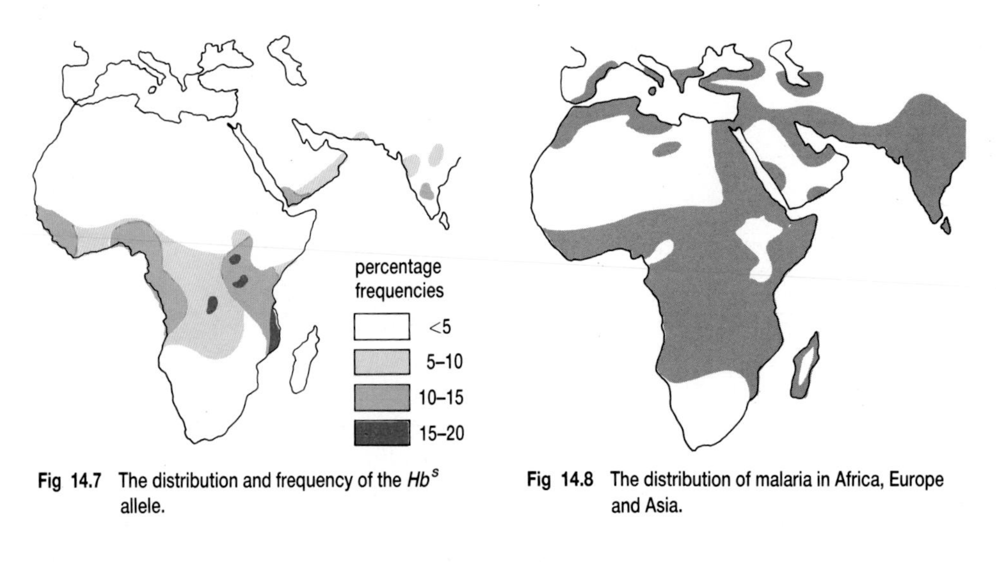

Fig 14.7 The distribution and frequency of the Hb^S allele.

Fig 14.8 The distribution of malaria in Africa, Europe and Asia.

QUESTIONS

14.1 The mutant alleles we have been looking at in the previous section are all very rare. What does that mean about their frequency in the gene pool?

14.2 For each of the disorders listed in Table 14.2 calculate the proportion of the alleles present in sufferers and carriers in our hypothetical sports crowd of 100 000.

14.3 **(a)** Why do you think that most of the genetic disorders discovered so far are due to dominant autosomal mutations?
(b) Why are X-linked recessive defects apparently relatively more common than autosomal recessive defects?

14.4 Explain why heterozygous advantage will act to maintain Hb^S alleles in human populations in areas where malaria is common.

14.3 CHROMOSOMAL DEFECTS

Abnormalities in chromosomes are usually so serious that their carriers are often unable to reproduce. Consequently nearly all the new cases are due to new 'mutations'. The syndromes produced are usually very complex since changes in the number or structure of the chromosomes will involve many genes. About 0.31 per cent of all live births are affected by some alteration in the autosomes, of which one-third are Down's syndrome, and about 0.2 per cent are affected by X chromosome abnormalities. Chromosomal abnormalities are detected using karyotype analysis (Box 14.1). Cells, usually white blood cells or fetal cells obtained by amniocentesis or chorionic villus sampling (see Box 15.2), are stimulated to undergo mitosis. A drug, colchicine, is then applied which inhibits the formation of the mitotic spindle. The result is a cell full of contracted chromosomes with nowhere to go. The cells and their contents are then stained and photographed under a microscope. The photograph is then cut up and the homologous chromosomes are arranged to form a karyotype.

BOX 14.1 Karyotype analysis

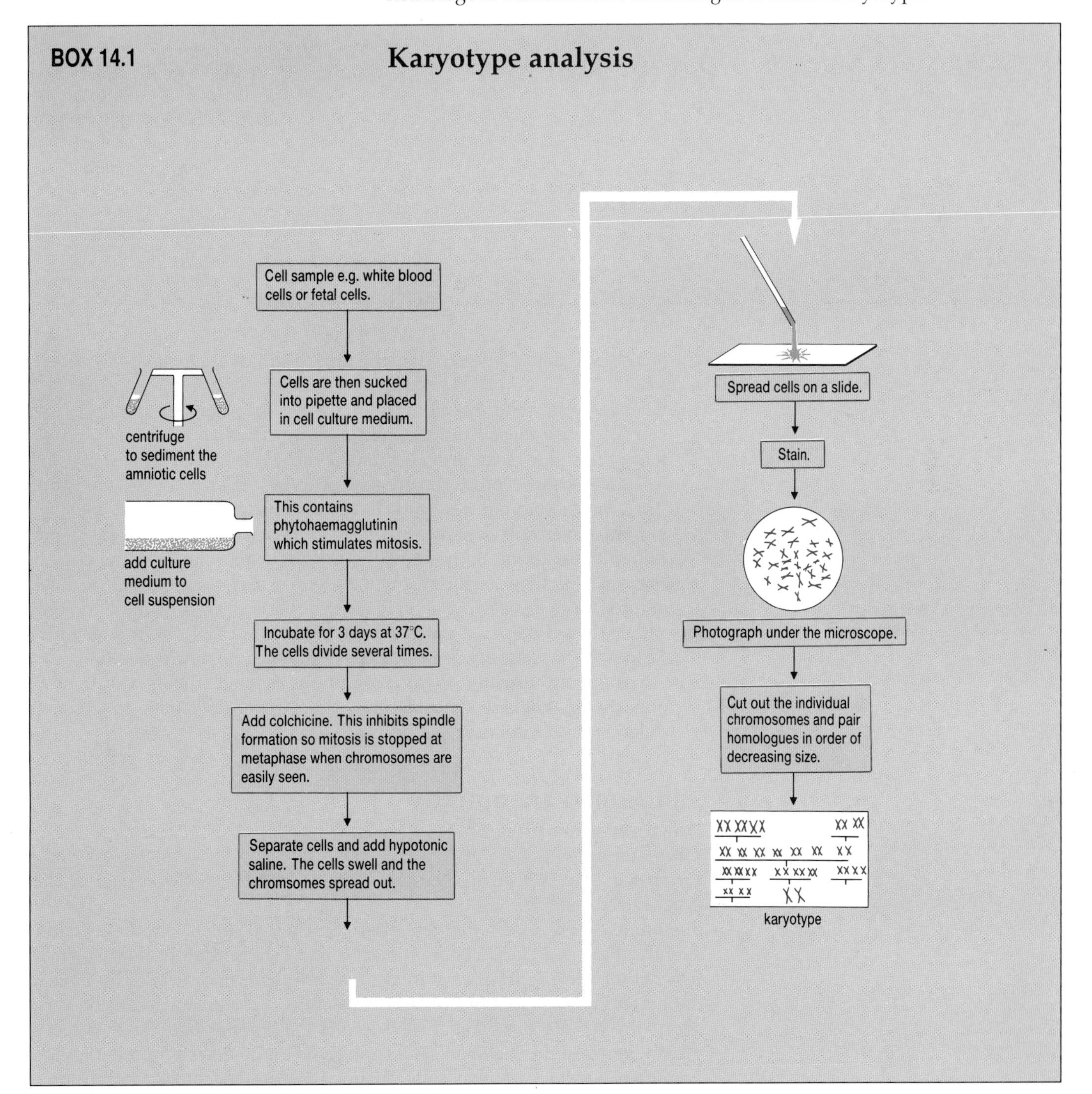

QUESTION

14.5 Fig 14.9 shows a normal human karyotype. Is this individual male or female?

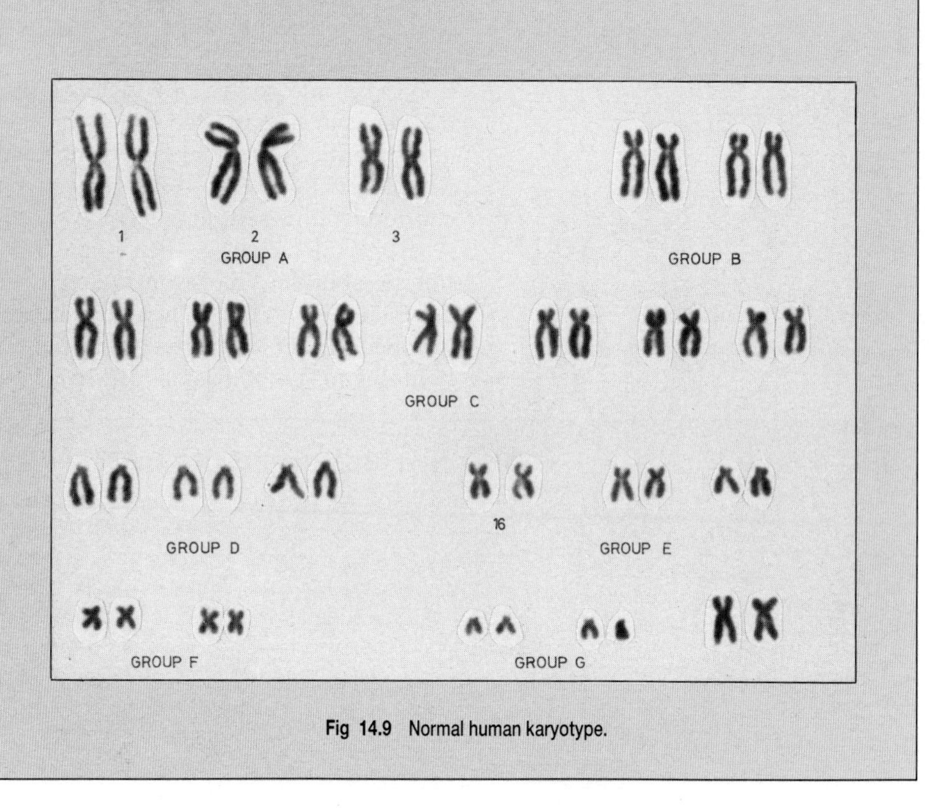

Fig 14.9 Normal human karyotype.

Chromosomal defects detected using this technique can take two basic forms.

1. Changes in the number of individual chromosomes (aneuploidy). You can recognise these from the suffix -**somy**. Thus an individual whose karyotype shows an extra chromosome 21 is called a **trisomy 21**. Trisomies can be produced by several types of error in cell division. The most common is called nondisjunction, where a pair of chromosomes fails to separate during gamete formation. This results in one cell containing both members of a particular chromosome pair. If this cell is then fertilised, the zygote will contain three of that chromosome instead of a pair. An individual possessing only one rather than a pair of a particular chromosome would be **monosomic**. Monosomy of autosomes in people is usually, if not always, lethal.
2. Alternatively, we can recognise conditions in which there are structural abnormalities in the chromosomes due to translocations, inversions, deletions and duplications (Fig 14.10).

Autosomal aneuploidy

Down's syndrome (Fig 14.11)

Cause: Trisomy 21

Frequency: 1 in 700 births, the frequency increasing with the age of the parents (see below).

Symptoms, treatment and detection: Epicanthal folds on the eyelids, slightly flattened face, poor muscle tone, short stature and mental retardation varying from very mild to severe. The children usually have a happy and extremely affectionate nature. Life expectancy used to be very short due to infections and heart defects. Antibiotics and heart surgery greatly increase the life span. The condition can be detected prenatally by karyotype analysis of fetal cells obtained by amniocentesis or chorionic

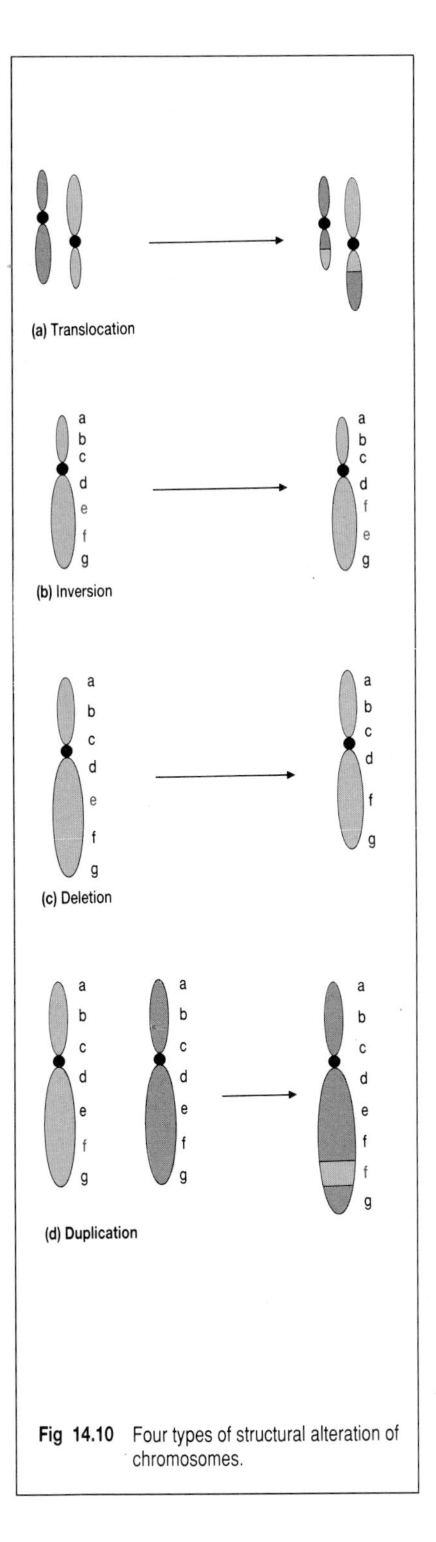

Fig 14.10 Four types of structural alteration of chromosomes.

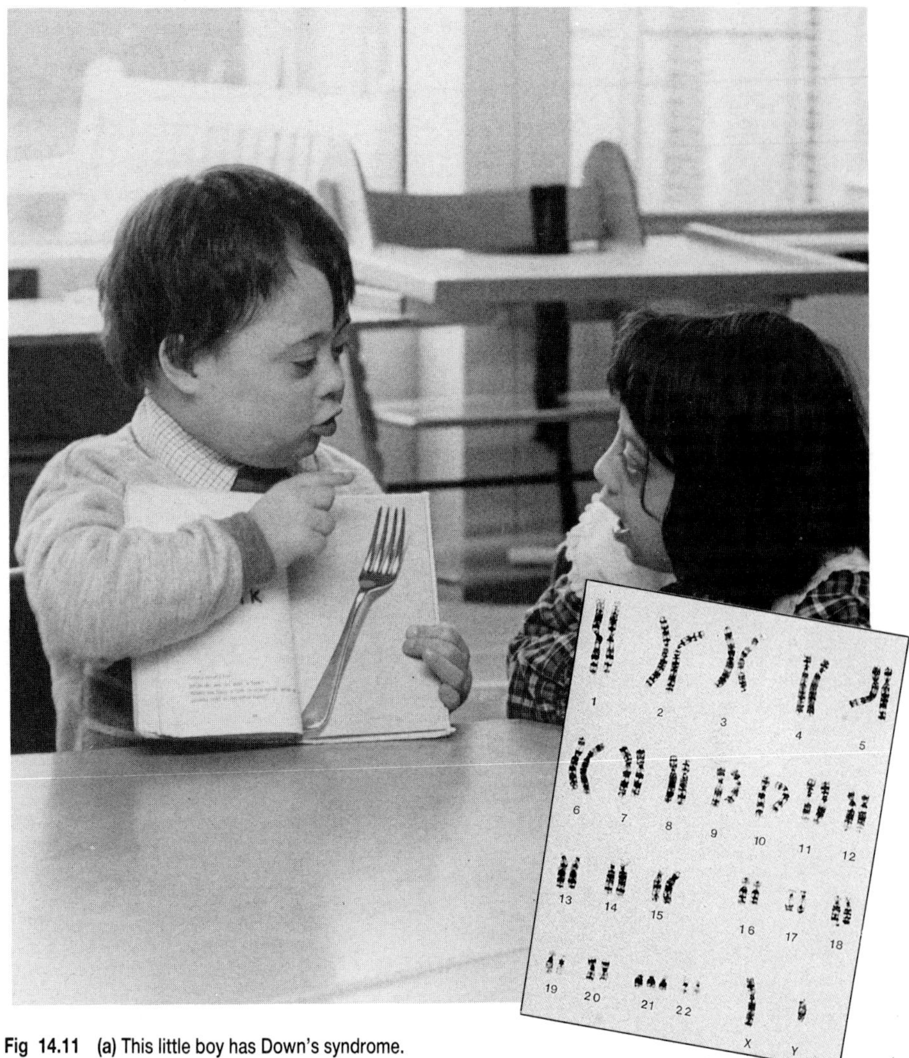

Fig 14.11 (a) This little boy has Down's syndrome.
(b) Karyotype showing the arrangement of chromosomes in a male with Down's syndrome.

villus sampling. Such an analysis is often offered routinely to women over the age of 40. Most cases of trisomy 21 arise *de novo,* presumably as a result of meiotic nondisjunction (Fig 14.13), so this form of Down's syndrome is not inherited.

The risk of having an affected child increases with maternal age (Fig 14.12), with mothers over 35 being particularly at risk. This may be the consequence of oocytes remaining in a 'resting stage' (dictyotene) from birth to just before ovulation. The many years of static dictyotene are thought to predispose the oocyte to the accumulation of genetic errors due to both internal factors, for example hormone levels, and external factors such as infection and ionising radiation.

Whilst the risk of having an affected child rises with maternal age, most (about 80 per cent) of Down's syndrome babies are born to women under the age of 35. There are two main reasons for this:

1. More women under 35 have babies than women over 35;
2. older women are usually offered routine screening for Down's syndrome and so can opt for abortion if they are carrying an affected fetus. Recently (1988) a blood test has been developed to screen all pregnant women to see if they are carrying a Down's syndrome baby. This test is expected to detect upto 60 per cent of Down's fetuses. A woman suspected of carrying a Down's baby would then be offered an aminocentesis. The test is likely to be offered to all women attending an antenatal clinic in Britain.

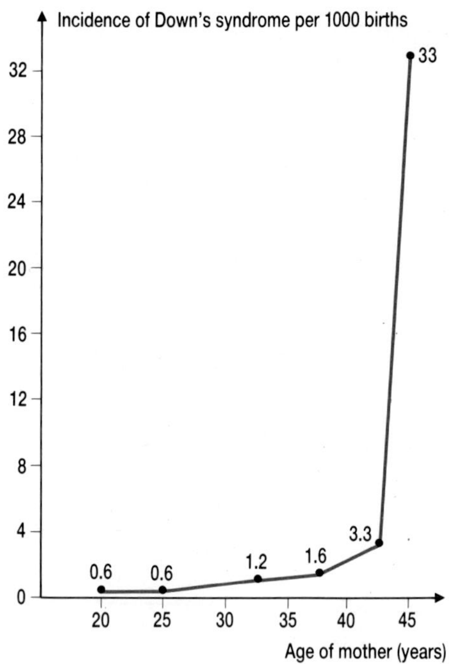

Fig 14.12 The relationship between the frequency of Down's syndrome and material age.

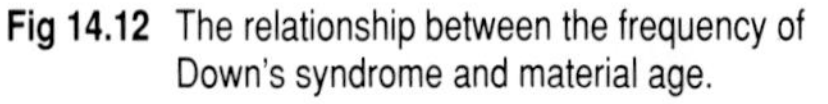

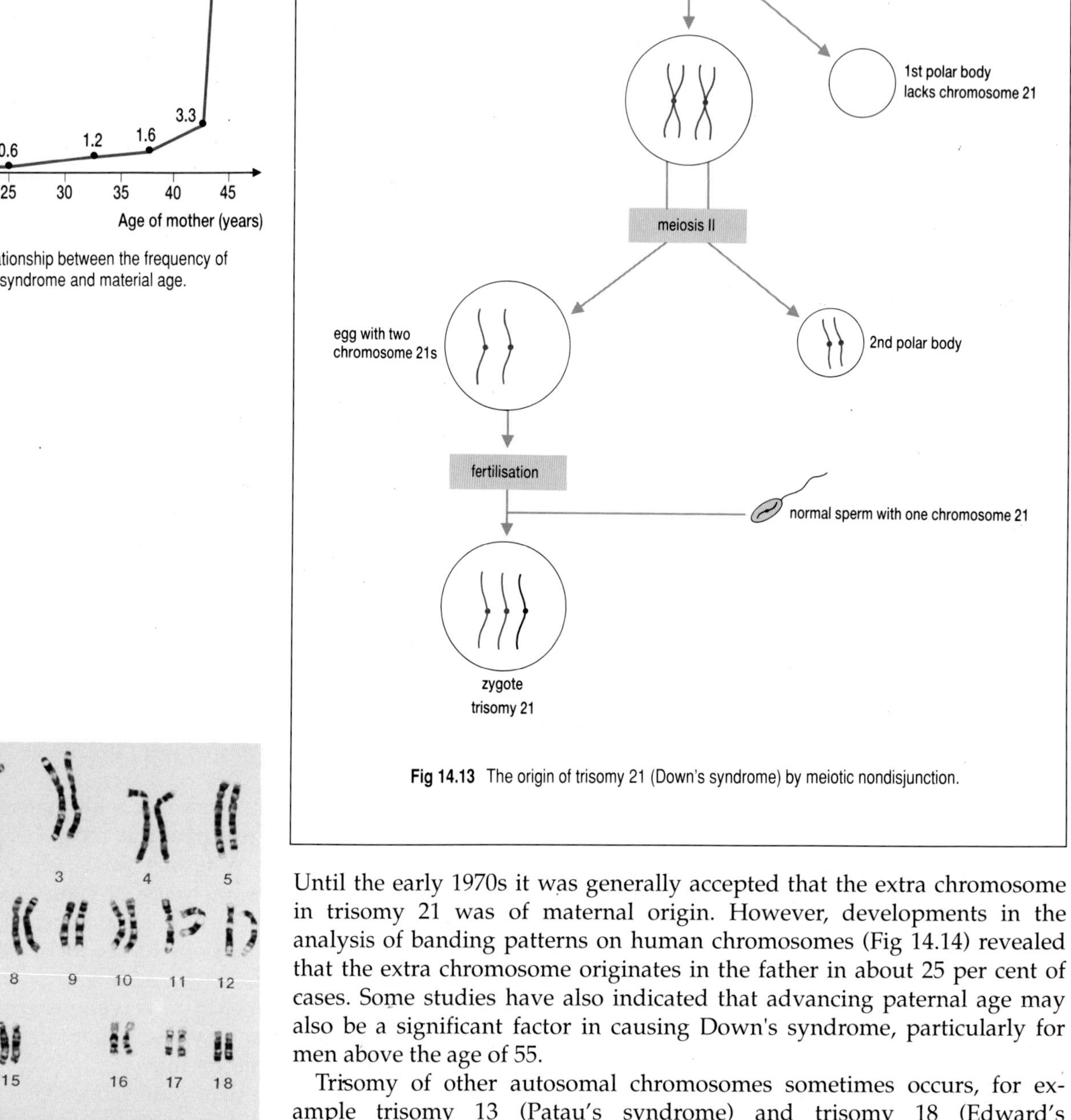

Fig 14.13 The origin of trisomy 21 (Down's syndrome) by meiotic nondisjunction.

Fig 14.14 Homologous pairs of human chromosomes showing characteristic banding patterns. These patterns can be used to track chromosomes between generations.

Until the early 1970s it was generally accepted that the extra chromosome in trisomy 21 was of maternal origin. However, developments in the analysis of banding patterns on human chromosomes (Fig 14.14) revealed that the extra chromosome originates in the father in about 25 per cent of cases. Some studies have also indicated that advancing paternal age may also be a significant factor in causing Down's syndrome, particularly for men above the age of 55.

Trisomy of other autosomal chromosomes sometimes occurs, for example trisomy 13 (Patau's syndrome) and trisomy 18 (Edward's syndrome). The effects of these are often more severe than Down's, and in the case of Patau's syndrome life expectancy is very short. Fortunately, both are very rare – 1 in 5000 and 1 in 3000 live births respectively.

Aneuploidy of the sex chromosomes

Nondisjunction of the sex chromosomes can produce a range of syndromes.

XO (Turner's syndrome). Found in about 1 in 3000 live births and common in about 20 per cent of spontaneous abortions. The characteristic picture is of a short girl, who fails to menstruate and who lacks secondary sex characteristics. There is often webbing of the neck. Some, but not all, sufferers may be mentally retarded. Not all patients develop these symptoms and many undoubtedly live perfectly normal lives.

YO Lethal. Possibly because the X chromosome carries the gene controlling blood clotting factors.

XXX (Triple X). A very rare condition. The affected women are usually perfectly normal.

XXY (Klinefelter's syndrome). Occurs in about 1 in 500 males. The only clinical sign in many patients are small testes which are unable to produce sperm. Long legs, breast development and a high voice may also occur. Otherwise many men look perfectly normal and probably live happy lives, totally unaware that they have the syndrome.

XYY Syndrome. These males are usually taller but otherwise perfectly normal. The association between this condition and aggression is based on spurious correlations involving members of the prison population.

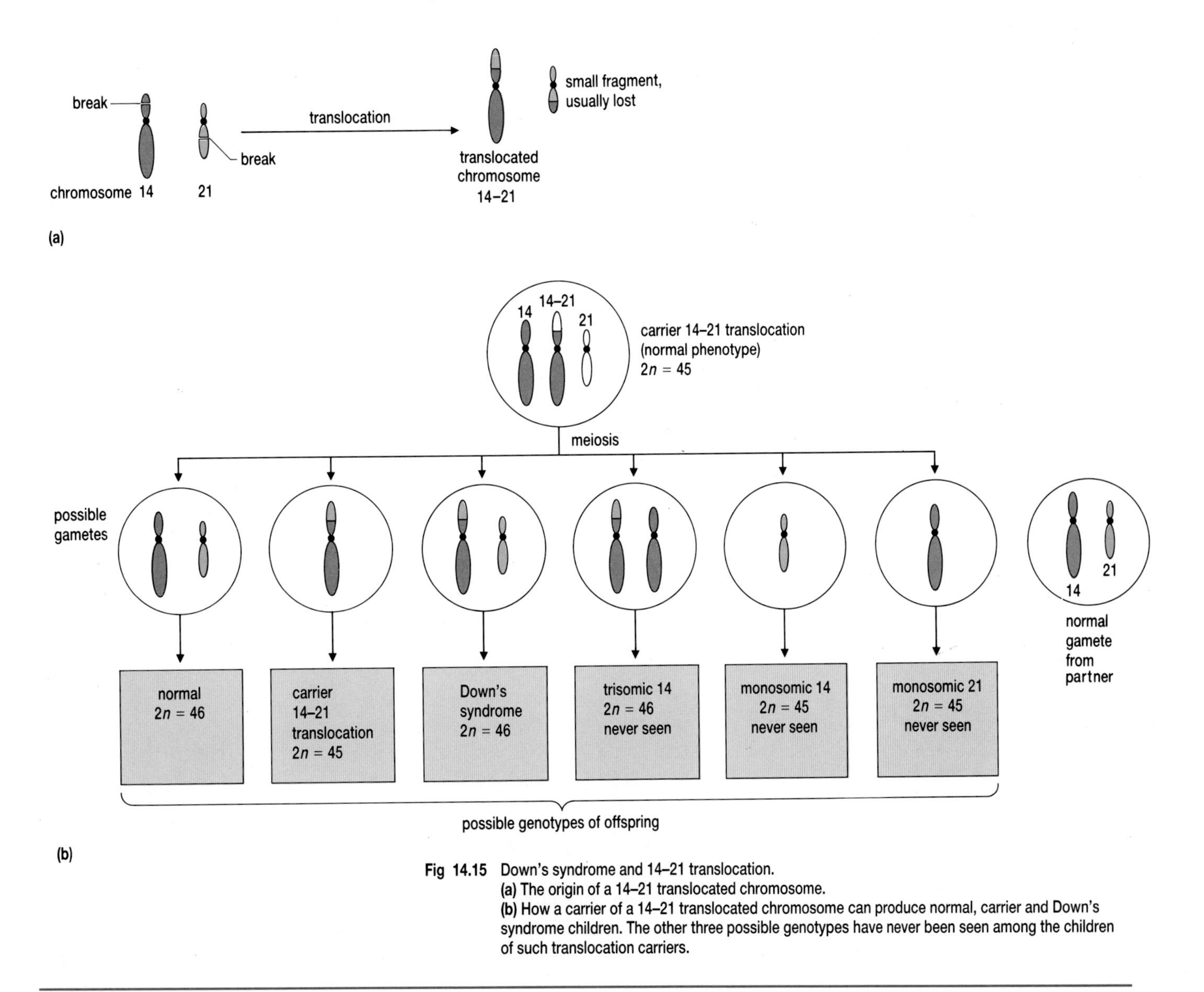

Fig 14.15 Down's syndrome and 14–21 translocation.
(a) The origin of a 14–21 translocated chromosome.
(b) How a carrier of a 14–21 translocated chromosome can produce normal, carrier and Down's syndrome children. The other three possible genotypes have never been seen among the children of such translocation carriers.

Interestingly, a Frenchman, Daniel Hugon, was actually awarded a lighter sentence for the murder of a prostitute because he was XYY, and an Australian charged with murder was acquitted altogether.

Structural abnormalities of chromosomes

Translocation involves the movement of a piece of one chromosome to another chromosome (Fig 14.10). During meiosis, therefore, one of the gametes may receive an extra piece of a particular chromosome. Should this gamete then be fertilised, the zygote will have an extra piece of the chromosome, as in trisomy. Down's syndrome may be caused by a translocation in about 5 per cent of cases. The translocated piece of chromosome 21 is usually transferred to chromosome 13, 14 or 15. Such translocations can produce an inherited form of Down's syndrome, with so called 'balanced carriers' transmitting the translocated chromosome to their offspring (Fig 14.15), but most cases of translocation arise *de novo*.

Deletions involve the loss of chromosome material. For example, one very rare lethal syndrome in man, cri-du-chat, involves a deletion in chromosome 5. Here the child carrying an affected chromosome mews like a cat, is mentally retarded and usually dies at a young age.

QUESTIONS

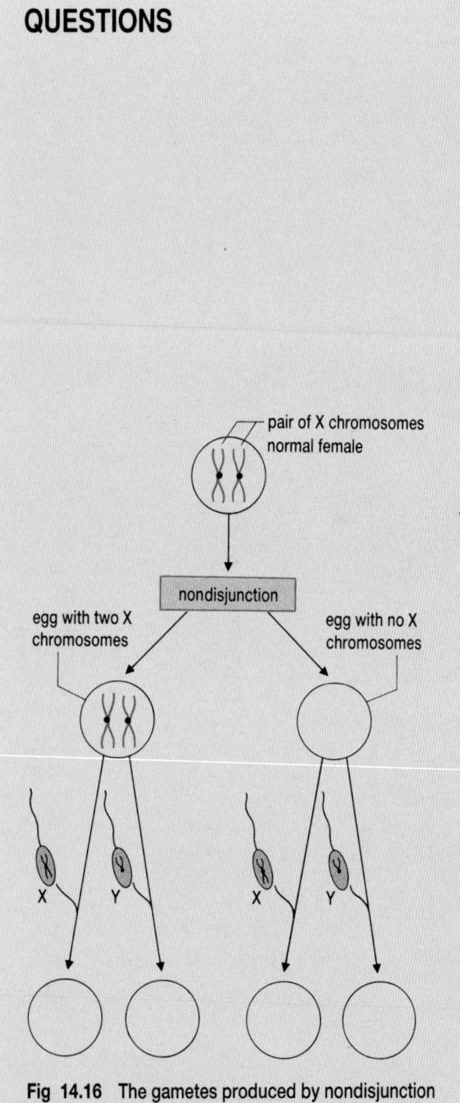

Fig 14.16 The gametes produced by nondisjunction of X chromosomes at meiosis I in females.

14.6 Fig 14.16 shows the gametes produced by nondisjunction of the sex chromosome at meiosis I in females. Copy and complete the diagram to show the genotypes and phenotypes of the offspring produced when the female gametes are fertilised by normal human sperm. What gametes and offspring would have been produced if nondisjunction had occurred at meiosis II?

14.7 **(a)** What gametes would be produced by nondisjunction during spermatogenesis?
(b) What would the genotypes and phenotypes of the offspring be if these abnormal sperm were to fertilise a normal egg?

14.8 Consider a woman with the triple X genotype married to a man with a normal complement of sex chromosomes. What proportion of their children would you expect to suffer from Klinefelter's syndrome?

14.9 Chromosome banding polymorphisms allow us to track the movement of chromosomes between generations since different 'morphs' act as chromosome markers. Consider the following marriage in which the parents are

female – 21^a21^b
male – 21^c21^d

where a, b, c and d stand for morphs of chromosome 21 which can be distinguished by means of variations in banding pattern. The following children, where 42A stands for the rest of the autosomes, are produced:

(a) 42A + 21^b21^d + XX
(b) 42A + $21^a21^c21^c$ + XY
(c) 42A + $21^b21^b21^c$ + XX
(d) 42A + 21^a21^d + XYY.

In each case :
(i) state the phenotype of the child
(ii) using diagrams, explain the events which gave rise to any abnormal condition
(iii) state in which parent the event took place.

(continued)

QUESTIONS

14.10 Colour blindness is an X-linked trait. A patient with Turner's syndrome is found to be colour blind. How can this be explained? Does this tell you whether the nondisjunction occurred in the mother or the father?

14.11 The parents of a Down's syndrome child want to have more children. What tests would you carry out and what advice would you give them?

14.4 WHAT DO WE MEAN BY GENETICALLY NORMAL?

In this last section, you should try to think about the distinction which is too often made between so-called genetically normal and abnormal people. If you went to a 20-year-old genetics textbook, a normal genotype would have been defined in terms of a 'wild type'. Such a genotype would be unvarying, have maximal fitness and would be found in all members of a population except for a rare unfit minority, that is, those possessing a genetic disease. However, we saw in Chapter 13 that human populations are genetically very variable. The human genotype has around 100 000 loci, of which about 10 per cent may be polymorphic. Even if you only had two different alleles at a locus, the number of ways of arranging a genotype at the 10 000 variable loci is enormous. If you consider the HLA system which provides practically all of us with a unique identity code, you will begin to realise the amount of genetic variation in human populations. Furthermore, there are now so many HLA types which can be associated with particular diseases, for example BW27 and ankylosing spondylitis, that most of us will probably have at least one predisposing HLA type but no disease, since the HLA type is a necessary but not sufficient precondition for the appearance of the disease.

If we look at the parents of a child with an inherited disease, say phenylketonuria, do they look any different from the rest of us? Of course not. It is the combination of these two people which has brought together two rare recessive alleles that has produced the disease. All of us are probably carrying about half a dozen recessive alleles which would cause severe disease in the homozygous state. The sobering thought that everyone is a 'carrier' of not just one but several genetic diseases should be tempered by the realisation that the chances of you meeting and having children with someone who is carrying the same rare alleles is minute. Consequently, few children are at risk, with the exception of groups like the white population of Great Britain who have high frequencies of some recessively inherited diseases, for example phenylketonuria and cystic fibrosis. Clearly, we cannot think about human populations consisting of individuals who all have the same healthy genotype (good light bulbs) and a few unfortunate people having rare mutations which make them unhealthy (dud light bulbs).

SUMMARY ASSIGNMENT

1. What do you understand by the term genetic disease? Produce a table to show the frequency, symptoms and methods of diagnosis of some genetic diseases.
2. Define nondisjunction and describe the common syndromes caused by nondisjunction of sex chromosomes and autosomes.
3. **(a)** Why do white Europeans have a high risk of having children who suffer from cystic fibrosis?
   (b) Can you suggest why the allele causing cystic fibrosis is not eliminated by natural selection?
4. 'We can speak of individuals with a genetic disease but there is no "normal" human genotype. We are all as normal as everyone else.' Do you agree with this point of view?

Chapter 15

GENETIC DETECTIVES

The classic methods of genetic analysis rely on the isolation of mutants and experimental crosses to unravel the patterns of inheritance for a particular trait. Just as a surgeon would probe the inner workings of a body with a scalpel, so geneticists dissect biological function with genetic variants, usually abnormal mutants. Enter the average genetics laboratory and you will find a collection of peculiar animals and plants with equally peculiar names like 'stumpy wing' and 'white eye' (both *Drosophila* mutants). These are essential to the work of an experimental geneticist. By crossing the organisms in the correct combinations the patterns of inheritance of particular traits can be unravelled. But can you carry out experimental crosses with people? Of course not. Our ethical code quite rightly prevents us from using people in the same way as we would use a fruit fly. Furthermore, if you think about the organisms that experimental geneticists use (bacteria, viruses, fruit flies and mice) you will realise that they all have short life cycles and produce many offspring. Consequently individual geneticists can see the results of many crosses during their working lifetime. By contrast, people have long life cycles and don't produce many children. Even if experimental genetics with people were ethically acceptable it is not a practical solution. In this chapter you are going to learn about some of the ways that geneticists get around these problems by becoming a genetic detective.

LEARNING OBJECTIVES

After completing the work in this chapter you will be able to:

1. interpret patterns of inheritance using pedigree analysis;
2. describe the techniques and uses of amniocentesis and chorionic villus sampling;
3. assess the risks and benefits of genetic screening.

15.1 PEDIGREE ANALYSIS

If geneticists cannot breed future generations of people, how can they unravel patterns of inheritance? The answer is to look back by collecting information about a particular family and their ancestors. This information is then used to construct a pedigree chart. This is a sort of family tree which contains not only the names of people but also information about their phenotypes. Using this chart it is possible to predict the risks of a particular couple having a child who might suffer from a particular genetic disorder. Notice we are talking about risk. There are very few absolute guarantees in this form of analysis. To understand how pedigree analysis works, you first need to learn the rules.

The rules of the game

Most of you will probably have seen diagrams like Fig 15.1 which shows the incidence of haemophilia in the royal families of Europe. But have you ever considered what all the symbols mean and why the symbols are connected up in the way they are? This is not just haphazard but follows a definite set of rules and conventions. These are laid out below. First the symbols.

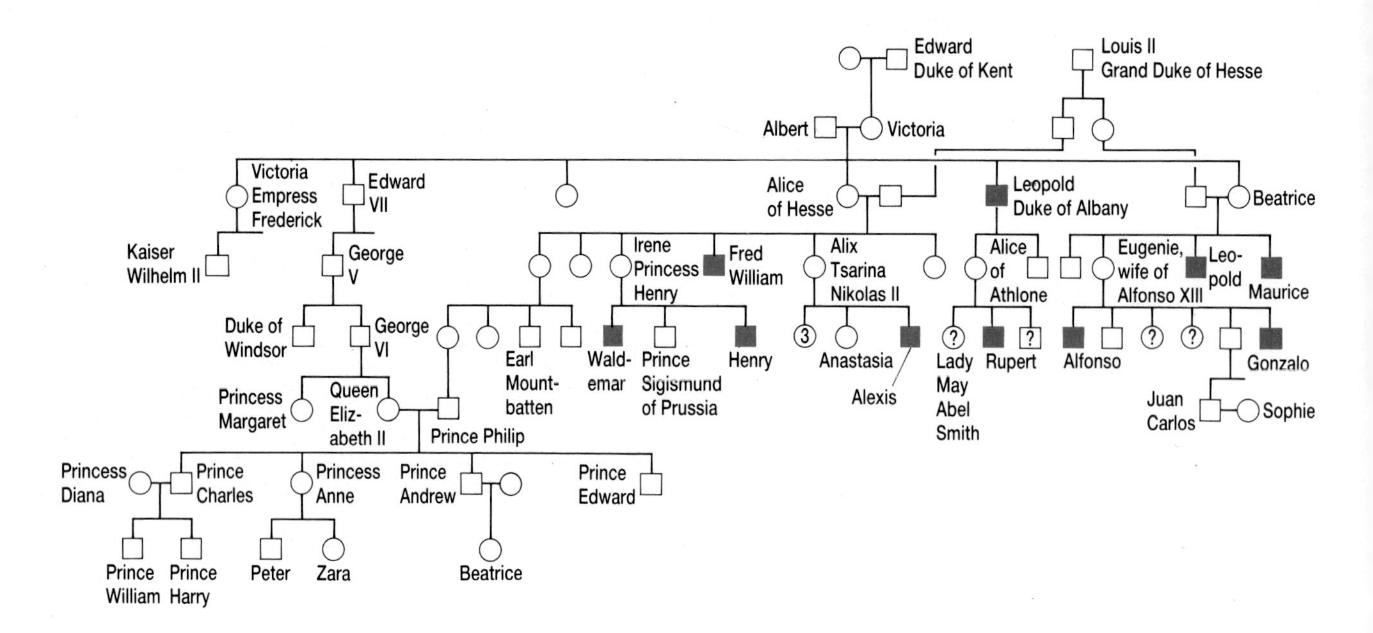

Fig 15.1 Haemophilia and the European royal families.

The symbols used in pedigree analysis. These are summarised in Fig 15.2. Two terms you will need to know:

1. **consanguinity** is the marriage of related individuals, for example cousins;
2. the **propositus** is the individual who first drew the geneticist's attention to a particular family.

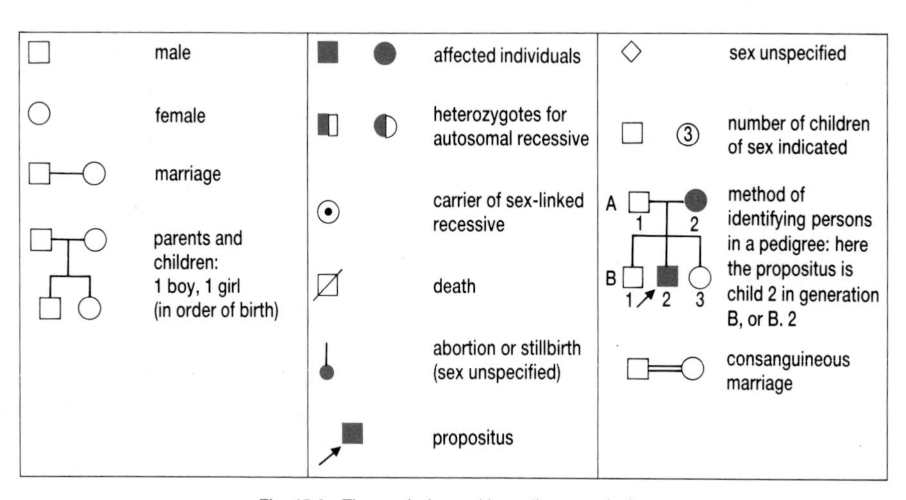

Fig 15.2 The symbols used in pedigree analysis.

The rules for drawing pedigrees. Individuals on the same line are from the same generation, for example generation B in Fig 15.3. Individuals in each generation are numbered in sequence – C.1, C.2 and so on. Brothers and sisters are attached to a common line. This line is then attached to the line joining the parents. C.1 and C.2 are brothers in Fig 15.3. The male B.3 has not reproduced asexually! It's just that his wife is irrelevant to the pedigree.

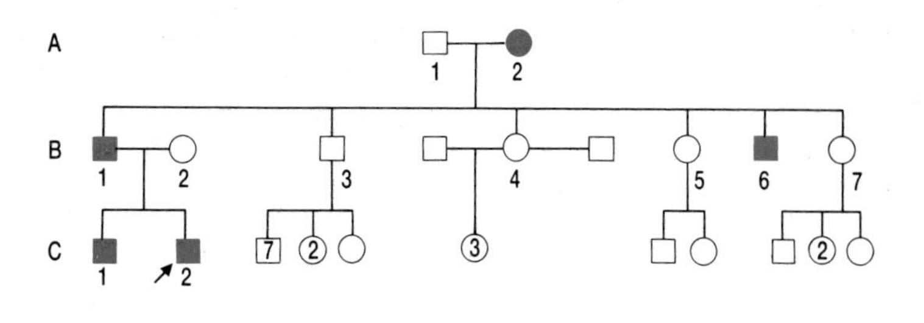

Fig 15.3 The rules for drawing pedigrees. See text for explanation.

Now you have a go.

QUESTION

15.1 Using Fig 15.3, answer the following questions.
- **(a)** Give the sex of individual B.7.
- **(b)** How many children has individual B.7 had, and in what order?
- **(c)** Identify the propositus.
- **(d)** Which female has a son and a daughter?
- **(e)** What do you think has happened to individual B.4?

Putting pedigrees to work

Using pedigree charts we can identify different patterns of inheritance. To illustrate these patterns, you will first be shown a hypothetical example so that you can identify the major points and then a real example where the interpretation may be a little more tricky. This really is a detective game where you have to hunt for clues. The trick is to develop the ability to spot them.

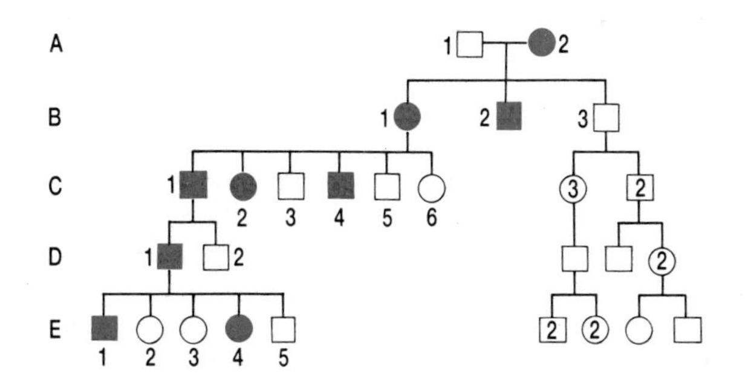

Fig 15.4 A pedigree showing a pattern of inheritance consistent with a trait inherited as an autosomal dominant.

Autosomal dominant. A sample pedigree illustrating the pattern of an autosomal dominant trait is shown in Fig 15.4. You can tell this is a trait determined by a dominant allele since:

- it appears in every generation;
- every affected individual has an affected parent;
- when one generation does not express the trait, the trait is lost and does not reappear in future generations (look at the children, grand-children and great-grandchildren of B.3).

How can we tell that it is an autosomal allele and not sex-linked? There are a number of clues. Firstly, equal numbers of males and females are affected. Compare this with Fig 15.1 which shows the inheritance of the sex-linked trait haemophilia. Secondly, the male C.1 passes the trait on to his sons. Since a man only passes his Y chromosome to his sons, the gene cannot be on the X chromosome. Thirdly, the affected male D.1 has got two unaffected daughters as well as an affected daughter. If the gene were an X-linked dominant then all his daughters would be affected, since he transmits his X chromosome to each of them.

QUESTION

15.2 What evidence is there from the pedigree shown in Fig 15.4 that this is not a trait caused by a Y-linked dominant gene? Will the children of D.2 have the trait?

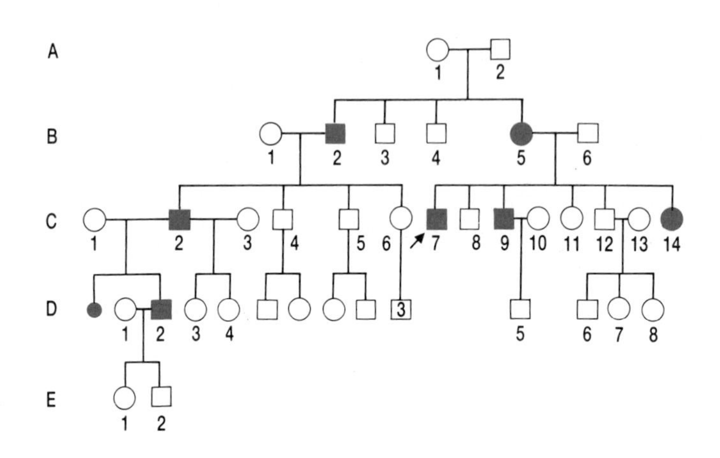

Fig 15.5 A pedigree illustrating the inheritance of an autosomal form of muscular dystrophy which shows incomplete penetrance.

Now look at Fig 15.5. This shows a pedigree for muscular dystrophy (an autosomal form, not the X-linked Duchenne muscular dystrophy). If you look carefully at the pedigree you will notice that the illness appears in every generation except for the first and the last. It looks like a dominant trait but the original parents (A.1 and A.2) and the children of D.1 and D.2 do not appear to have the disease. This is where the biological skill of the geneticist comes in. The effect of a gene which causes a disease can vary because people vary. In particular, the age at which a disease develops and the degree to which it develops will be affected by a host of other genetic and environmental factors. When an individual carrying the relevant gene is so mildly affected that the disease passes unnoticed we say that gene shows **incomplete penetrance**. In Fig 15.5 one of the two original parents must have had the disease, but they were both dead by the time the pedigree was drawn up. If one of them only developed the disease very mildly then their children might not have noticed. So this pedigree chart shows a pattern of inheritance consistent with a dominant allele showing incomplete penetrance.

QUESTIONS

15.3 Look at Fig 15.5. Which individual first drew the geneticist's attention to this family?

15.4 Explain why E.1 and E.2 have not developed muscular dystrophy. Would you expect them to develop the disease? What advice would you give them about having children?

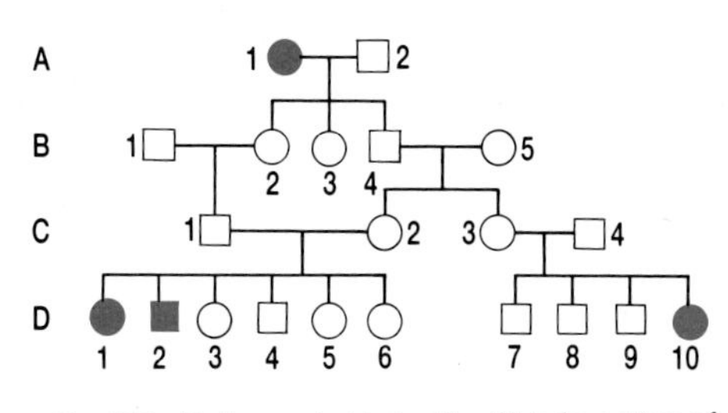

Fig 15.6 Pedigree of a kindred in which the pattern of inheritance indicates an autosomal recessive trait.

Autosomal recessive inheritance. Fig 15.6 shows a pedigree chart for a **kindred** (a group of related people) in which the pattern of inheritance indicates an autosomal recessive trait. The recessive nature of the trait is suggested by:

- the rarity of the trait;
- skipping of generations;
- the observation that D.1 and D.2 are the children of a **consanguineous marriage**, here between cousins.

It is not unusual to see the appearance of a recessive trait linked to a consanguineous marriage. The reason is that since most recessive traits are rare, the chances of two people being heterozygous for it are greater if they are related than if they are unrelated since they may both have inherited the recessive allele from a common ancestor. We shall return to consanguinity later.

QUESTIONS

15.5 What evidence is there in Fig 15.6 that this is a recessive autosomal trait and not a recessive sex-linked trait?

15.6 Explain the appearance of the trait in D.10. Could you have predicted this?

15.7 Fig 15.7 shows the pedigree of a kindred with cystic fibrosis. Summarise the evidence that cystic fibrosis is due to a recessive autosomal allele.

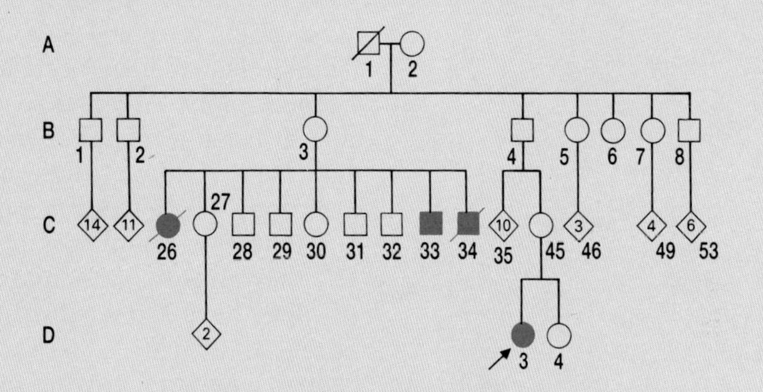

Fig 15.7 The pedigree of a kindred with cystic fibrosis.

15.8 A European woman living in Eastwood gave birth to a daughter who had a rare form of a recessive trait called thalassaemia, which is generally only found in people from the Indian subcontinent and the Mediterranean. A geneticist discovered that she and her husband were distantly related through the son of an army captain who had married an Indian princess in 1785. How does this discovery explain the baby being born with thalassaemia?

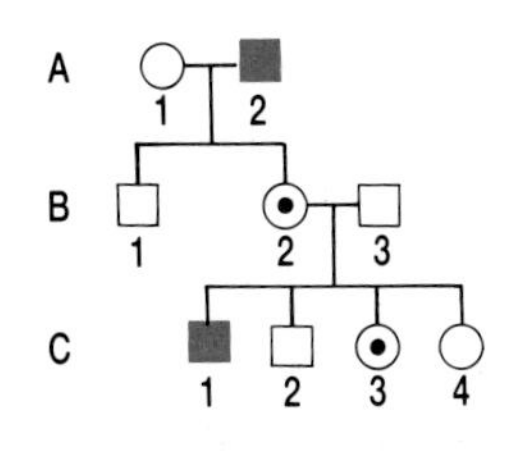

Fig 15.8 A pedigree showing the inheritance of an X-linked recessive trait.

Sex linked traits. Fig 15.8 shows the characteristic pattern of inheritance for an **X-linked recessive trait**. Important clues here are:

- many more males than females are affected. If the recessive gene is very rare almost all observed cases will be males;
- usually none of the offspring of an affected male will be affected, but all his daughters will carry the allele, so half of their sons, on average, should be affected;
- None of the sons of an affected male will be affected.

Fig 15.9 shows the pattern of inheritance for an X-linked dominant trait.

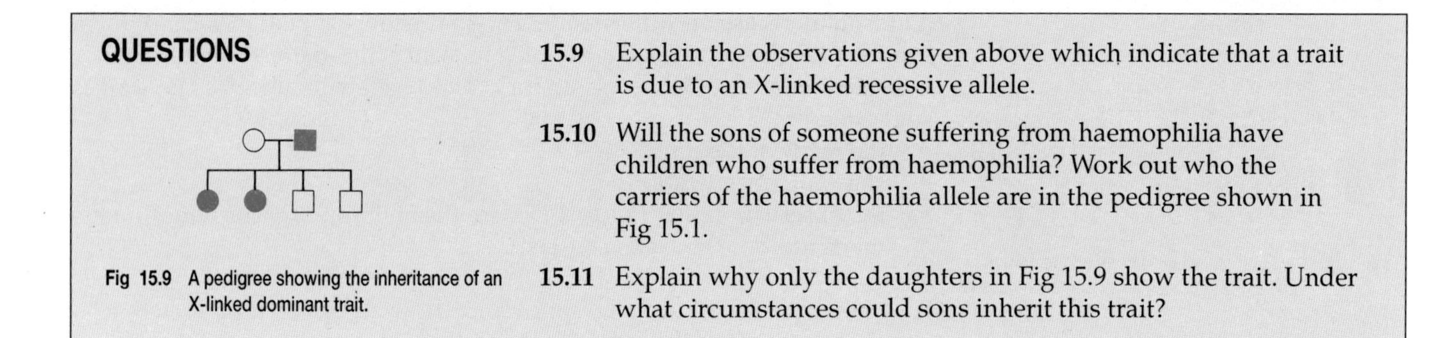

QUESTIONS

Fig 15.9 A pedigree showing the inheritance of an X-linked dominant trait.

15.9 Explain the observations given above which indicate that a trait is due to an X-linked recessive allele.

15.10 Will the sons of someone suffering from haemophilia have children who suffer from haemophilia? Work out who the carriers of the haemophilia allele are in the pedigree shown in Fig 15.1.

15.11 Explain why only the daughters in Fig 15.9 show the trait. Under what circumstances could sons inherit this trait?

15.2 GENETIC COUNSELLING

The theory above is all very well, but what practical use is it to a genetic counsellor? In this section we will examine how a genetic counsellor can use pedigree analysis to advise people.

Why visit a genetic counsellor?

Most major hospitals have a genetic counselling department. The people who do this job have been specially trained not only as geneticists but also as counsellors. A genetic counsellor has, above all, to be sympathetic and able to listen. Most people would visit a counsellor if they were worried that their children might inherit some genetic disorder, say muscular dystrophy. These worries might have arisen because:

- they have relatives with the disease;
- they already have children with some disorder;
- they have had a baby die or been still born.

People want to know if the illness or death is due to some inherited abnormality and what the chances are of having a child or another child with the same defect. The job of the counsellor is to assess that risk. Notice we are talking about risks and probabilities. The answer that the counsellor will come up with will be in the form of a probability, say 1 in 4. The counsellor then has to communicate this risk factor to their clients in a manner in which they can understand. The decision to have children is then up to the parents.

What will a genetic counsellor do?

Obviously the first thing is an accurate diagnosis of the propositus. For example, imagine a couple have had a child who is blind or deaf. There could be a number of causes, the definition of which will depend on an accurate diagnosis. This problem is particularly challenging with disorders like essential hypertension (high blood pressure) where there may be a genetic predisposition towards developing a disease. Clearly diagnosis is not easy, but assume that we are only dealing with clear-cut disorders which are easy to define medically.

Having decided the propositus has a particular disorder, the genetic counsellor now has to collect information about the phenotypes of other members of the family. This again is no easy task. How much do you know about your great-grandparents or your great-aunts? Perhaps there are some cousins overseas or a member of the family who has disappeared. The counsellor will probably have to conduct interviews with a number of members of the family. Let us assume that enough information is collected to produce a pedigree chart. Analysis of this chart suggests that the cause of the problem is genetic. What happens next?

Assessing the risk

Let us look at the risks associated with different patterns of inheritance of major gene defects like cystic fibrosis and PKU.

Autosomal dominant. If the gene always penetrates, then the situation is relatively straightforward. If you have an autosomal dominant defect you are likely to be heterozygous for that defect since the homozygous condition is nearly always lethal. What is the probability that your child will also have the defect? This is a monohybrid cross, as shown in Fig 15.10 and on average half of your children will inherit the allele and the defect.

parents	Huntington's chorea		×	'normal'	
genotype	*Hh*			*hh*	
gametes	(*H*) or (*h*)			(*h*)	
children	*Hh*	*Hh*		*hh*	*hh*
probability	1/4 +	1/4		1/4 +	1/4
	1/2 Huntington's disease			1/2 'normal'	

Fig 15.10 Assessing the risk: Huntington's chorea – an autosomal dominant trait.

Stop and think about this for a moment. This situation is like tossing a coin. If the coin comes up heads, the child has the defect. If the coin comes up tails then the child does not have the defect. What we are saying is that if we toss the coin enough times, say 1000, then provided the coin is unbiased we should have 500 tails and 500 heads. But toss the coin four times and you might get four tails, that is, four normal children, or you might be unlucky and get four heads, that is, four abnormal children. In the long run (1000 tosses) we expect half-and-half; in the short run (four tosses) you cannot say you will definitely get half-and-half. You can now, perhaps, see the problem for both the parents and the counsellor. The counsellor is talking in terms of probability and chance, not making a definite prediction. The parents have to decide 'do we risk it?'

Anyway, predicting the risk at least was fairly simple, but what about a real family with a history of Huntington's chorea? Would you run the risk of having children if you knew that one of your parents had this disease? But this is an unrealistic situation since you probably won't know if either of your parents have got the disease until you are in your 30s. But you know that your mother's father (your grandfather) developed the disease when he was 52, so the probability that you have the allele is $\frac{1}{4}$ $(\frac{1}{2} \times \frac{1}{2})$. What do you do? Well you could wait to see if your mother develops the disease, which increases the odds that you will have got the allele. But how long do you wait before having children? Huntington's chorea may not appear until your mother is well into her 50s, by which time you could be in your middle 30s the very time at which the risks of having a Down's syndrome child are also increasing rapidly. Again the choice is a difficult one.

Autosomal recessive. Analysis of a pedigree diagram has indicated that one member of a couple who want to have a child, the man, has a recessive trait in his family, say cystic fibrosis. His wife is totally unrelated to him. What advice can a genetic counsellor give?

This is a real challenge. From Table 14.2 we can see that the frequency of individuals who are heterozygous for the cystic fibrosis allele in the general population (carriers) is 1 in 22, that is, $\frac{1}{22}$. This then is the probability that the man's wife is a heterozygote. By looking at the husband's pedigree chart the counsellor will be able to calculate the probability that he is a carrier. Assuming they are both carriers then, on average, one-quarter of their children will develop cystic fibrosis (that is, a monohybrid cross with complete dominance, see Fig 15.11). By combining all these probabilities the counsellor will be able to give some estimate of the risk of having a child with cystic fibrosis. This will probably be very small!

Let us look at another situation. Assume that the couple are related and that they have a history of cystic fibrosis in their family. This increases the risk because of consanguinity. The effects of consanguinity are well illustrated by Fig 15.12. The degree of closeness of the relationship between a husband and wife will affect the probability of them being heterozygous for the same gene. One way of measuring this probability is to calculate the coefficient of consanguinity or inbreeding (*F*). **The coefficient of inbreeding of an individual is the probability that this individual receives at a given locus two genes that are identical by descent that is, are copied from a single gene carried by a common ancestor.** The method of calculation is shown in Box 15.1.

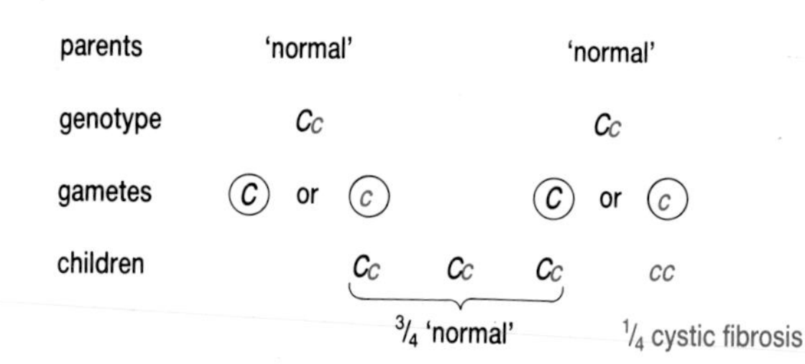

Fig 15.11 Assessing the risk: cystic fibrosis – an autosomal recessive trait.

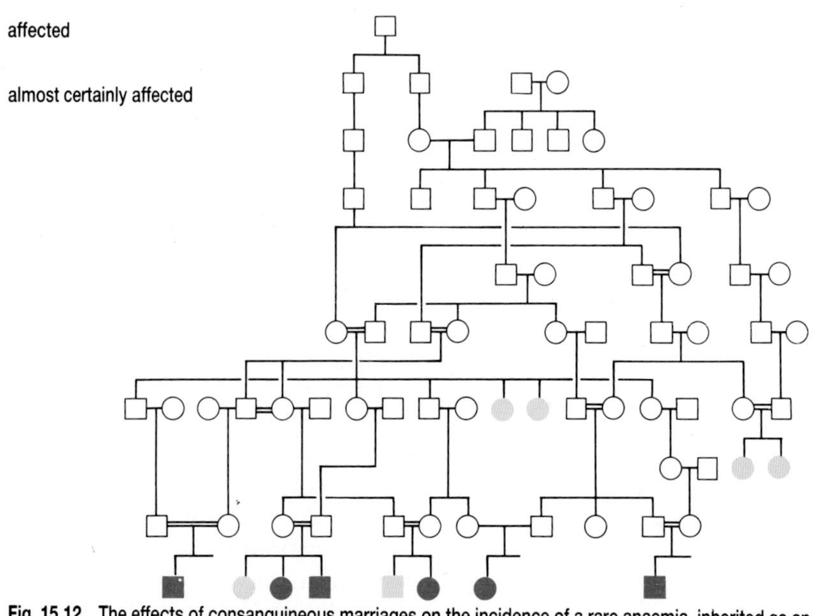

Fig 15.12 The effects of consanguineous marriages on the incidence of a rare anaemia, inherited as an autosomal recessive in an Amish family.

BOX 15.1

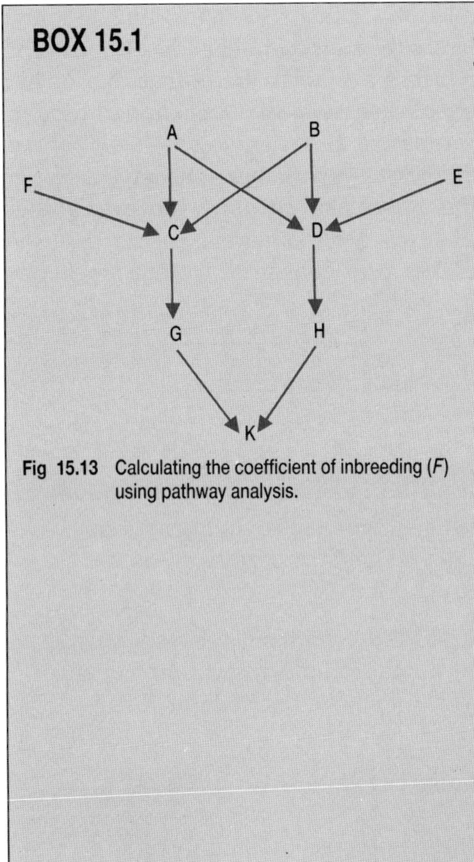

Fig 15.13 Calculating the coefficient of inbreeding (F) using pathway analysis.

Calculating the coefficient of inbreeding (F) – Pathway analysis

This method allows you to calculate the coefficient of inbreeding for an individual with a known pedigree. We might do such an analysis if, for example, there was history of cystic fibrosis in a family. By looking at the pedigree the genetic counsellor is able to work out that both ancestors, A and B, must have been carriers. The couple, G and H, want to know what the probability is that a child of theirs will have cystic fibrosis.

Look carefully at Fig 15.13. Here we have a child (K) whose parents (G and H) are first cousins. Path analysis involves tracing the arrows in a pedigree from the child back to itself through each ancestor common to both parents. In our example, A and B are the two ancestors common to both parents, G and H. So in our example the two paths are:

$$K \rightarrow G \rightarrow C \rightarrow A \rightarrow D \rightarrow H \rightarrow K = 6 \text{ steps}$$
$$\text{and } K \rightarrow G \rightarrow C \rightarrow B \rightarrow D \rightarrow H \rightarrow K = 6 \text{ steps}$$

The contribution each path makes to F is $(\frac{1}{2})^n$, where n = either the number of steps in the path minus one if an individual appears twice in the same path, or just the number of steps if the individual under consideration appears only once in each path.

Since in our example K appears twice in each path, $n = 6 - 1 = 5$. So in our pedigree the contribution made by each path to $F = (\frac{1}{2})^5 = \frac{1}{32}$. F equals the sum of the two paths $= \frac{1}{32} + \frac{1}{32} = \frac{1}{16}$. This then is the probability that the child will inherit a copy of the allele which causes cystic fibrosis from each parent. By contrast, the probability of two people drawn at random from the general population having a child with cystic fibrosis is $\frac{1}{22} \times \frac{1}{22} \times \frac{1}{4} = \frac{1}{1936}$.

QUESTIONS

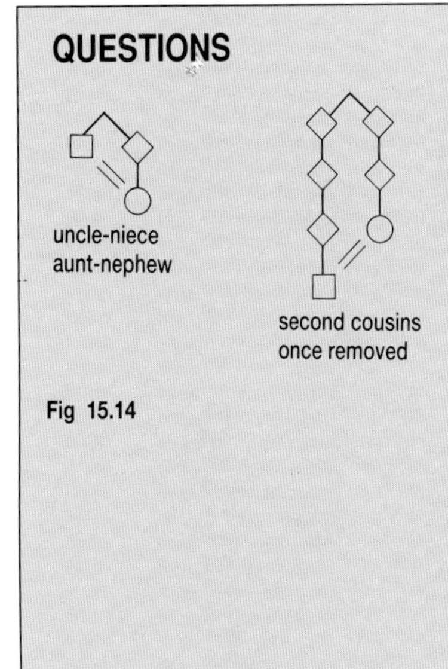

Fig 15.14

15.12 What is the probability that a couple who have had one child with cystic fibrosis will have another child who also suffers with this disease?

15.13 Calculate the coefficients of inbreeding for children of the marriages shown in Fig 15.14. What do you notice about the coefficient as the distance between the relatives increases?

15.14 'Over half the states in the United States of America forbid marriages between uncles and nieces, aunts and nephews and first cousins. In Japan first cousin marriage is encouraged, whilst in Andra Pradesh (India) certain castes favour uncle-niece marriages.' Comment on these social conventions in the light of the data given in Table 15.1 and the work you have done in this section.

Table 15.1 Data showing the frequency of mortality among young children born to consanguineous marriages in Hiroshima, Japan.

Parental relationship	Number of births	Deaths between 1 and 8 years: Number dead	Percentage of dead
first cousins	326	15	0.0460
second cousins	139	3	0.0218
unrelated	544	8	0.0417

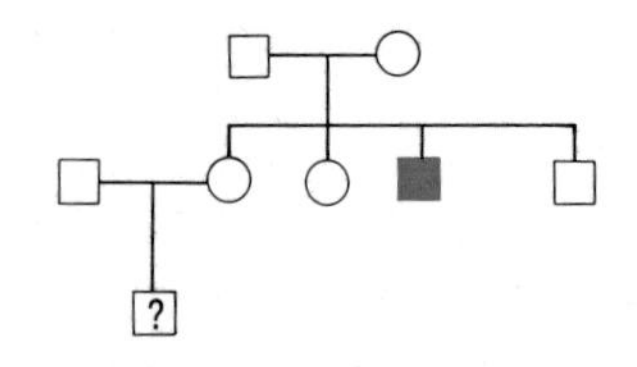

Fig 15.15 Assessing the risk: Duchenne muscular dystrophy – an X-linked recessive trait.

Sex-linked inheritance. Here the dilemma is that a woman wants to know if she might be a carrier who could pass the trait on to her sons. Look at the pedigree chart shown in Fig 15.15. Here a woman finds that one of her brothers has developed a sex-linked disease, say Duchenne muscular dystrophy. This means her mother must be a carrier, so she has a 50 per cent chance of being a carrier too. If she is a carrier, her sons have a 50 per cent chance of developing Duchenne muscular dystrophy, however all her daughters will be normal. So if she knows her mother is a carrier then she can predict that her sons have a 25 per cent chance of having Duchenne muscular dystrophy.

QUESTIONS

15.15 What is the probability that a woman whose brother has Duchenne muscular dystrophy will give birth to a daughter who is a carrier?

15.16 Calculate the probability of having four children all of the same sex. Imagine you could examine all the families living in your country which have four children. What proportion of these would you expect to have four children of the same sex?

15.17 Do you think society has a right to stop someone whose family has a history of Huntington's chorea from having children even though there is no evidence that they have the disease? Justify your answer.

15.18 **(a)** Analyse the pedigrees shown in Fig 15.16 for mode of inheritance.
(b) For each pedigree, give the genotypes of the numbered individuals. You will need to assign letters to the genes.

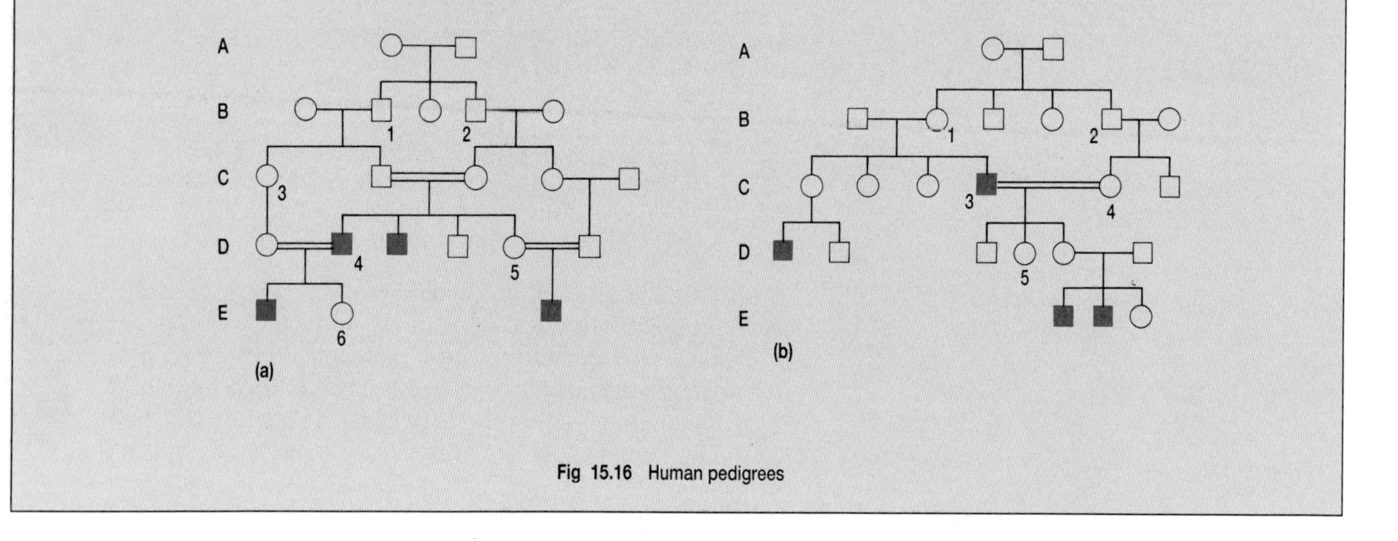

Fig 15.16 Human pedigrees

15.3 GENETIC SCREENING

Whilst pedigree analysis allows us to chart the pattern of disease in a family, it really only offers the prospect of assigning risk to a particular event. Can't we be more certain and actually state that a fetus has cystic fibrosis or Duchenne muscular dystrophy? Even better, why can't we routinely screen everybody to see if they are a carrier of one of these rare recessive alleles which, when in the homozygous condition, can cause genetic disease?

Genetic screening is the systematic testing of fetuses, newborn children or individuals of any age for the purpose of ascertaining potential genetic handicaps in them or in their progeny that may require treatment.

Prenatal screening

Most of you will probably have heard of amniocentesis which is now performed routinely in our hospitals. Details of the technique are given in Box 15.2. You are probably aware that the technique is widely used to detect chromosome abnormalities, for example Down's syndrome, in those pregnancies considered at risk. However, a number of other disorders can be detected using the technique coupled with biochemical tests and/or recombinant DNA techniques described in Chapter 14.

BOX 15.2

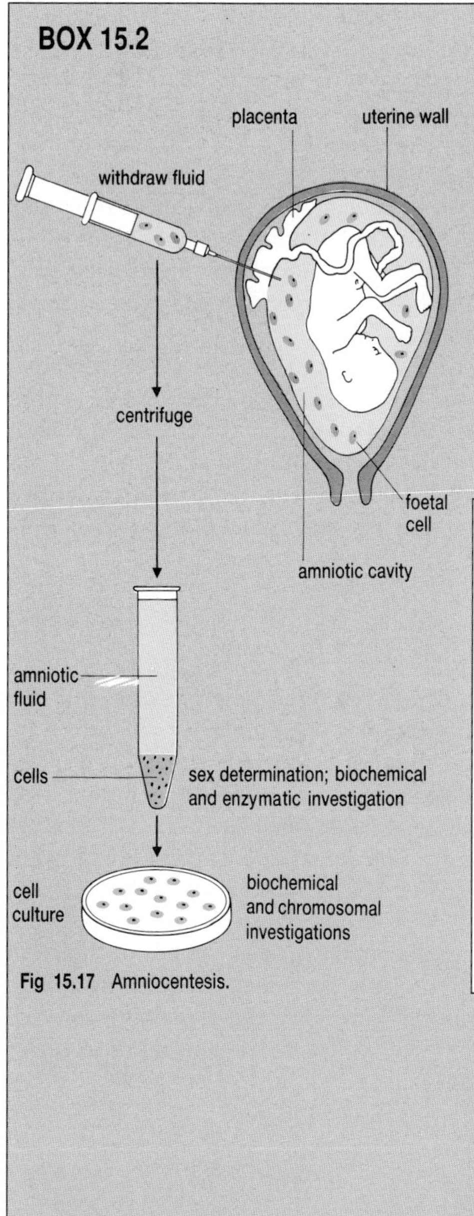

Fig 15.17 Amniocentesis.

Amniocentesis and chorionic villus sampling

Amniocentesis is a technique for the prenatal diagnosis of congenital abnormalities (defects present at birth). These may be due to genetic, environmental or a combination of these two factors. A hypodermic syringe is inserted into the amnion with the help of an ultrasound scanner which shows the position of the fetus. A sample of 10–15 cm^3 of the amniotic fluid surrounding the fetus is then withdrawn. The fluid contains cells which can be cultured in the laboratory (see Box 14.1). The cells can then be subjected to a number of tests, as shown in Fig 15.17. The test is usually given between weeks 13 and 16 of the pregnancy.

A new prenatal testing procedure for obtaining fetal cells called chorionic villus sampling (see Fig 15.18) is currently undergoing trials to ascertain the risks associated with it. This technique has two advantages over amniocentensis.

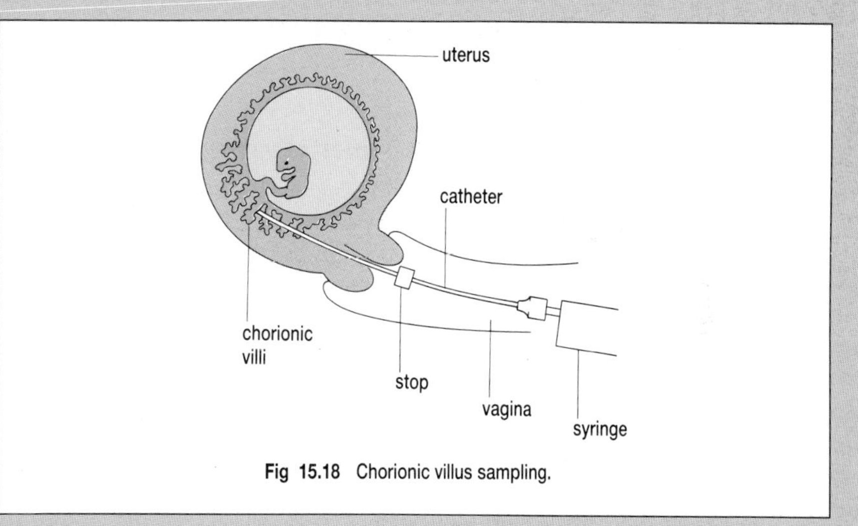

Fig 15.18 Chorionic villus sampling.

1. It can be performed earlier in pregnancy and so allows earlier diagnosis of genetic disease.
2. Since the cells of the chorionic villi are actively dividing, karyotype analysis can be performed without having to grow the cells in tissue culture first. This reduces the time required to obtain results from weeks to days.

Both of these advantages reduce the amount of stress on anxious parents in high risk groups.

However, you should not think that such prenatal screening is the answer to our genetic counsellor's prayers. Let us consider just amniocentesis, though many of the comments made below will also apply to chorionic villus sampling. The practicalities of the techniques present problems. For example, many biochemical tests and all karyotyping require the successful culture of fetal cells. The earlier in pregnancy the sample is taken the more easily the cells can be grown in culture. But the commonest detectable abnormality using amniocentesis is spina bifida and the tests for this give the best results if performed later in pregnancy. Unless you are going to give several amniocenteses during a pregnancy, each of which has a risk of causing spontaneous abortion or damaging the baby, then you have to reach a compromise on the timing of the amniocentesis based on the competing requirements of different techniques. In other words, you still have to decide which of a number of possible conditions a mother is most at risk from and time the amniocentesis accordingly. Furthermore, since the range of tests which can be performed is still limited and you may not be able to perform the test at the optimum time, you can still make an incorrect diagnosis. Amniocentesis does not guarantee a perfect baby.

Still, you might think it advisable for every woman to have at least one amniocentesis during each pregnancy just in case. Certainly, pregnant women are often given ultrasound tests routinely, so why not amniocentesis? Firstly, there is a definite risk of spontaneous abortion, still birth or damage to the baby associated with amniocentesis, probably of the order of 1–2 per cent. Unless the risk of a mother having a baby which is ill greatly exceeds the risks associated with the amniocentesis then she should not have the test. Secondly, it is simply not cost effective to screen those mothers who are not in high risk groups.

Similarly, the answer to the question 'why don't we test every newborn baby for every single genetic disorder we can?' is, again, that it will not be cost effective and it is not necessary. The risk of people outside certain high risk groups of developing a genetic illness is so minute that it is simply not worthwhile screening them. Given limited resources it is better to identify high risk groups and concentrate on them. An example will make this clear.

Identifying carriers

Tay-Sachs disease occurs in individuals who are homozygous for an autosomal recessive allele causing a defect in an enzyme called hexosaminidase A. The disease is particularly distressing and the children usually die by three to five years of age. Although there is no known treatment for the disease, the heterozygotes can be detected using a relatively simple blood test and homozygous fetuses following amniocentesis. The disease is particularly common among Jews of Central and Eastern Europe so it would seem sensible to screen all married couples in this group to ascertain heterozygotes for Tay-Sachs disease. Where both members of a couple are heterozygous then they can have amniocentesis to test for the presence of an affected fetus, and this information offers the opportunity to terminate the pregnancy and try again for a normal baby. So a marriage between two heterozygotes each carrying a copy of the Tay-Sachs allele is an example of a high risk group for whom prenatal diagnosis should be available.

A brave new world?

Imagine, though, we were able to develop prenatal tests which have no risks associated with them. Wouldn't it then be worthwhile testing every pregnant woman. You still have the problem of cost effectiveness but perhaps we also need to ask a more fundamental question. Why are we doing all this testing? The answer is too often a eugenic one. The idea that we can somehow breed genetic abnormality out of the human race.

There is undoubtedly a view point that if we had more testing, screening of babies and genetic counselling we could drastically reduce the incidence of genetic abnormality in our population. Certainly, genetically based diseases are expensive to treat and in the United States of America and Great Britain upto 30 per cent of the children in pediatric wards may have illnesses which are due to some defect present at birth. However the idea that we could reduce this number by more extensive screening is false. We have already seen that amniocentesis can only detect a limited range of abnormalities. Furthermore, of the 3 per cent of pregnancies which result in an ill baby, only 0.5–1 per cent of these children have some form of genetic disease. This includes both chromosomal abnormalities, the vast majority of which are not inherited, and all those traits which show non-Mendelian patterns of inheritance, for example heart disease. The vast majority of these unfortunate children are not suffering from genetic disease at all but have been born with illnesses due to infection (cytomegalovirus, rubella and so on); birth trauma, for example lack of oxygen; or environmental effects, for example a mother smoking or drinking alcohol during pregnancy.

Certainly the techniques of recombinant DNA technology can be used to detect fetuses with genotypes which could produce diseased phenotypes, but such tests will not stop children from being born who are suffering from a genetic disease. You cannot screen every pregnancy. Even if you screened every pregnancy where there was a history of a genetic disease in the family this would not make a great impact since so many of the alleles involved occur around the mutation frequency. This means that these alleles will be fed into the population as fast as they are lost. For example, 20 per cent of all cases of Duchenne muscular dystrophy are due to new mutations.

Clearly, then, it is not enough to simply detect fetuses which have a genotype which may produce a diseased phenotype. We must find out why this phenotype develops. What has gone wrong with, say, the gene which in its mutated form causes Duchenne muscular dystrophy? What does this gene do in people who are not suffering from the disease? How is it regulated? Using recombinant DNA technology we may be able to start answering questions like these, which in turn holds out the prospect of treating, even finding cures for, inherited diseases like Duchenne muscular dystrophy.

In the next few years, especially with the advent of tests based on recombinant DNA, we will be able to detect a whole range of new genetic abnormalities prenatally. This raises an important moral dilemma. How serious does a defect have to be to decide upon selective abortion? In some parts of the world girl fetuses are regularly aborted. If a fetus is detected with allele B27 in the HLA system, which is known to carry a risk of upto a few per cent of getting the disease ankylosing spondylitis, should the fetus be aborted? What should we do about fetuses which have genetic diseases that are untreatable but which nonetheless would live happy and fulfilling lives if they were allowed to be born? Should males who are red–green colour blind be aborted or those fetuses which might suffer from essential hypertension later in life? Where do you draw the line? Biologists have and will make enormous strides in diagnosing and treating genetic diseases, but with such discoveries goes the social responsibility about what to do with them. You really cannot divorce science and society.

QUESTIONS

15.19 Explain why all babies in Britain are tested at birth for phenylketonuria but not for Tay-Sachs disease.

15.20 If techniques were available, would it be cost effective to screen all fetuses for
(a) cystic fibrosis
(b) phenylketonuria
(c) galactosaemia.
If the results were positive what advice would you give the parents in each case?

SUMMARY ASSIGNMENT

1. Summarise how pedigree analysis can be used by genetic counsellors to assess the risk of a couple having a child with a genetic disorder.
2. Compare and contrast prenatal screening based on amniocentesis and chorionic villus sampling.
3. A civil servant in the Department of Health has been asked to investigate the possibility of setting up a screening programme for a particular genetic disorder. What criteria would you use to evaluate whether such a programme is worthwhile? Discuss your criteria with your friends.
4. 'If we enable the weak and the deformed to live and propagate their kind, we face the prospect of a genetic twilight. But if we let them die or suffer when we can save or help them, we face the certainty of a moral twilight.' (Theodosius Dobzhansky). Discuss this statement from both a genetic and a moral point of view.

Appendix A

FURTHER READING

General

Ayala, F.J. & Kiger Jr, J.A. (1984) *Modern Genetics,* 2nd ed. Benjamin/ Cummings.
Suzuki, D.T., Griffiths, A.F.J., Miller, J.H. & Lewontin, R.C. (1986) *An introduction to genetic analysis.* Freeman.

Theme 1

Burnett, L. (1986) *Essential Genetics.* Cambridge University Press (1, 2, 3).
Jones, R.N. & Karp, A, (1986) *Introducing genetics.* John Murray (1, 2, 3).

Theme 2

Allard, R.W. (1960) *Principles of plant breeding.* Wiley (5, 6).
Bowman, J.C. (1984) *An introduction to animal breeding,* 2nd ed. Edward Arnold *Studies in Biology* **46** (5, 6, 7).
Dalton, D.C. (1985) *An introduction to practical animal breeding,* 2nd ed. Collins (5, 6, 7).
Day, P.R. (1974) *Genetics of host parasite interactions.* Freeman (8).
Dixon, G.R. (1984) *Plant Pathogens and their control in horticulture.* Macmillan (8).
Dixon, R.A. (1985) *Plant Cell Culture: a practical approach.* IRL Press (7).
Falconer, D.S. (1981) *Introduction to Quantitative Genetics.* Oliver Boyd (4).
Gordon, I. (1983) *Controlled Breeding of Farm Animals.* Pergamon Press (5, 6, 7).
Hammond Jr, J., Bowman, J.C. & Robinson, T.J. (revisers) (1983) *Hammond's Farm Animals,* 5th ed. Edward Arnold (4).
Land, J.B. & Land, R.B. (1983) *Food Chains to Biotechnology.* Nelson (5, 6, 7).
Manners, J.G. (1982) *Principles of Plant Pathology.* Cambridge University Press (8).
Simmonds, N.W. (1979) *Principles of Crop Improvement.* Longman (6, 7).
Tudge, C (1988) *Food Crops for the Future.* Blackwell (6, 7).

Theme 3

Connor, S. & Kingman, S. (1988) *The Search for the Virus.* Penguin (9).
Hardy, K. (1986) *Bacterial plasmids,* 2nd ed. Van Nostrand Reinhold (10).
Lin, E.C.C., Goldstein, R. & Syvanen, M. (1984) *Bacterial, plasmids and phages.* Harvard University Press (9 and 10).
Metcalfe, J. & Baumberg, S. (1988) *An overview of genetic mechanisms in bacterial cells. Journal of Biological Education* **22**: 23–30 (10).

Theme 4

AFRC (1988) *Biotechnology in Agriculture.* Available from National Centre for School Biotechnology, Dept. of Microbiology, University of Reading, London Road, Reading RG1 5AQ (12).
Bains, W. (1988) *Genetics engineering for almost everybody.* Penguin (11 and 12).
Brown, T.A. (1986) *Gene cloning, and introduction.* Van Nostrand Reinhold (11).
Cherfas, J. (1982) *Man made life.* Blackwells. (11 and 12).
Oliver, S.G. & Ward, J.M. (1985) *A dictionary of Genetic Engineering* Cambridge University Press (11).
Watson, J.D., Tooze, J. & Kurtz, D.T. (1983) *Recombinant DNA, a short course.* Scientific American Books (11).

Theme 5

Bodmer, W.F. & Cavilli Sforza, L.L. (1976) *Genetics, evolution and man.* Freeman (13).
Clarke, C.A. (1987) *Human Genetics and Medicine,* 3rd ed. Edward Arnold, *New Studies in Biology* (14).
Harper, P.S. (1984) *Practical genetic counselling,* 2nd ed. Wright, Bristol (14).
Holton, J.B. (1985) *An introduction to Inherited Metabolic Diseases* (14).
Lewontin, R. (1982) *Human diversity.* Scientific American Books (13).
Macleod, A. & Sikora, K. (1984) *Molecular biology and human disease.* Blackwell Scientific Publications (14).
Singer, S. (1985) *Human Genetics.* Freeman (13 &14).
Weatherall, D. (1985) *The new genetics and clinical practice,* 2nd ed. Oxford University Press (14).

Appendix B

ANSWERS TO IN-TEXT QUESTIONS

Chapter 1

1.1 Plants (ii) and (iii) since they are homozygous at all loci. Consequently they can only produce one type of gamete.

1.2 A clone. Not necessarily – they could be heterozygous.

1.3 To ensure that the response of the mice is the result of the administration of the drug and not the result of genetic differences between the mice.

1.4 This individual is heterozygous with the genotype Pgm^{100}/Pgm^{108}.

1.5 You could choose a wide variety of traits including the following blood groups, tongue rolling in humans, sex, protein polymorphism, melanism in moths.

1.6 Number of possible genotypes = $n(n + 1)/2$ where n = number of alleles.

1.7 **(a)** Different triplets of bases can code for the same amino acid, e.g. AAA and AAG both code for phenylalanine. Clearly a mutation which changes the triplet AAA to AAG would have no effect on the charge on the protein.
(b) Even where a mutation results in the substitution of one amino acid by another, if the charge on the amino acids is the same then there will be no alteration in the overall charge of the protein.

Chapter 2

2.1 Heterozygous individuals are able to synthesise enough functional HAO to catalyze the conversion of all the homogentisic acid to maleyacetoacetic acid.

2.2 Phenylketonuria – E_1; albinism – E_2; tyrosinosis – E_4.

2.3 **(a)** *h*;
(b) no.

2.4 **(a)** All the F_1 plants will be *CcPp* and will therefore produce the functional form of both the enzymes needed for the synthesis of the purple pigment.
(b) Expected F_2 phenotypic ratio of 9 purple : 7 white.

2.5 Exactly the same as in question 2.4.

2.6 *(i)* agouti; *(ii)* agouti; *(iii)* black; *(iv)* albino.

2.7 All the F_1 birds will have the genotype *IiCc* and will be white. Crossing these F_1 birds gives an expected F_2 phenotypic ratio of 13 white : 3 coloured.

2.8 The F_2 phenotypic ratio is approximately 12 white : 3 black : 1 brown which is consistent with a dominant epistatic interaction. So the white colouration is due to the presence of a dominant allele, call it *W*, at the locus which is epistatic to the locus controlling fur colour. The fur colour locus has two alleles: *B* which codes for black fur and *b* which codes for brown fur. *B* is dominant to *b*. Pure-bred brown cats have the genotype *WWbb*, whilst the F_1 cats will all have the genotype *WwBb* and so will be white. Of the F_2 cats we would expect $\frac{12}{16}$ to have at least one *W* allele so they will be white. Of the remaining $\frac{4}{16}$ with the *ww* genotype at the epistatic locus, $\frac{3}{16}$ will have at least one *B* allele and will be black whilst $\frac{1}{16}$ will be wwbb and so will be brown.

2.9 **(a)** *(i)* 3 walnut : 1 rose. *(ii)* 9 walnut : 3 rose : 3 pea : 1 single. *(iii)* 3 walnut : 3 rose : 1 pea : 1 single. *(iv)* 1 walnut : 1 rose : 1 pea : 1 single.

2.9 **(b)** Single = *rrpp*; walnut = *RrPp*.

2.10 **(a)** 5 white : 3 purple.
(b) 1 purple : 1 white.
(c) 6 purple : 2 white.
(d) 3 white : 1 purple.

2.11 The phenotypic ratio in the F_2 generation suggests that this is an example of dominant epistasis rather like the white Leghorn chickens. Call the epistatic locus *L* and assume there are two alleles of this gene – *L* and *l* – where *L* blocks the formation of malvidin and is dominant to *l*. Call the malvidin locus *M* and again assume that there are two possible alleles, *M* and *m*, where *M* codes for the production of malvidin and is dominant to the *m* allele. The parent which produces malvidin must then be *llMM*. The parent which lacks malvidin could have three possible genotypes; *LLmm*, *LLMM* or *llmm*. The 13 :3 F_2 ratio suggests the first possibility. (Try the others and see what ratios you get in the F_1 and F_2 generations.) Crossing these two parents, *llMM* and *LLmm*, will give an F_1 generation which all have the genotype *LlMm* and so will not produce malvidin. The expected phenotypes of the F_2 produced by crossing the F_1 are 13 non-producers : 3 malvidin producers.

Chapter 3

3.1 **(a)** Each *L* allele will contribute $\frac{24}{4}$ = 6 cm.
(b) *(i)* 15 cm; *(ii)* 19.5 cm; *(iii)* 15 cm.

3.2 **(b)** As the number of gene pairs controlling cob length increases the number of phenotypic classes in the F_2 generation increases and the histograms become more 'bell'-shaped.
(c) The effects of the environment would be to smooth out the steps in the histograms so making them more like the 'bell'-shaped normal distribution curve associated with quantitative characters.

3.3 **(a)** We would expect about $\frac{1}{64}$ of the F_2 offspring (about 2%) to resemble one of the original parents. If you look at the number of F_2 individuals which are 19 cm in length less than 1% are similar to F_2. This suggests four loci are involved in determining cob length.
(b) 9 – 10 cm.

3.4 **(a)** The white kernelled parent will have the genotype $r_1r_1r_2r_2r_3r_3$ whilst the light red kernelled parent will have the genotype $R_1R_1r_2r_2r_3r_3$. So all the F_1 generation will have the genotype $R_1r_1r_2r_2r_3r_3$. Since they only have one *R* allele they will synthesise very little of the red pigment and so they will be intermediate between the two parental varieties. Crossing F_1 plants will produce an F_2 generation with the following expected genotypes and phenotypes:

Genotypes	Expected frequency	Phenotype
$R_1R_1r_2r_2r_3r_3$	$\frac{1}{4}$	Light red
$R_1r_1r_2r_2r_3r_3$	$\frac{2}{4}$	Intermediate
$r_1r_1r_2r_2r_3r_3$	$\frac{1}{4}$	White

These predictions are consistent with the actual results observed in the experimental crosses.

3.5 F_1 will all yield 7 g. The F_2 will have the following phenotypes: 4 g ($\frac{1}{64}$); 5 g ($\frac{6}{64}$); 6 g ($\frac{15}{64}$); 7 g ($\frac{20}{64}$); 8 g ($\frac{15}{64}$); 9 g ($\frac{6}{64}$); 10 g ($\frac{1}{64}$)

3.6 **(a)** Long fruited parent = $L_1L_1L_2L_2L_3L_3$; Short fruited parent = $S_1S_1S_2S_2S_3S_3$. $F_1 = L_1S_1L_2S_2L_3S_3$; have fruits 9 cm long.

(b) Any genotype which contains two *L* alleles and four *S* alleles.

3.7 **(a)** $\frac{1}{4}$; **(b)** $\frac{1}{8}$; **(c)** $\frac{1}{6}$.

3.8 Of the 560 F_2 plants 36 resemble the white seeded parent – $\frac{1}{16}$ of the F_2 generation. This suggests that at least three gene pairs are involved in determining seed colour in oats.

3.9 **(a)** Since both parental lines are pure-breeding they must be homozygous at all loci. Consequently the variation in the cob length of the parents must be the result of environmental variation because each set of parents is genetically identical. The F_1, though heterozygous at all loci, are genetically identical so that any variation will again be the result of environmental variation. Provided that the parents and the F_1 plants are raised in similar environments we would expect the F_1 generation to show the same amount of variation in cob length as each parent. However in the F_2 generation produced by crossing the F_1 generation we would expect a number of different geneotypes to segregate. Consequently the F_2 plants will not be genetically identical. So now both genetic and environmental variation will contribute to the overall phenotypic variation of the F_2 generation. Hence we would expect this generation to show greater variation than either the F_1 or the parental generation.

(b) See the summary on page 26.

(c) Continuous variation.

Chapter 4

4.1 **(a)** Mean = 28.7 dm^3; variance = 226.5.

(b) Increase the variance.

(c) Either reduce the number of cows which produce the extreme amount of milk or increase the number of cows which produce close to the mean amount of milk.

4.2

Characteristic	Broad-sense Heritability	Narrow-sense Heritability
No. of bristles	0.61	0.52
Thorax length	0.49	0.43
Ovary size	0.65	0.30
No. of eggs	0.62	0.18

4.3 Broad-sense heritability = 0.36; narrow-sense heritability = 0.24.

4.4 You will need to plot a graph similar to Fig 4.4 and then calculate the slope of the regression line that you fit to the data. You should get a value of about 0.6.

4.5 You need to measure the cranial capacity of a large number of parents and their [illegible] ph like Fig 4.4. The slope of the [illegible] give you the heritability of t[illegible] ır tutor your method of deter[illegible]

4.6 This means that 60% of the [illegible] ;g mass in a particular flock [illegible] particular time can be att[illegible] n-ponent of genetic variance.

4.7 Heritability measurements only relate to the population in which they were measured. You cannot therefore use heritability measurements to argue that differences in mean IQ scores between two different populations, in this case black and white Americans, are due to genetic differences between them.

4.8 Heritability of large body mass = 0.42; heritability of low body mass = 0.46.

4.9 Heritability will decline with time as the amount of genetic variance in the selected line will decrease as selection proceeds. So the top term in equation (4) will decrease at a faster rate than the bottom term.

4.10 Changing the environment, for example adding more fertiliser, may improve conditions for growth. Increasing V_G, i.e. adding genetic variation, may result in an increased response to selection for a given selection differential.

4.11 Generally speaking the larger V_G the greater the response to selection for a given selection differential.

4.12 The higher the heritability of a trait the greater the response to selection. A heritability of 0.6 suggests that the prospects for selecting for an increase in egg mass are good.

Chapter 5

5.1 **(a)** *AaBbCcDd*; **(b)** 16.

5.2 $q = 0.0027$, $p = 0.9973$. In a randomly mated population the frequency of homozygous recessives = $q^2 = 7 \times 10^{-6} \approx 10^{-5}$. In an inbred population the frequency of homozygous recessives = $q^2 + Fpq$. So (a) with $F = 0.5$, frequency of homozygous recessives = $10^{-5} + (0.5 \times 0.0027 \times 0.9973) = 10^{-5} + 1.35 \times 10^{-3} \approx 10^{-3}$.
So when $F = 0.5$ we would expect homozygous recessives to be about 100 times more common than in a randomly mated population. With (b) $F = 0.0625$ the frequency of homozygous recessives will be about 10 times greater in the inbred population compared with the randomly mated outbred population.

5.3 Pure-bred pedigree dogs will tend to show inbreeding depression. This may take the form of a number of conditions, depending upon the breed of dog, but all may need expensive veterinary care in the future. By contrast a mongrel will show hybrid vigour.

5.4 You should have produced a diagram similar to Fig 5.10. Such information is useful because it tells the selective breeder the ease with which crosses to introduce new genes into a plant can be made.

5.5 Backcrossing is a breeding procedure which increases the proportion of genes in the progeny derived from one parent, say P_1, whilst reducing the proportion of genes derived from the other parent P_2. This is achieved by crossing the progeny with plants having the desired parental gentoype. See Fig 5.11 for a possible scheme. The proportion of genes derived from the wild parent halves with an increasing number of backcrosses. After 5 or 6 backcrosses any further reduction in the proportion of genes derived from the wild parent will be so small as to be not worth the cost involved in carrying out further backcrossing.

5.6 A cultivated variety will contain the essential background genes which adapt a cultivated plant to a particular environment. It is highly unlikely that a wild plant will contain the correct mix of background genes. So using a cultivated variety to introduce a new gene will reduce the number of backcrosses needed and so reduce the cost.

5.7 **(a)** Pollen.

(b) Outbreeder since the turnips readily cross-pollinate with closely related plants.

(c) Cabbage, rape and charlock.

(d) Yes. The progeny produced by crossing members of the primary gene pool are usually fertile.

(e) Bearing in mind the date, 1822, Cobbett will know nothing about the mechanism of heredity and nothing about chromosomes. So you will need to explain these before you go on to discuss gamete formation. Remember turnips are flowering plants and they produce gametes by mitosis and not meiosis. Finally you will need to explain the events which occur during fertilisation and the subsequent development of the plant and how introgression of the rape genes proved disastrous for the turnip harvest.

5.8 **(a)** You would need to use a disruptive selection procedure.

(b) First you would need to estimate the heritability of high and low gluten content of wheat flour. Then using equation (5) you could estimate the selection response for a given selection differential.

Chapter 6

6.1 **(a)** Number of pure breeding lines = 2^n, where n = number of loci.

(b) The effect of increasing the number of alleles at each locus will be to increase the number of pure breeding lines.

(c) Since each plant will contain 1000s of genes, each of which will probably have several alleles, the plant breeder will be able to produce a vast number of pure-bred lines, far more than can ever be realistically evaluated. So the plant breeder needs to reduce the number of lines under consideration to manageable proportions by a programme of intense selection during the first few generations after the initial cross. This raises the plant breeder's nightmare – discarding a pure-bred line during this early stage of a selective breeding programme which could be of value later.

6.2 Discuss your table with your tutor to ensure you have summarised all the points.

6.3 An outbreeder will be heterozyous at most loci, e.g. *AaBbCc*, whilst an inbreeder will be homozygous, e.g. *aaBBcc* or *AAbbcc* and so on. The outbreeder will be able to produce gametes with eight possible genotypes; the inbreeder will produce gametes which have identical genotypes. Consequently the F_1 generation produced by crossing two different inbred lines, say *AABBCC* and *aabbcc*, will all have the same genotype, e.g. *AaBbCc*, and will not be genetically variable. By contrast the F_1 produced by crossing two outbred plants will be highly variable. Now think about crossing the F_1 plants produced from the initial cross between the two inbred parents. The F_1 plants are heterozygous at all loci so they will be able to produce gametes with eight different genotypes, and consequently the F_2 generation will be highly variable.

6.4 See Fig 5.3. There is some disagreement over this amongst geneticists but the most likely explanation is that during their evolution deleterious alleles have been almost wholly eliminated from the gene pools of inbreeders.

Chapter 7

7.1 **(a)** S_1S_3, S_1S_4, S_2S_3, S_2S_4 **(b)** S_1S_3, S_2S_3 **(c)** S_1S_2, S_2S_3.

7.2 **(a)** S_1S_3, S_2S_3, S_1S_4, S_2S_4 **(b)** none **(c)** none **(d)** S_1S_2, S_1S_3, S_2S_3, S_3S_3.

7.3 **(a)** This is an example of gametophytic incompatibility so you will need to explain the mechanism involved.

(b) No progeny as this is now a sporophytic incompatibility system and all the pollen would have been phenotypically S_1.

(c) Prevent inbreeding.

(d) Allows crosses to be set up without the need for expensive hand emasculation to prevent self fertilisation. This cuts costs and so increases the number of crosses that can be made.

7.4 Take anthers from a pure-breeding plant (i.e. *WxWx*) which contains amylose in its starch and use them to grow monoploid plants. Test the cells of the monoploid plants for the presence of amylose. Those plants which do not contain amylose have grown from a pollen grain which contains a *wx* allele produced by mutation from the *Wx* allele. If 1 out of say 10 000 plants so tested did not contain amylose then the mutation rate for *Wx* to *wx* would be 10^{-4} per generation.

7.5 **(a)** Autotetraploids often show increased vigour whist some polyploids may be sterile which is useful for producing seedless fruit. Allopolyploids are used to combine the useful characteristics from two different species. You should give a specific example of each use.

(b) The original hybrid between turnip and black mustard must have been $2n = 18$. Since brown mustard has 36 chromosomes a doubling of chromosome number must have occurred. Such a plant is called an amphidiploid.

(c) See Fig 7.13.

Chapter 8

8.1 **(a)** Species A shows a discontinuous pattern of resistance suggesting that resistance in this species is under the control of one or two major genes. Resistance in species B shows continuous variation suggesting polygenic control.

(b) The variation in resistance within each major class is the result of environmental variation.

8.2 1. Cross pure-breeding, warfarin-resistant rats with pure-breeding, warfarin-susceptible rats.
All the F1 progeny should be warfarin resistant.
2. Cross the F1 progeny with pure breeding, warfarin-susceptible rats. If warfarin-resistance is due to a single dominant allele at one locus then we would expect $\frac{1}{2}$ the progeny of this test cross to be warfarin-resistant and $\frac{1}{2}$ to be warfarin-susceptible.

8.3 Mutation continually produces warfarin resistance alleles albeit at a slow rate. However these only occur at a low frequency in the gene pool because rats which carry the resistance alleles have an increased demand for vitamin K and so are at a selective disadvantage relative to rats who do not carry resistance alleles. However when warfarin is used to poison rats those rats which have resistance alleles are now at a selective advantage relative to those rats which do not. Consequently the warfarin-resistant rats will now leave more offspring than the warfarin-susceptible rats and so the frequency of the warfarin-resistance alleles will increase in the gene pool. Eventually nearly all the rats will carry at least one resistance allele and the population will be warfarin-resistant.

8.4 Your model should show that both homozygotes are at a selective disadvantage relative to the heterozygotes but for different reasons. Consequently whilst the frequency of the resistance allele increases rapidly to begin with, and the frequency of the susceptible allele decreases, there comes a point when there is no further change in the relative frequency of the two alleles. This is because the heterozygotes contain one copy of each allele.

8.5 Since the heterozygotes are at a selective advantage over both homozygotes so there will always be a reservoir of susceptible alleles in the heterozygote portion of the population. Since this genotype will come to dominate the population the frequency of the susceptible allele will actually remain quite high, approaching 50%.

8.6 **(a)** Once the selection pressure of warfarin has been removed rats with resistance alleles will once again be at a selective disadvantage compared to those individuals who do not have resistance alleles because of their increased demand for vitamin K.

(b) Under normal circumstances individuals who carry resistance alleles will be at a selective disadvantage compared with individuals who do not carry resistance alleles. It is only when the population is challenged by a pesticide that the resistant individuals begin to enjoy a selective advantage.

8.7 A variety of poisons are now used but you would expect rat populations to become resistant to them. Your local EHO will be able to tell you what strategies are used, if any, to overcome this problem.

8.8 Frequency of $Rw^s = 0.707$ and of $Rw^R = 0.293$. The frequency of each genotype is $Rw^SRw^S= 0.5$; $Rw^SRw^R = 0.414$; $Rw^RRw^R = 0.086$.

8.9 Narrow-spectrum insecticides kill only one or a few related species of insect whilst broad spectrum insecticides kill a wide range of unrelated species.

8.10 There are a lot of points you will need to make and your tutor will guide you, but the four essential ones are:
1. broad-spectrum insecticides kill a wide range of insects including the pest's predators, leading to ecological instability;
2. you only need to spray when the pests are causing economic damage;
3. too much spraying with the same insecticide will lead to the rapid development of pesticide resistance;
4. It is better to use a range of narrow-spectrum insecticides rather than one broad-spectrum insecticide.

8.11 **(a)** Insects have very short generation times, breed rapidly and leave many offspring. So resistant individuals can produce many offspring in a very short space of time and consequently the frequency of resistance alleles in the gene pool will increase rapidly.

(b) Annuals have a short generation time and leave many offspring compared to other plants. The fact that they are self-fertile will mean that they are inbred and so will be homozygous for any resistance alleles. As a result there will be no reservoir of susceptible alleles in the heterozygotes. These two factors together will lead to a rapid increase of resistance alleles in a gene pool once application of a herbicide begins.

8.12 Heavy metals, like mercury and copper, are general inhibitors of enzymes and so will affect a large number of different metabolic processes in the cell. For a fungus to be resistant to a heavy metal would mean that it would have a large number of resistance alleles. By contrast a fungicide which affects only a single enzyme can be rendered useless by a single mutation.

8.13 Your model should show that:
1. two loci are involved in governing virulence of *M.lini*;
2. one locus governs the ability of *M.lini* to attack Ottawa, the other locus governs the ability to attack Bombay;
3. the virulence allele at each locus is dominant to the non-virulence allele.

8.14 The cultivars show a low level of general resistance to all physiologic races of the pathogen. They show general resistance to five of the physiologic races and specific resistance to three of them.

8.15 Specific resistance is under major gene control so even a single mutation will have a large effect on the organism's phenotype, perhaps preventing the production of a toxic compound which inhibits the growth of a particular physiologic race of the pathogen. By contrast general resistance is under polygenic control so a single gene mutation will only have a very small effect on an organism's phenotype.

8.16 Old cultivated varieties, perhaps stored in a gene bank, land races or even wild relatives of cultivated maize. The method used will depend upon the source of the genes. See Chapters 5 and 6 for details.

8.17 Plants grown from seed will be genetically variable whilst plants grown from cuttings will be genetically identical. The variation strategy is a cheap option because only some of the plants will succumb to local pests so there will be little need to use expensive pesticide sprays. In addition, providing genetically variable material gives local growers the opportunity to produce cultivars which are best suited to local growing conditions.

8.18 To answer this question successfully you will need to use all your knowledge gained from the first two themes of this book and apply some of the basic Mendelian genetics that you should already know.

(a) *(i)* Both strains will be homozygous at most if not all gene loci.

(ii) Let the genotype of strain A = *RR*. Let the genotype of strain S = *rr*. Crossing the two strains gives an F_1 generation with the genotype *Rr*. Since these are all resistant this suggests that *R* is dominant to *r*. This is confirmed by the test cross between the F_1 and the S strain giving a 1:1 ratio of susceptible to resistant plants. (You will need to show details of this cross.) Furthermore the test cross also confirms that the difference between the two strains is due to a single gene. If, for example, two genes, each with two alleles, had been involved then only $\frac{1}{4}$ of the progeny of the test cross would have been susceptible.

(iii) All the plants would be resistant since half will have the genotype *RR* and the other half will have the genotype *Rr*. (Again you sould show details of the cross, i.e. *Rr* × *RR*).

(b) *(i)* (In answering these questions it is important that you set out the details of the crosses correctly. Ask your tutor for guidance.)
You should show that : F_1 = *Qqrr* (all resistant) and B = $\frac{1}{2}$ *Qqrr* : $\frac{1}{2}$ *qqrr* (50% resistant : 50% suceptible).

(ii) A. All the F_1 would be resistant with the genotype *RR*.
B. All the F_1 would be resistant with the genotype *QqRr*.

(iii) You will need to give details of the crosses between the F_1 in each case, i.e. *RR* × *RR*; *QqRr* × *QqRr*. The F_2 from the first cross will all be resistant. However $\frac{1}{16}$ of the F_2 from the second cross would be expected to be susceptible because they have the genotype *qqrr*.

(c) *(i)* See section 8.3 page 101.

(ii) You need to consider two aspects of the biology of the fungus relative to the maize plant. Firstly the fungus has a shorter life cycle and will leave more offspring than a maize plant. How will this affect the rate at which the frequency of the virulence alleles increases in the fungus gene pool compared with the resistance alleles in the maize gene pool? Secondly the fungus is haploid whilst maize is diploid. This means that selection can only act against susceptible alleles in the gene pool of maize when they are in the homozygous condition. Literally, susceptible alleles can hide from the effects of selection when they are in heterozygotes. Consequently the frequency of susceptible alleles will fall only slowly and so the frequency of the dominant alleles for resistance will increase slowly. Contrast this situation with that operating in the fungus gene pool where there are no heterozygotes because the fungus is haploid and so selection can act on all the alleles in the gene pool in every generation. So not only does the fungus have shorter generations than maize but selection can remove non-virulent alleles from the fungal gene pool rapidly. Consequently we would expect the frequency of virulence alleles to rise faster in the fungal gene pool compared with the rate of increase in the frequency alleles in the maize gene pool.

(d) This is an example of additive gene interaction. Let each *G* or *H* allele add 1 unit of resistance to the maize phenotype, i.e. a plant with resistance level 2 has 2 units of resistance and contains two resistance alleles. So strain D has resistance level 4 = 4 units of resistance = four resistance alleles, i.e. its genotype is *GGHH*. The fully susceptible strain, *gghh*, has no resistance alleles and therefore has no units of resistance, i.e. it is resistance level 0. The F_1 generation produced by crossing the two strains all have the genotype *GgHh* = 2 resistance alleles = 2 units of resistance, i.e. resistance level 2 as shown in Fig (a). Now crossthe F_1 and compare the proportion of the F_2 phenotypes with each resistance level (4, 3, 2, 1, 0) with Fig 2. Do they match up?

Chapter 9

9.1 Nucleic acid, either DNA or RNA, forms a central core surrounded by a protein coat called a capsid.

9.2 TMV is 10 nm wide; Adenovirus is 35.7 nm in diameter; T4 phage is approximately 100 nm long.

9.3 A model system is one which captures the essential features of a whole group of organisms. In this case phages capture the essential features of virus life cycles and viral genetics. The importance of phages to geneticists is summarised in section 9.1, page 112.

9.4 There is no simple answer to this question. Discuss your point of view with your friends and your tutor.

Chapter 10

10.1 There are a number of points that you should make including:
1. in both eukaryotes and prokaryotes recombination involves pairing of homologous DNA sequences followed by breaking and subsequent rejoining of DNA;
2. in prokaryotes a linear fragment of DNA must be recombined into a closed circle of DNA if it is not to be lost from a cell. This means that successful recombination in prokaryotes must usually involve an even number of cross overs;
3. in eukaryotes recombination occurs primarily during gamete formation. This is not the case in prokaryotes;
4. the effect of crossing over in both eukaryotes and prokaryotes is to increase genetic variation.

10.2 Resistance to antibiotics. This means that bacteria which contain *R* factors will be at a selective advantage in antibiotic-rich environments, e.g. hospitals or your intestines if you are taking antibiotics.

10.3 If the sequence of bases which constitutes a gene is disrupted by the insertion of a piece of foreign DNA, e.g. a transposon, then the information in the gene can no longer be read correctly resulting in, for example, a polypeptide which does not contain the correct sequence of amino acids. Such a polypeptide will not function correctly leading to a change in the phenotype of the organism, i.e. a mutation.

10.4 See Figs 10.3, 10.4, 10.11, 10.12, 10.14. In all bacteria conjugation involves the formation of a conjugation tube and the subsequent transfer of DNA, usually a plasmid, from donor to recipient cell. Hfr is different in that it involves the directed transfer of chromosomal DNA alongwith the plasmid DNA.

10.5 This is the classic example of transformation of one type of bacterial cell with DNA from a different type of bacterial cell. The R strain bacteria acquire DNA from the heat-killed S strain bacteria (The heating process lyses the S strain cells and fragments their DNA) producing partially diploid R cells. If one of the pieces of DNA acquired by an R strain cell carries virulence gene and if that piece of DNA becomes recombined into the R strain cell's genome then the virulence gene will be expressed, i.e. the R strain has been transformed into an S strain.

10.6 The use of antibiotics creates an enormous selection pressure in favour of those bacterial cells which carry resistance genes. Usually such cells occur at a low frequency in a bacterial population presumably because there is some metabolic cost associated with producing the product encoded by the resistance gene. However in an antibiotic-rich environment resistant cells will be at a selective advantage and they will increase rapidly in numbers with a corresponding increase in the frequency of the antibiotic-resistance allele in the bacterial gene pool. Consequently antibiotics should only be used when absolutely necessary and never as growth promoters.

10.7 Again you could make a number of points but the following are the main ones:
1. only prescribe antibiotics when absolutely necessary;
2. use narrow-spectrum rather than broad-spectrum antibiotics;
3. ensure that the dose is large enough and the treatment goes on for long enough to ensure that all the bacteria are killed;

10.8 You will need to explain the dangers of antibiotic - resistance and how such resistance can threaten the lives of people when they are infected by antibiotic - resistant bacteria. Real examples are generally more persuasive than just theory.

10.9 Hospital sewage will tend to contain faeces and urine from people who are being treated with antibiotics. Consequently there will be selection pressure in favour of antibiotic-resistant strains of bacteria. Furthermore many of those bacteria will have come from the guts of people who are being treated with antibiotics. The commonest of these gut bacteria is *E. coli*, a member of the Enterobacteriaceae, and this group of bacteria is known for its ability to pass R plasmids to each other by conjugation. So hospital sewage represents an antibiotic-rich environment compared with normal domestic sewage, which is inhabited by bacteria which have the ability to transfer antibiotic-resistance plasmids to each other. Under these circumstances you would expect a thriving population of antibiotic-resistant cells.

Chapter 11

11.1 **(a)** i **(b)** iii, v, vi **(c)** ii, iv.

11.2 **(a)** GCTGCCGATGGT **(b)** GCUGCCGAUGGU.

11.3 Prokaryotic genes never contain introns and exons, they occur in operons consisting of several genes under the control of a single promoter. Eukaryotic genes often consist of introns and exons, they do not occur in operons and each gene has its own promoter.

11.4 TCGA.

11.5 Resistance to the antibiotics tetracycline and ampicillin.

11.6 You are only expected to know the basic outline of how this is achieved. Details of amounts, temperatures and incubation times are not needed.
1. In separate test tubes mix the restriction enzyme with the mammalian DNA and pBR322. The effect is to produce, in each tube, DNA with complementary sticky ends.
2. Mix the contents of the two test tubes and incubate. During incubation annealing will occur producing, among other things, recombinant DNA molecules.
3. Add DNA ligase to the incubation mixture to seal the nicks in the recombinant DNA molecules.

11.7 See Fig 11.11 which shows the intact plasmid. Now look at Fig 11.15 and imagine that the foreign DNA was removed. This is how the cut plasmid would look.

11.8 **(a)** 573.
(b) It may contain a number of introns in addition to the coding exons.
(c) Some amino acids are coded for by more than one triplet. For example the amino acid arginine may be represented as TCC in the probe but in the actual gene it may be encoded as TCT.

11.9 Only a tiny number of the restriction fragments produced by digesting DNA isolated from cells will contain the gene you are interested in. Consequently the chance of recombining one of these fragments into a vector molecule and subseqently isolating a bacterial clone which is carrying it is very small. By contrast a set of restriction fragments produced by the cDNA method or by DNA synthesis will all carry the gene you are interested in. Consequently the chances of making the correct recombinant DNA molecule containing the relevant genetic information and subsequently isolating a bacterial clone expressing that information is much greater.

11.10 **(a)** 7.5×10^5 fragments. A purified sample of DNA will be produced by extracting the DNA from many cells. If, for example, we used 10^6 cells to produce our DNA sample then we could expect $(7.5 \times 10^5) \times 10^6 = 7.5 \times 10^{11}$ restriction fragments. In practice you would never get as many fragments as this because the restriction enzyme would not be given enough time to cut the DNA at all the restriction sites.
(b) 1. pBR322, like most plasmids, cannot carry the large DNA fragments necessary to successfully make a gene library for a genome as large as the human one.

2. The sheer number of DNA fragments produced by shotgunning would involve maintaining an enormous number of bacterial cells to hold all the DNA fragments. This would be time consuming and expensive. Maintaining and screening an equivalent number of λ Charon phage particles is much easier and cheaper.

Chapter 12

12.1 The insulin gene is a split gene. Bacteria do not possess the necessary enzymes to cut and splice the primary mRNA transcript to produce a functional mRNA molecule.

12.2 The pre-pro-insulin polypetide has to undergo post-translational modification to produce functional insulin. Again the bacterial cells do not have the enzymes necessary to do this.

12.3 The amino acid sequences of the two polypeptide chains.

12.4 In the presence of lactose.

12.5 Discuss and present your examples to your tutor.

12.6 You should include the following points:
1. malaria is a disease which affects a lot of people so there is a large market;
2. the market is increasing as a result of (i) the increase in insecticide resistance among the mosquitoes which carry the malarial parasite; (ii) high birth rates in countries where malaria is endemic. You should also include some background information about the malarial parasite's life cycle, how it is transmitted and the present methods of treating and preventing malaria.

12.7 Reduce the costs to farmers by reducing the need to apply fertilisers and spray crops with expensive pesticides. Prevent the ecological damage caused by contaminating water supplies with nitrates, e.g. eutrophication, and the problems caused when pesticides get into food chains.

12.8 The sodium hydroxide will break the hydrogen bonds which hold the complementary DNA strands together so making the DNA on the filter single-stranded. The gene probe can then bind to the single-stranded DNA wherever it encounters a complementary DNA sequence. If the gene probe is made radioactive, for example by incorporating ^{32}P, then the presence of the probe on the filter can be detected using autoradiography.

12.9 Gel (a) corresponds to an individual who is homozygous for the normal haemoglobin allele; gel (b) to an individual who is homozygous for the sickle-cell allele; gel (c) represents a carrier of the sickle-cell allele.

12.10 The phenotypes of individuals who are homozygous for a dominant allele and those who are heterozygous are often indistinguishable. However if the dominant and recessive alleles produce a change in the restriction sites leading to restriction fragment length polymorphism, as in the case with sickle-cell anaemia, then heterozygotes can be detected provided a suitable gene probe is available.

12.11 The probability of having an affected child if you have Huntington's chorea is 0.5. However you will probably not develop the symptoms of Huntington's chorea until you are in your 40s by which time you may already have had children. So a person with a history of Huntington's chorea in their family may opt to have a test to see if they are carrying the Huntington's allele, before thay have children, to ensure that they do not pass the allele on to their children.

Chapter 13

13.1 Alleles.

13.2 **(a)** Two. **(b)** Form 1 – 13%; Form 2 – 36%; Form 3 – 0%; Form $\frac{1}{2}$ – 43%; Form $\frac{2}{3}$ – 5%; Form $\frac{1}{3}$ – 3%.

13.3 Electrophoresis separates proteins on the basis of their net charge and size which is determined by the amino acids which make up the protein. If the amino acid sequence is changed, as a result of a mutation, then the electrical properties of the protein may also be changed and this can be detected by electrophoresis.

13.4 **(a)** The heterozygosity of each locus is simply the number of heterozygotes divided by 100. For example in the case of acid phosphatase = $\frac{52}{100}$ = 0.52.
(b) 0.067.

13.5 **(a)** Identical twins are produced when the two cells which result from the first cell divsion of the fertilised egg become separated. Since this is a mitotic division each cell contains exactly the same genetic information. So identical twins are the result of asexual repro-duction.
(b) Populations of organisms produced by asexual reproduction are genetically identical. Populations of organisms produced by sexual reproduction are genetically variable. Consequently sexually reproducing populations contain more genetic variability on which selection can act.

13.6 AB.

13.7 **(a)**

N	A	A	A
N	N	A	A
N	A	N	A
N	N	N	N

(b) Red blood cells from blood group O individuals do not carry AB antigens on their surface. Consequently they will not agglutinate with any agglutinins which may be present in the blood of another individual. The amount of anti-A and anti-B agglutinogen present in the quantity of blood normally given in a blood transfusion is very small and so has a negligible effect. Individuals who are blood group AB do not have either anti-A or anti-B agglutinogens in their blood so they can receive red blood cells from an individual with any other ABO blood group without running the risk of agglutination occurring. Again the amount of anti-A or anti-B agglutinogen which may be present in the blood being given to the AB person will be so small as to have a negligible effect.

13.8 The father is $I^A i$, the mother is $I^B i$.

13.9 **(a)** 8.
(b) *(i)* 8. *(ii)* 0 – red blood cells do not carry HLA antigens.
(c) HLA-A = 210; HLA-B = 820; HLA-C = 36; HLA-D = 78.

(d) 4.84×10^8.
(e) About 1 in 500 million.

13.10 2.

13.11 (a) (i) $\frac{1}{4}$. (ii) $\frac{1}{2}$. (iii) $\frac{1}{4}$.
(b) Since the HLA alleles are inherited as a block , this means that the probability of members of the same family having the same haplotype is much higher.
(c) Blood group.

13.12 You can analyse this data in a number of ways. You could try plotting triallelic diagrams for each gene and then superimposing them. Do you get two separate distinct groups of alleles with different frequencies as you would expect if these two populations were genetic races? Are there particular genes where the allele frequencies are very different in the two populations? Are there particular genes where the allele frequencies are very similar? Are there more similarities than differences? Discuss your conclusions with your friends and tutor.

Chapter 14

14.1 Their frequency will be low.

14.2 Albinism: 5 sufferers (s), 1391 carriers (c); phenylketonuria: 45s, 1250c; alkaptonuria: <1s, 199c; cystic fibrosis: 50s, 4545c.

14.3 Most deleterious alleles occur at around the mutation frequency, i.e. they are very rare. Consequently you are far more likely to find such an allele in a single copy in a genotype, i.e. in a heterozygote. An individual carrying a single copy of a dominant deleterious allele will show symptoms of the disease whilst an individual heterozygous for a deleterious recessive allele will show no symptoms of the disease. It is only in the very rare circumstances of two individuals who are both heterozygous for the same rare recessive allele having a child who is homozygous for the recessive allele that the disease will appear. Even for a fairly common deleterious recessive allele like the one involved in cystic fibrosis the incidence of the actual disease is low.

14.4 Individuals who are homozygous for the sickle-cell allele will suffer from sickle-cell disease and, without medical help, will probably die before they can have children reducing the frequency of the sickle-cell allele in the population. In areas where malaria is endemic individuals who are homozygous for the normal haemoglobin allele will suffer more from the disease than heterozygous individuals. Consequently heterozygous individuals will be at a selective advantage. So selection against the sickle-cell allele in homozygotes is just counterbalanced by selection in favour of the sickle-cell allele in heterozygotes so maintaining the allele at a fairly high frequency in spite of its lethality in the homozygous condition.

14.5 Female.

14.6 (a) Reading from left to right the genotypes of the zygotes will be XXX (Triple X syndrome), XXY (Klinefelter's syndrome), XO (Turner's syndrome), YO (lethal).
(b) Equal numbers of XXX, XXY, XO, YO.

14.7 Equal numbers of Klinefelter's syndrome and Turner's syndrome if nondisjunction occurs at meiosis I. 1 Triple X : 2 Turner's syndrome : 1 Klinefelter's syndrome if nondisjunction occurs at meiosis II in both spermatocytes. 1 normal boy : 1 normal girl : 1 Turners syndrome : 1 Klinefelter's syndrome if nondisjunction only occurred in one spermatocyte.

14.8 $\frac{1}{4}$.

14.9 (a) Normal female.
(b) Female trisomy 21 (Down's syndrome) where nondisjunction occurred in the mother.
(c) Male trisomy 21 (Down's syndrome) where nondisjunction of chromosome 21 occurred in the father.
(d) Male with XYY syndrome caused by nondisjunction of the Y chromosomes in the father at meiosis II.

14.10 A normal female is XX so even if she carries a colour blindness allele on one X chromsome the normal allele on the other chromosome compensates. However a female with Turner's syndrome is XO so if she is carrying a colour blindness allele on her single X chromosome then she will be colour blind. You cannot decide from this information where the nondisjunction occurred.

14.11 Both parents should have a karyotype analysis done to ensure neither is carrying a 14–21 translocation. Otherwise there is no reason for them not to have another baby. The mother should have an amniocentesis or CVS followed by a karyotype analysis on the baby if she is in a high risk group, e.g. above the age of 40.

Chapter 15

15.1 (a) Female.
(b) 1 boy followed by three girls.
(c) C.2.
(d) B.5.
(e) She has married twice and has three daughters from her first marriage.

15.2 You would expect all the sons of the fathers who are affected to also have the trait but D.2 is not affected. No.

15.3 C.7.

15.4 There are two possibilities:
1. neither of the children has inherited the allele which causes this form of muscular dystrophy from their father;
2. the disease has not yet developed. Given that this allele apparently shows incomplete penetrance and that both of the children stood a 50% chance of inheriting the allele from their father they should consult a genetic counsellor before they have children.

15.5 The trait affects equal numbers of males and females whilst X-linked traits affect males more than females.

15.6 D.3 and D.4 must both be carriers of the recessive allele. Given that such alleles are rare (see Chapter 14) there is no way that you could have predicted this outcome.

15.7 The trait is rare, it affects equal numbers of males and females and it skips generations.

15.8 The baby has parents who are related, i.e. this is a consangineous marriage. Such marriages increase the probability of bringing together two carriers of a recessive allele. In this case the most likely explanation is that the distant common ancestor of our two parents inherited a single copy of the recessive thalassaemia allele, probably from his mother the – Indian princess. This allele has then passed down through their separate families to each of them.

15.9 Since males are XY they only need one copy of the allele to show the trait whilst females will have to be homozygous for the allele to show the trait. A son inherits the Y chromosome from his father whilst his daughters will inherit the X chromosome carrying the trait making them carriers. A daughter will only be affected if her mother is a carrier and her father has the trait. Since such alleles are rare the probability of this occurring is small so we would expect none of the children of an affected male to develop the trait.

15.10 No. Victoria, Alice of Hesse, Beatrice, Irene, Alix, Alice of Athlone.

15.11 This trait is the result of an X-linked dominant allele. The sons would only develop the trait if their mother had it.

15.12 $\frac{1}{4}$.

15.13 **(a)** $\frac{1}{8}$. **(b)** $\frac{1}{64}$. They decrease.

15.14 Marriage between first cousins does increase, slightly, the risk of mortality among children born in such marriages. We might expect the percentage mortality to increase further the more closely related the parents. However, provided both parents are not carriers of the same rare recessive allele, there seems no reason to prevent marriages between close relatives. Remember a consanguineous marriage only increases the slight risk of having a child with a genetic illness – it does not mean that the children will have a genetic disease.

15.15 0.25.

15.16 $\frac{1}{16}$ and $\frac{2}{5}$ respectively.

15.17 This is for you to decide. Discuss your point of view with your friends and tutor or even better organise a debate between those who think the answer is yes and those who think the answer is no.

15.18 **(a)** Autosomal recessive trait. Let the dominant allele = A and the recessive allele = a. Then 1 = Aa; 2 = Aa; 3 = Aa; 4 = aa; 5 = Aa; 6 = Aa.

(b) X-linked recessive trait. Let X = the normal chromosome and X^a = X chromosome with the recessive allele. Then 1 = XX^a; 2 = XY; 3 = X^aY; 4 = XX; 5 = XX^a.

15.19 Tay-Sachs disease only occurs at a high frequency among the Jews of central and Eastern Europe. So unless our British parents originally came from this particular group of people the chances of then having a baby with Tay-Sachs disease are so small that the test is simply not worth doing. By contrast phenylketonuria is a disease which is relatively common in all groups therefore it is worthwhile testing all babies at birth.

15.20 Such tests are only worthwhile if the risks associated with the test are much less than the risk of developing the disease. Provided we have a test which is completely safe then the only factors which will prevent testing all fetuses is money, time and of course parental choice. Under these circumstances the two common disorders, cystic fibrosis and phenylketonuria, seem worth testing for. However it is possible to detect phenylketonurics at birth with a well established, cheap test and it would seem reasonable to carry on with this practice. So that just leaves us with cystic fibrosis. Galactosaemia is so rare that a prenatal screening programme for all would seem unjustified though it should be offered to those individuals with a history of galactosaemia in their family. The decision of what to do with the results of such tests is more difficult. Both phenylketonuria and galactosaemia are treatable diseases and provided the sufferers are identified in time they lead perfectly normal lives. People with cystic fibrosis cannot at present be cured but the use of drug therapy and physiotherapy has greatly extended their lives. The decision as to what to do with the information from such tests must therefore be left to the parents who will need support from appropriate and understanding counselling.

Index